VARIORUM COLLECTED STUDIES SERIES

The New Science of Geology

Professor Martin J.S. Rudwick

Martin J.S. Rudwick

The New Science of Geology

Studies in the Earth Sciences in the Age of Revolution

Routledge
Taylor & Francis Group
LONDON AND NEW YORK

First published 2004 by Ashgate Publishing

2 Park Square, Milton Park, Abingdon, Oxfordshire OX14 4RN
711 Third Avenue, New York, NY 10017

Routledge is an imprint of the Taylor & Francis Group, an informa business

First issued in paperback 2018

British Library Cataloguing in Publication Data
Rudwick, Martin J.S. (Martin John Spencer), 1932–
   The new science of geology : studies in the earth sciences in the age of
   revolution. – (Variorum collected studies series)
   1. Geology – History
   I. Title
   551'.09

Library of Congress Cataloging-in-Publication Data
Rudwick, M. J. S.
   The new science of geology: studies in the earth sciences in the age of
   revolution / Martin J. S. Rudwick.
      p. cm. – (Variorum collected studies)
   Includes bibliographical references and index.
   ISBN 0–86078–958–6 (acid-free paper)
   1. Earth sciences. I. Title. II. Series: Collected studies.

   QE26. 3 R83 2004
   550–dc22   2003025799

ISBN 13: 978-0-86078-958-1 (hbk)
ISBN 13: 978-1-138-38250-3 (pbk)

VARIORUM COLLECTED STUDIES SERIES CS789

CONTENTS

This volume contains xviii + 316 pages

INTRODUCTION

This volume reproduces some of my published articles on the history of the natural sciences. I have used three criteria in making the selection. First, all these articles focus on the late eighteenth and early nineteenth centuries – or on what political historians often call "the age of revolution" – although a few range over a longer timespan. Second, all of them deal with the area of natural knowledge that at just this time was being transformed into the science of "geology". And third, these articles seem to have stood the test of time, in that they continue to be cited by historians and geologists and to be used in teaching, and I continue to be asked for offprints of them. This brief introduction summarizes their contents and explains the relations between them; it is followed by brief notes on the circumstances in which they were written and published, and a bibliography giving full details of all the books and articles mentioned. Most of the articles are reproduced here exactly as they were published, with their original pagination, to facilitate citation etc.; a few have had to be reset to ensure legibility, but are unchanged in wording (except that a few citations have been brought up to date).

In one way or another, all these articles are related to the historical problem that has been at the heart of my research, ever since I turned myself in mid-career from a geologist into a historian. I have tried to understand how the science of geology developed from antecedents that were significantly different in their practical methods and cognitive goals, and how it came to be recognized as a new *kind* of natural science. Geology as a new science was constituted around the idea that the natural world itself had had a *history*: it had to be understood not only – as in the physical sciences – in terms of unchanging natural laws that could be observed in action in the present, but also in terms of a "deep" past that was reliably knowable, even though it was pre-human, not directly observable, and only fragmentarily preserved. This new conception of a thoroughly historicized natural world first became visible in the science of geology, during – perhaps not coincidentally – the political age of revolution.

The articles are identified by Roman numerals, and are arranged in three groups. The first group deals with general issues and syntheses. The history of geology is commonly misconceived in terms of just one overriding theme, the expansion of the earth's putative timescale from a few thousand years in the seventeenth century to several billions by the end of the twentieth. "Geologists' time: a brief history" (I) surveys that *longue durée*, but replaces the usual polarities with continuities. It suggests that the early textual science of "chronology" (with Ussher and his Creation date of 4004 BC as its most famous exponent) was not the reactionary opponent of The Progress of Science, as much modern mythmaking would have it. Instead it was the direct conceptual forerunner of the natural science of "geochronology", which now provides earth scientists with an indispensable temporal framework for their research. "The shape and meaning of earth-history" (II) covers the same three centuries with more explicit attention to the shifting relations between scientific and religious beliefs. Again, the suggestion is that the usual polarities are deeply misleading; that on the specific issue of the history of the earth – what scientists now call "geohistory" – there was no proper ground for conflict; and that the modern arguments between religious fundamentalists and their atheistic counterparts are profoundly misconceived.

"Minerals, strata and fossils" (III) suggests how, during the period from about 1750 to 1850, ways of studying the earth were radically reconstituted. Sciences of descriptive "natural history" – of mineral and fossil specimens, of the spatial features of physical geography, and of three-dimensional rock structures – were all transformed into a unified and deeply historicized "geology". This development is analysed in more detail in "The emergence of a new science" (IV), which is an extended review, both appreciative and critical, of Rachel Laudan's important book *From Mineralogy to Geology*. It evaluates her historical interpretation in relation to my own; in particular I support her case for the rehabilitation of Abraham Werner's much maligned science of "geognosy", while remaining sceptical about her argument that it was truly geohistorical. Finally in this first group of articles, "The emergence of a visual language for geological science 1760–1840" (V) surveys the varied kinds of imagery that were used during this formative period to express the vitally important *visual* component of the new science; it traces how they coalesced into the standardized visual or graphical "language" that geologists now take for granted. It is worth mentioning that this article, with its plea for serious attention to be given to images as well as texts in the history of science, was well received at the time by scientists with historical interests, but it was almost ignored by historians – although it was published in one of our major periodicals – until, several years later, it started being treated retrospectively as a "classic"! Such has been the change in historical fashions in the past quarter-century.

The second group of articles focusses on the crucial role of Georges Cuvier in the construction of a new science of geology around the turn of the nineteenth century; they can be read in conjunction with my interpretative anthology *Georges Cuvier*. But first, "Jean André de Luc and nature's chronology" (VI) deals with an older naturalist whose ideas about reconstructing nature's own history, and about its relation to the far briefer span of human history, were significant in moulding Cuvier's approach. In particular, de Luc argued effectively that certain physical features could be used as "nature's chronometers", enabling the earth's putative last "revolution" (later identified as the Pleistocene glaciations) to be assigned an approximate date, while treating still earlier geohistory as immeasurably longer in duration. "Cuvier and Brongniart, William Smith, and the reconstruction of geohistory" (VII) describes how Cuvier and his mineralogist collaborator Alexandre Brongniart first worked out how to transform "stratigraphical" piles of rock formations – as mapped by the English surveyor William Smith – into nature's "documents" for tracing the sequence of events in nature's history. Cuvier and Brongniart's joint field survey of what they termed the "Paris Basin" became the most widely admired model for turning stratigraphy into geohistory.

The following two articles focus, in complementary ways, on the indoor working methods of Cuvier himself, which formed the foundation for all his inferences about geohistory. "*Researches on Fossil Bones*: Georges Cuvier and the collecting of international allies" (VIII), which has not been published previously in English, traces how he built up a network of collaborators – at the height of Napoleon's wars – to supply him with specimens for his greatest work, or at least with "proxies" for them in the form of pictures on paper. "Georges Cuvier's paper museum of fossil bones" (IX) follows the trajectory of this indoor scientific practice all the way from his specimens (or their proxies) to his publications, with particular attention to his artistic activity and that of his engravers. Cuvier's plates of engravings, as much as or even more than his text, enabled him to claim persuasively that the fossil species of "quadrupeds" (mammals and reptiles) were all distinct from living species, and all extinct. This in turn focussed attention on the natural "catastrophes" in geohistory that alone seemed adequate to explain such massive episodes of extinction.

The third and last group of articles moves forwards to the second quarter of the nineteenth century, and to the time when Charles Lyell was beginning to occupy a pivotal position in geology not unlike that of Cuvier in earlier years. "Encounters with Adam, or at least the hyaenas" (X) analyses how geologists working in the Cuvierian tradition developed a new pictorial genre that depicted reconstructed scenes from the distant past history of the earth. It covers briefly some of the same ground as my book on *Scenes from Deep Time*, but in a different way. It uses a retrospective mode to trace how this genre – now taken for granted in countless computer simulations of dinosaurs – was first constructed, and how it came to be accepted as respectably scientific rather than merely speculative. "A year in the life of Adam Sedgwick and company, geologists" (XI) surveys how geology was being practised throughout Europe in 1835. It uses the famous Cambridge geologist (who that year turned fifty) as a focus for showing how well established the science had become. In particular it analyses the social composition and geographical distribution of the members of the Geological Society – the first body of its kind in the world – and of its French counterpart and other institutions.

"Travel, travel, travel: geological fieldwork in the 1830s" (XII), which has not been published previously, uses Lyell's first long field expedition on the Continent (initially in the company of Roderick Murchison) as an example of geological field practices around that time. It aims to clarify the relation between geologists' first-hand encounter with the phenomena of importance in their science, and the arenas of scientific debate (such as the Geological Society) in which they expounded their interpretations of what they had seen. "The group construction of scientific knowledge" (XIII) uses the famous Devonian controversy, which raged from 1834 to about 1842, to enlarge on this analysis of the social interactions of geologists in relation to their experiences in the field. It traces how the complex dynamics of the argument – at once social and cognitive – led not to the triumph of either of the initial explanations of the puzzle, nor to any mere compromise between them, but to a novel synthesis that was unexpected by either party (and one that has endured to the present day, in all its essentials). This article offers a summary of some aspects of the detailed account given in my book on *The Great Devonian Controversy*.

Finally, "The glacial theory" (XIV) traces the course of the heated arguments – coincidentally, over much the same years as the Devonian controversy – about the geologically recent past that Lyell called the "Pleistocene" period. It uses Albert Carozzi's English edition of Louis Agassiz's *Studies on Glaciers* as a peg on which to hang a review of the formulation and reception of the glacial theory. Most geologists were highly sceptical about Agassiz's claim that there had been a drastic Ice Age, with widespread glaciers and ice-sheets in Europe and north America, around the time that human beings first made their appearance. The idea was utterly unanticipated and the theory seemed extravagantly speculative, and anyway less satisfactory than the then current idea of a "geological deluge" (a mega-tsunami, as it were), which was identified by some geologists – but certainly not by all – as being dimly recorded in the story of Noah's Flood. In time, however, a more moderate form of the glacial theory came to be widely accepted; it underlined the radically contingent character of geohistory.

A companion volume, entitled *Lyell and Darwin, Geologists*, will offer a further selection of my articles, focussed more specifically on Lyell's work and that of his most famous disciple. Charles Darwin first made his name in the scientific world as a Lyellian geologist, and his early geological work was intimately linked with his later research on the origin of species and other biological problems.

MARTIN J.S. RUDWICK

Cambridge
July 2003

PUBLISHER'S NOTE

The articles in this volume, as in all others in the Variorum Collected Studies Series, have not been given a new, continuous pagination. In order to avoid confusion, and to facilitate their use where these same studies have been referred to elsewhere, the original pagination has been maintained wherever possible.

Each article has been given a Roman number in order of appearance, as listed in the Contents. This number is repeated on each page and is quoted in the index entries.

NOTES ON THE ARTICLES

"Geologists' time: a brief history" (I) was commissioned for *The Story of Time* (1999), which was published by the National Maritime Museum in London in conjunction with its millennial exhibition on the same theme. "The shape and meaning of earth history" (II) was written for a conference in Madison, Wisconsin, in 1981, and published in *God and Nature* (1986); this collection of historical essays has been influential in breaking down the misleading stereotype of an intrinsic and perennial conflict between science and religion. "Minerals, strata and fossils" (III) was my contribution to *Cultures of Natural History* (1996), a volume that has likewise helped to rescue the concept of natural history from scientists' scorn and historians' neglect. "The emergence of a new science" (IV) was published in *Minerva* (1990) as an essay-review of Rachel Laudan's *From Mineralogy to Geology* (1987). "The emergence of a visual language for geological science" (V) was my main contribution to the conference held in 1975 by the British Society for the History of Science to mark the centenary of Lyell's death; it was published in *History of Science* (1976).

"Jean-André de Luc and nature's chronology" (VI) was written for a conference organized in London by the Geological Society to mark the millennium; it was published in *The Age of the Earth* (2001), a volume that combines historical essays with reviews of recent scientific work. The next three articles were all related to research for my book on *Georges Cuvier* (1997). "Cuvier and Brongniart, William Smith, and the reconstruction of geohistory" (VII) was written for *De la Géologie à son Histoire* (1997), a Festschrift in honour of the late François Ellenberger, the doyen of francophone historians of geology; it is reprinted here from my English translation (which, fortuitously, was published in *Earth Sciences History* before the original French text). "*Researches on Fossil Bones:* Georges Cuvier and the collecting of international allies" (VIII) was written for the conference held in Paris in 1993 to mark the bicentenary of the Muséum d'Histoire Naturelle, the world's foremost institution for the natural history sciences in the age of Cuvier; it was published in French in *Le Muséum au Premier Siècle* (1997), but is published here for the first time in English. "Georges Cuvier's paper museum of fossil bones" (IX) was written for the conference on "Drawing from nature" organized by the Society for the History of Natural History, and published in its *Archives of Natural History* (2000).

"Encounters with Adam, or at least the hyaenas" (X) was my contribution to *History, Humanity and Evolution* (1988), a Festschrift for the distinguished Darwinian historian John C. Greene; it summarizes some aspects of the argument set out later, with many more examples, in my book on *Scenes from Deep Time* (1992). "A year in the life of Adam Sedgwick and company, geologists" (XI) was given as the sixth Ramsbottom Lecture to the Society for the History of Natural History, at a meeting in Cambridge in 1985 to mark the bicentenary of Sedgwick's birth, and was published in its *Archives* (1988). "Travel, travel, travel: geological fieldwork in the 1830s" (XII) was written for a conference on British culture in that decade, held in 1994 at Trinity College, Cambridge, to mark the bicentenary of the birth of Sedgwick's polymathic Trinity colleague William Whewell; it has not been published previously (it is printed here as it was circulated to the conference participants, in an informal style and without erudite footnotes). "The group construction of scientific knowledge" (XIII) was given in Jerusalem in 1982 to the first Israel Colloquium for the History, Philosophy and Sociology of Science, and published in *The Kaleidoscope of Science* (1985); it embodied some reflections on the case-study that I published in detail in *The Great Devonian Controversy* (1985). Finally, "The glacial theory" (XIV) was published in *History of Science* (1970), prompted by Albert V. Carozzi's edition (1967) of Agassiz's *Études sur les Glaciers.*

BIBLIOGRAPHY

Blanckaert, Claude, Claudine Cohen, Pietro Corsi and Jean-Louis Fischer (eds.). 1997. *Le Muséum au Premier Siècle de son Histoire*. Paris: Muséum Nationale d'Histoire Naturelle.

Carozzi, Albert V. (ed.). 1967. *Studies on Glaciers by Louis Agassiz*. New York and London: Hafner.

Gohau, Gabriel, and Goulven Laurent (eds.). 1997. *De la Géologie à son Histoire*. Paris: Société Géologique de France.

Jardine, N., J.A. Secord and E.C. Spary (eds.). 1996. *Cultures of Natural History*. Cambridge and New York: Cambridge University Press.

Laudan, Rachel. 1987. *From Mineralogy to Geology: the Foundations of a Science, 1650–1830*. Chicago and London: University of Chicago Press.

Lewis, C.L.E. and S.J. Knell (eds.). 2001. *The Age of the Earth: from 4004 BC to AD 2002*. London: The Geological Society [Special Publication 190].

Lindberg, David C. and Ronald L. Numbers (eds.). 1986. *God and Nature: Historical Essays on the Encounter between Christianity and Science*. Berkeley and Los Angeles: University of California Press.

Lippincott, Kristen (ed.). 1999. *The Story of Time*. London: Merrell Holberton.

Moore, James R. (ed.). 1988. *History, Humanity and Evolution*. Cambridge and New York: Cambridge University Press.

Rudwick, Martin J.S. 1970. The glacial theory. *History of Science* 8: 136–57. [ARTICLE XIV]

———. 1976. The emergence of a visual language for geological science 1760–1840. *History of Science* 14: 149–95. [ARTICLE V]

———. 1985. The group construction of scientific knowledge: gentlemen-specialists and the Devonian controversy. *In* Ullmann-Margalit, *The Kaleidoscope of Science*, 193–217. [ARTICLE XIII]

———. 1985. *The Great Devonian Controversy: The Shaping of Scientific Knowledge among Gentlemanly Specialists*. Chicago and London: University of Chicago Press.

———. 1986. The shape and meaning of earth-history. *In* Lindberg and Numbers, *God and Nature*, 296–321. [ARTICLE II]

———. 1988. A year in the life of Adam Sedgwick and company, geologists. *Archives of Natural History* 15: 243–68. [ARTICLE XI]

———. 1988. Encounters with Adam, or at least the hyaenas: nineteenth-century visual representations of the deep past. *In* Moore, *History, Humanity and Evolution*, 231–51. [ARTICLE X]

———. 1990. The emergence of a new science. *Minerva* 28: 386–97. [ARTICLE IV]

———. 1992. *Scenes from Deep Time: Early Pictorial Representations of the Prehistoric World.* Chicago and London: University of Chicago Press.

———. 1996. Minerals, strata and fossils. *In* Jardine et al., *Cultures of Natural History*, 266–86. [ARTICLE III]

———. 1996. Cuvier and Brongniart, William Smith, and the reconstruction of geohistory. *Earth Sciences History* 15: 25–36. [ARTICLE VII]

———. 1997. Smith, Cuvier et Brongniart, et la reconstruction de la géohistoire. *In* Gohau and Laurent, *De la Géologie à son Histoire*, 119–28.

———. 1997. *Recherches sur les Ossements Fossiles*: Georges Cuvier et la collecte des alliés internationaux. *In* Blanckaert et al., *Le Muséum au Premier Siècle*, 591–606. [text in English, ARTICLE VIII]

———. 1997. *Georges Cuvier, Fossil Bones, and Geological Catastrophes: New Translations and Interpretations of the Primary Texts.* Chicago and London: University of Chicago Press.

———. 1999. Geologists' time: a brief history. *In* Lippincott, *The Story of Time*, 250–53. [ARTICLE I]

———. 2000. Georges Cuvier's paper museum of fossil bones. *Archives of Natural History* 27: 51–68. [ARTICLE IX]

———. 2001. Jean André de Luc and nature's chronology. *In* Lewis and Knell, *The Age of the Earth*, 51–60. [ARTICLE VI]

Ullmann-Margalit, Edna (ed.). 1985. *The Kaleidoscope of Science.* Dordrecht and Boston: D. Reidel.

ACKNOWLEDGEMENTS

Grateful acknowledgement is made to the following persons, institutions and publishers for their kind permission to reproduce the papers included in this volume: University of California Press, Berkeley, California (article II); Cambridge University Press, Cambridge (articles III, X); Kluwer Academic Publishers BV, Dordrecht (articles IV, XIII); Science History Publications Ltd, Cambridge (articles V, XIV); Geological Society Publishing House, Bath (article VI); *Earth Sciences History*, Morgantown, West Virginia (article VII); Society for the History of Natural History, London (articles IX, XI).

I

Geologists' Time: A Brief History

'Time we may comprehend; 'tis but five days older than ourselves'

Sir Thomas Browne

In explaining that the Earth was just five days older than man, Sir Thomas Browne expressed a belief that would have been taken for granted in the seventeenth century, even by the best informed. The universe – and time itself – had, so far, lasted no more than a few thousand years; and apart from a brief prelude to set the scene for them, human beings had been on stage throughout. James Ussher, Browne's older contemporary, claimed that the moment at which time began could be dated precisely, to a specific Saturday evening in the autumn of the year 4004 BC. Far from being scolded as an obscurantist or derided for his naivety, the scholarly Irish archbishop was admired for giving quantitative precision to the accepted account of the world's history. Other and equally learned 'chronologers' disputed Ussher's particular date, but they all shared the scholarly methods by which he had reached it. A short timescale for the universe, for the Earth and for human beings was simply taken for granted.

In the modern world a quite different time scale is equally taken for granted, at least among scientists. Astronomers and cosmologists date the 'big bang' in terms of billions of years, and do so as nonchalantly as they talk of the literally inconceivable distances to the furthest galaxies. Geologists and other Earth scientists deal with time in almost equally vast quantities, but they also date a plethora of past events with a casual confidence and a sense of relative precision that recall Ussher's efforts. Multicellular life first exploded into diversity in the early Cambrian seas of 540 million years ago; the mass extinction at the 'K/T boundary' (probably the result of an asteroid's impact), took place 65 million years ago; the last retreat of the Pleistocene ice-sheets happened swiftly some 10,000 years ago. Such dates, progressively refined and with known margins of error, are the everyday currency of twentieth-century geologists.

Ussher and his colleagues practised the seventeenth-century science of chronology; modern geologists practise the twentieth-century science of geochronology. The similarity of terms points to shared concepts and even methods. Both groups have been at the forefront of intellectual life in their respective centuries.

In fact, far from being diametrically opposed, what Ussher and other chronologers were trying to do was the direct lineal ancestor of what Earth scientists do in the modern world. This essay describes very briefly how the conclusions of the first group evolved, by the kind of learning process that characterizes all scientific work, into those of the second.

Ussher, and many other chronologers all across Europe, aimed at correlating the calendars of all civilizations into a single universal time-line for world history. Since the correlations were often uncertain, the so-called 'Julian Period' was widely used as the standard of reference. This deliberately artificial timeline, which was independent of any particular culture or religious tradition, was chosen because its starting point was further back in the past than anyone imagined Creation to have been. Ussher, for example, dated the beginning of time at 710 Julian, which meant that any earlier dates had been as it were in 'virtual' time. Although its use did not outlast the seventeenth century, the Julian Period was a crucial innovation, because it established the idea of a 'neutral' time-line that stood, as it were, outside the flux of historical events and the diversity of human cultures.

On to this time-line, chronologers tried to plot the important events in the histories of all the ancient cultures they knew: Jewish, Greek and Roman, but also Egyptian, Indian and Chinese cultures (as far as they were understood at the time). In cases of inconsistency, it is hardly surprising – given the cultural basis of Christendom – that biblical data tended to be privileged above others: Ussher, for example, dismissed as 'mythical' the ancient Egyptian claims to dynasties extending far back before his date for Creation itself. However, such debates obliged chronologers to judge all the records in terms of their relative reliability and, for the first time, to distinguish such categories as 'myth' and 'legend'. So, in effect, the practice of textual criticism undermined any naive literalism about any of the sources – even the biblical texts.

The chronologers were always frustrated by the paucity of reliable records for the earliest periods. Traced backwards, most histories ran out into the mythical times of heroes and demigods relatively quickly. Even the biblical texts became disappointingly thin, with little more than genealogical lists of who begat whom. So when, about the same time as the end of Ussher's life, naturalists began to draw attention to fossils – many of which were easy to recognise as organic in origin – these natural objects were soon recruited as evidence that could supplement the textual sources for chronology. Fossils were referred to as 'witnesses' to events for which the textual evidence was scanty or obscure. For example, fossils that looked like living marine shellfish were found widely on dry land, far from the sea, so they were treated as nature's 'documents'. They were evidence of the greater extent of the sea at some remote period. In the 1660s, Robert Hooke, the employee of the new Royal Society in London, even suggested that it might be possible to 'raise a Chronology' from fossils – but he meant that they could supplement textual evidence by throwing light on the obscure earliest periods of history. Neither he nor his contemporaries had any clear notion that fossils might be witnesses of a history dating from before any written records whatever.

Naturally enough, fossil shells were initially assumed to be evidence for the one and only recorded event that seemed large enough to account for them: Noah's Flood, which was thought to be matched by similar stories in other ancient cultures. Yet the more closely fossils were studied, the less plausible this explanation became, on any literal reading of Genesis. Fossils often seemed too well preserved to have been swept hundreds of miles in a turbulent flood. Moreover they were found in thick piles of finely layered rocks, which could hardly have been deposited in any brief event. Around the end of the seventeenth century the London physician John Woodward, who later endowed a chair at Cambridge to promote the study of fossils, argued that the piles of rock, with their fossils, had been deposited out of a kind of global soup, in order of their specific gravity. But this entailed modifying the Flood story out of all recognition, and later naturalists quietly dropped the equation with Genesis. They retained, however, the idea that the pile of rocks could be read as the record of a sequence of events, or, in effect, as a history based on natural evidence.

In the early eighteenth century, naturalists came to suspect increasingly that most of the rocks and their fossils must date from long before the few millennia documented by the chronologers. At the same time orthodox beliefs in the created status of everything in the world, from atoms to human beings, were challenged by 'deistic' or even atheistic claims that the world – including its human inhabitants – was uncreated and eternal. Such 'eternalist' views, which reached back to Aristotle, were the often covert rival to the traditional short time scale. Yet these rivals shared one crucial assumption, which distances both from any modern conception of cosmic time and history. Both took it for granted that human beings had been present throughout. A world without a human presence seemed meaningless, indeed literally unthinkable, to those on both sides of the argument.

What was novel in the eighteenth century, therefore, was the growing suspicion, among naturalists who studied rocks and fossils, that the world's timescale might be far longer than the work of the chronologers suggested – but not that it was eternal – and that most of this long history might have been non-human, with the whole of human history crammed into just its final phase. These hunches – at first they were no more than that – were, above all, the product of an increasing focus towards studying rocks and fossils in the field, and not merely indoors in museums.

However, the very long but finite time scale involved could not be quantified with any confidence. The leading French naturalist Georges Leclerc, comte de Buffon, tried to put figures on it, by extrapolating the results of his experiments on the cooling of model globes. In the 1770s, he calculated that some 74,000 years had elapsed since the Earth originated as an incandescent body thrown off from the Sun. But any such figure depended, of course, on the validity of that theory of the Earth's origin, which, at that point, was deemed both speculative and controversial. In any event, Buffon himself suspected his figure was much too low to account for the piles of rock formations and, privately, he thought some three million years was more likely.

Although such a figure may seem absurdly inadequate to modern geologists, even a million years entailed a huge stretching of the imagination, to spans of time that were as inconceivable – literally – as the cosmic distances estimated by astronomers around the same time. When, in the 1780s, the Scottish philosopher James Hutton claimed that the earth had 'no vestige of a beginning, no prospect of an end', it was his blatant eternalism that drew criticism. His implicitly vast sense of time was by then almost a commonplace among naturalists, even if it was still unfamiliar to the wider public.

Ironically, it was the story in Genesis that provided the conceptual model for enlarging human history into a far longer 'geohistory', to use the modern term. In the 1770s, Buffon defined seven successive 'epochs' or significant moments in the Earth's history. In doing this, he offered, in effect, an updated and secularized version of the seven 'days' of the Creation story. By defining his last epoch as the first appearance of human beings – and no longer as God's Sabbath rest! – he made explicit what other naturalists already suspected: that most of geohistory had been pre-human history. This conclusion seemed to be reinforced by the continuing failure to find any signs of human life, either bones or artefacts, in any but the most recent deposits.

More specifically, and again ironically, the idea of calibrating nature's history with a quantified chronometry arose from a concern to defend the historicity of the biblical Flood, by using natural evidence to confirm the conclusions of the chronologers. The Genevan naturalist Jean-André de Luc claimed on the basis of his fieldwork that there had been a drastic change in geography in the infancy of the human race – the present continents had risen from the ocean floor and the former continents had sunk below the waves. He matched this event with a rather loose interpretation of the story in Genesis, and he tried to date it by extrapolating the known rates of various observable natural processes back into the past. By the 1790s, he was calling these processes 'nature's chronometers', deliberately recalling John Harrison's recent invention, the great 'high-tech' achievement of the century.

River deltas, for example, were growing at rates that could be estimated from historical records. As they were of finite size, they could be used to calculate the approximate date at which they had originated. De Luc concluded, from several independent 'chronometers', that nature's great 'revolution' had happened only a few thousand years ago. The physical evidence was, therefore, compatible with the textual evidence assembled by chronologers. (Modern geologists would recognize much of de Luc's evidence as marking the end of the last glacial period, indeed only a few thousand years ago.)

In the early nineteenth century, geologists – as they may now be called without anachronism – assumed, like de Luc, that the Earth's still earlier history was far longer, but unquantifiable. Leaving the magnitude of the timescale aside, they therefore concentrated on clarifying the sequences of rock formations. The discovery by the English mineral surveyor William Smith that some formations had what he called 'characteristic' fossils, by which they could be traced across wide tracts of country, greatly aided this new study of the sequences of rock formations – or stratigraphy, as it was later called. But it was only gradually that geologists, unlike Smith himself, began routinely to treat such sequences of formations as evidence for geohistory, and it was still longer before they revived earlier attempts to quantify its timescale.

In the 1830s, for example, the London geologist Charles Lyell tried to quantify the dates of each of the more recent sets of rock formations by calculating the percentages of still living molluscan species found among their respective fossil shells. But this method depended on the validity of his theory that faunas and floras have changed continuously over time, by the appearance of new species and the extinction of old ones at a statistically uniform rate. Lyell's contemporary John Phillips pursued a more fruitful approach, exploiting the new results of stratigraphical research in many parts of the world. In the 1840s, he tried to estimate the maximum thickness of sediments deposited in each of what were now defined as successive 'periods' of geohistory, and then to match them with current rates of deposition, as far as those could be measured. Although, like Lyell's method, this depended on an assumption of uniform rates through time, it yielded figures that most geologists found much more plausible: a total of some 100 million years for the whole known fossil record, starting in what had recently been defined as the 'Cambrian' period.

In the 1860s, the Scottish physicist William Thomson (much later ennobled as Lord Kelvin) asserted, on the quite independent grounds of the Sun's supposed rate of cooling, that the Earth's solid crust was no more than 98 million years old; and he put this forward with all the arrogance of a physicist lording it over mere geologists (the precision of his figure was spurious, since he conceded a margin of error ranging from 20 to 400). In fact, however, most geologists were initially content with Kelvin's estimate, which matched their own quite well. But it was incompatible with the far longer timescale demanded both by Lyell's extreme version of uniformity and by Darwin's new theory of evolution by natural selection. By the end of the nineteenth century, however, Kelvin and his followers had refined their calculations and tightened the screws on the geologists. Their new estimates, allowing only some 20 or even merely 10 million years for the whole record of the rocks, were resisted not only by the followers of Lyell and Darwin, but by a much broader range of geologists. They argued that the physicists must have made some mistake, because the geological evidence, however hard it might be to quantify, seemed to demand substantially more time. They were reassured, for example, when the Irish physicist John Joly offered an estimate of about 90 million years – not far from Phillips's figure – based on the inferred rate of accumulation of salt in the world's oceans.

The discovery of radioactivity at the turn of the century upset both apple carts. Since this strange phenomenon occurred in rocks, its potential as a new kind of 'natural chronometer' was quickly appreciated: once the rates of radioactive decay had been measured, analysis of its products could, in principle, yield quantitative ages for rocks. Even the early results, uncertain though they were, suggested a timescale far in excess of what geologists expected and, of course, still further in excess of what Kelvin and his colleagues had allowed. The geologists, having been bitten once by the physicists, declared themselves twice shy: if the physicists could swing suddenly from one extreme to another, the geologists thought it prudent to be sceptical about both.

After the First World War, however, geologists gradually came to appreciate that their own distinctive evidence did, in fact, support the longer timescale now offered by the physicists. This reconciliation was greatly facilitated by the emergence of a new breed of scientist who, unlike Kelvin, understood geology as well as physics. The Englishman Arthur Holmes, for example, summarized the research on radioactivity and the age of the earth during the second decade of this century and, after the end of the First World War, Holmes became a leading advocate of the new synthesis. By the late 1930s, most scientists were agreed that the Earth must be at least a couple of billion years old.

More unexpectedly, it had also become apparent that the entire fossil record, from the Cambrian period to the present, was only a small fraction of geohistory. The 'Precambrian', in which almost no fossils had been found, was no longer treated tacitly as a relatively brief prelude, and became instead the bulk of the story. This new perspective was just the first sign of a radical change in the character of what was now called 'geochronology'. Instead of merely trying to estimate the total age of the Earth, geologists were beginning to use the new 'radiometric' methods for truly geochronological purposes: to put dates (however approximate they might be) on many successive events in geohistory and to calibrate the stratigraphical sequence into quantified periods of time. During the years following the Second World War, a crucial technical development made this far more reliable: the mass spectrometer greatly accelerated the conversion of rock samples into accurate and consistent dates.

By the end of the twentieth century, the radiometric time scale had come to be taken for granted by geologists and other Earth scientists. It provided a dimension of time that lay, as it were, outside the flux of terrestrial events, thereby allowing them to be dated with a steadily increasing confidence and precision. This geochronology offered a framework for the more interpretative work of under-standing the causal relations of events and their roles in geohistory. Geohistory had turned out to be unimaginably lengthy and complex. The history of our species had been reduced to a brief final phase, while the history recorded in textual records had become an even smaller sliver of time. Yet the achievements of modern geochronology should not blind us to the continuity that links it back to the more modest history offered three or four centuries ago by the science of chronology. The timescale is now almost inconceivably greater in magnitude, but the ideal of a quantified history and the underlying passion for precision remain the same. At the start of a third Christian millennium – a date first established by the old chronologers – it is right and proper to acknowledge that the spirit of Ussher and his colleagues is alive and well among modern geologists.

II

The Shape and Meaning
of Earth History

Archbishop James Ussher, a scholarly Irishman of the seventeenth century, has become unjustly famous as the supposed author of the claim that the creation of the world could be dated to the year 4004 B.C. More than two centuries later, when the classic histories of science and religion were being written in terms of warlike conflicts,[1] similar claims were still being made by some culturally conservative groups in Western societies. Today, routine estimates of the earth's antiquity in terms of several billion years, made by geologists on the basis of radioactivity in rocks, contrast strikingly with the assertions of the small but vocal group of creationists, who retain something like Ussher's calculation in their body of alternative knowledge. It is not surprising, therefore, that this piece of history is still commonly seen as a story of continuous conflict between Christianity and science, in which Christians—with the exception of the so-called fundamentalists—have compromised their traditional beliefs in the face of the triumphant march of scientific knowledge.

This kind of scientific triumphalism is long overdue for critical reappraisal. Its claims to serious attention have been thoroughly demolished in other areas of the history of science, but it survives as an anomaly in the historical treatment of the relation of science to religious belief.[2] This may be because the historians' own attitudes are conditioned by the immature age at which religious beliefs and practices are abandoned by many, though not all, intellectuals in modern Western societies. This common experience may explain why many historians of science seem incapable of giving the religious beliefs of past cultures the same intelligent and empathic respect that they now routinely accord to even the strangest scientific beliefs of the past. Traditional interpretations in terms of conflict and compromise do more, however, than fail to treat religious beliefs seriously. These well-worn categories also encourage the reification of science and religion into contrasting bundles of abstract propositions. The crucial third term, society, is either ignored[3] or else invoked in the most naive form, the contents of religion and science being attributed to two similarly polarized social groups, namely religious believers and scientists.

In contrast with such simplistic treatments I suggest that specific episodes of conflict should be regarded as stories about the interaction of rival *cosmologies*—using that term, as I shall throughout this essay, in its anthropological sense.[4] In other words, they are episodes in which people on both sides appealed to some aspect of nature, such as the origin and history of the earth, in order to support and justify their attempts to propagate their own view of the *meaning* of personal and social life and of the conduct appropriate to that life, whether that meaning was formulated in religious terms or not. In historical studies there should be no place for sweeping generalizations that contrast the progressive outlook of scientists with the reactionary attitudes of Christians. We must expect to find that scientists of varying degrees of originality and competence and Christians of varying degrees of insight and orthodoxy were prominent on *both* sides of many controversies. Furthermore, there is the obvious fact that one and the same individual has frequently been both Christian and scientist (a term I use for convenience, although it is highly anachronistic before the mid-nineteenth century).

In the history of controversies about the age of the earth, few modern scholarly studies of particular episodes take the religious dimension seriously and at the same time give full attention to the social uses to which rival viewpoints were put. What I shall do in this brief essay, therefore, is simply to sketch the main outlines of the changing interpretations of earth history since the Middle Ages and to suggest how rival religious and social meanings have been expressed through those interpretations. In particular, I shall argue that the quantitative figures that have been given at various times for the age of the earth are far less significant than the qualitative patterns that have been discerned in, or attributed to, the whole history of the earth, of life, and of mankind (see fig. 12.1). I shall first summarize the tradition of biblical chronology that formed the temporal component of the geocentric picture of the cosmos. Second, I shall describe the new enterprise that was termed the "theory of the earth" and its relation to the rise of critical methods in biblical interpretation. And

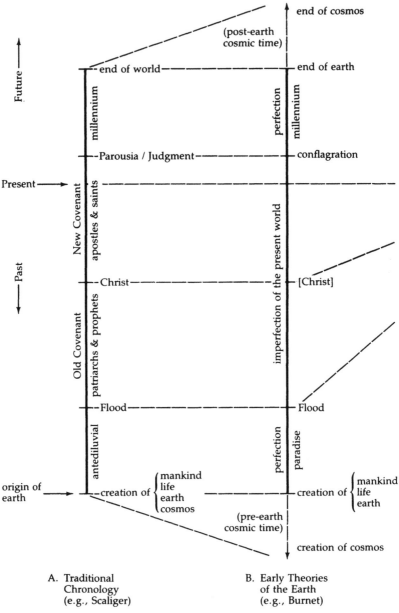

A. Traditional
Chronology
(e.g., Scaliger)

B. Early Theories
of the Earth
(e.g., Burnet)

Fig. 12.1. Earth histories of the mind. Diagrammatic representations of four successive ways in which the shape of earth history and its most significant events were conceived. Time is shown flowing upward (as in classical geological diagrams). *Quantitative* estimates or calculations of the magnitudes of time involved are deliberately ignored; the span of time from the origin of

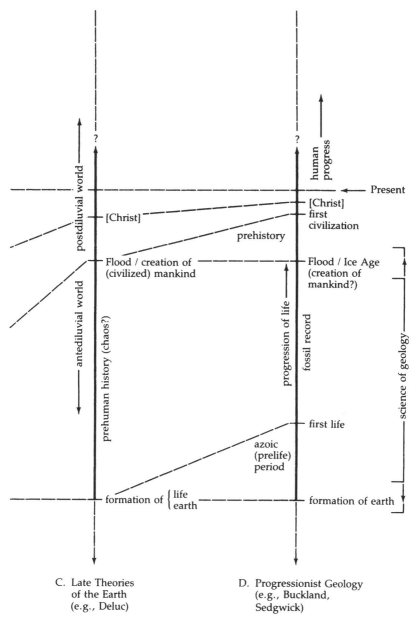

C. Late Theories
   of the Earth
   (e.g., Deluc)

D. Progressionist Geology
   (e.g., Buckland,
   Sedgwick)

the earth to the present is represented (arbitrarily) as uniform in all four diagrams, in order to highlight the qualitative changes in the conceptualized pattern of events. Note that eternalistic conceptions of earth history cannot be adequately depicted in this kind of visual representation.

third, I shall outline the way that a new science of geology, with deliberately limited cognitive goals, pushed attributions of cosmological meanings out of the scientific study of the earth and relegated them to the social and cognitive margins of science.

THE GEOCENTRIC COSMOS

The spatial aspect of the picture of the cosmos culturally dominant around the end of the Middle Ages in the West is comparatively familiar. The universe was conceived as a bounded system, with the earth lying immobile at the center of an ascending hierarchy of ceaselessly revolving celestial spheres. The abode of mankind was central in position, but furthest from the divine presence and least in glory. This image embodied and expressed a cosmology that related mankind both to the environment and to the transcendence of God. It also reflected, and justified as natural, an analogous social order of stable hierarchical equilibrium. The temporal pattern that would have matched this spatial structure was one of merely cyclic change, without truly historical development. Up to a point this image of cyclic change was indeed incorporated into the personal and social life of Christendom, as for example in the public celebration of the cycle of the church's year. But a concept of merely cyclic change could all too easily be extrapolated into an image of a cosmos that was eternal, self-maintaining, and—above all—uncreated. This eternalism, which, like the accepted spatial picture, derived principally from Aristotle, was therefore suspect to those who saw themselves as the guardians of the order of nature and society. Conversely, an eternalistic picture of the cosmos was a valuable resource in the hands of any critics of that social order, the more so if the spatial bounds of the conventional cosmos were dissolved into boundless infinity.

Those who sought to maintain the stable order of society, as justified by the order of nature, therefore united a limited acknowledgment of cyclic change with an affirmation of the finite spatial limits of the cosmos and the underlying directional pattern of its history. In this image of world history the cosmos was bounded as much temporally as spatially: it had a clear beginning in Creation, and God's action would bring it to an equally decisive End. It was also structured as much temporally as spatially within these limits (see fig. 1A). It had a unique midpoint in the events surrounding the life of Christ, which divided the old relationship or covenant between mankind and God from the new. On either side of that divide, the particularities of patriarchs and prophets, apostles and saints, were framed by past and future events of global significance, namely the Flood and the Parousia (or Second Coming of Christ). Beyond that frame lay only the swift prelude of the Creation, Paradise, and the Fall in the past, and the expected culmination of the final Judgment and (more controversially) the Millennium in the future. This whole temporal structure imbued the past, present, and future of the cosmos with human meaning: it attributed order to the often chaotic flux of human lives and gave social action a transcendent context and justification.

This picture of earth history was derived, of course, principally from the Bible, the diverse components of which were interpreted in terms of an underlying unity of narrative history. Spiritual methods of interpretation, such as allegory and typology, were built on a substrate of literal historical meaning. The narrative story discerned in the Bible was not simply a religious way of looking at the history of the world. For most members of Western societies it *was* the history of the world, at least in outline. Other secular events, above all the life of society and its constituent persons, received meaning by being seen in their appropriate place within the narrative structure. The brief narratives recorded in the opening passages of Genesis, the first book of the Bible, posed few problems of credibility. It seemed plausible to regard human history as virtually coextensive with earth history, and that with cosmic history; without mankind the earth and the cosmos would have seemed to lack meaning and purpose. The closely integrated Creation of cosmos, earth, and life, the swift sequence of Paradise and the Fall, and the great global action of the Flood together formed an intelligible prelude to the main historical narrative of redemption embodied in the Bible. As Sir Thomas Browne put it, "time we may comprehend; 'tis but five days older than ourselves" (*Religio Medici*, 1635).

CHRONICLES OF THE WORLD

The construction of a single narrative story of cosmic history received new impetus during the period of the Renaissance and Reformation. Improved methods of textual scholarship were used to eliminate apparent discrepancies within the Bible itself. The interest of humanist scholars in the texts and monuments of classical antiquity, later broadened to include those of other ancient civilizations, provided a new wealth of historical information that needed to be integrated into the biblical narrative. A tradition of scholarly chronology grew up, in which that narrative was given quantitative precision by the dating of its events. The scattered calendrical information in the Bible was collated and compared with, for example, astronomical calculations on historically recorded eclipses and conjunctions. The resultant outline of datable history was then enriched by fitting into it whatever was known about the history of the world from nonbiblical sources. The result of this scholarly activity was the production of an outline of world history that had pretensions to being universal and was no longer confined to biblical sources; an example of this genre is J. J. Scaliger's *De Emendatione Temporum* (1583). At first there was little to suggest any marked deviation from traditional estimates of the scale of cosmic history, even back to the Creation. All reliable calculations, despite minor differences of scholarly opinion, could be plotted without strain within a few thousand years.[5] Ussher's famous figure of 4004 B.C., which he did not originate, was only one of many rival scholarly estimates of this kind; it has become the best known—in the English-speaking world—only through its adoption in the marginal notes of some editions of the seventeenth-century King James translation of the Bible.

This scholarly consensus was disturbed, however, by ancient records of Egyptian dynasties that implied a higher antiquity; and these were followed in the seventeenth century by reports of alleged records of Chinese civilization that pushed the history of mankind still further back in time. Moreover, problems of another kind emerged in the wake of European exploration of other continents. It became increasingly difficult to integrate what was becoming known about the distribution of exotic animals and plants—and, above all, of human beings—into the brief narrative framework of the Bible. It was not clear, for example, how the descendants of Noah and of the animal inmates of the ark could have had time to repopulate all the scattered continents after the Flood. Such problems gave rise to an extensive scholarly literature in the sixteenth and seventeenth centuries.[6] Its significance lies in the fact that these problems were used as resources for *either* the defense *or* the criticism of the established view of a temporally finite world. The consensus among scholars held that documented human history could not be carried back in any civilization more than a few thousand years; texts that implied a higher antiquity were regarded as either legendary or fraudulent. Those who claimed on the contrary that such documents were reliable used them to attack the authenticity of the biblical narrative in the interests of alternative, and generally anti-Christian, interpretations of the world. In other words, speculations about a vast antiquity for mankind, about the possible existence of human beings before Adam, and about the inadequacy of the ark as an explanation of organic distribution were never just enlightened and disinterested scholarly inquiries. They were put forward to support specific cosmologies, as were the more literalistic interpretations to which they were opposed.

It would be wholly anachronistic to contrast the opinions of scholarly chronologists with those of contemporary scientists. Natural philosophers, as the latter generally called themselves, also worked within the taken-for-granted assumptions of their societies. For example, even the recognition of fossils as organic remains was highly problematical; the most obviously organic fossils were precisely those that could most readily be attributed to a single large-scale event.[7] The most appropriate recorded event was of course the Flood, the historicity of which seemed to be confirmed by analogous stories in other ancient literatures. For example, the Danish scholar Niels Stensen (1638–1686), better known as Steno, interpreted the rock structure of Tuscany in Italy in terms of a temporal sequence of events; but he inferred that the fossil-bearing strata must have been deposited during the Flood, and he thought they could not be much older than the monuments of the ancient Etruscan civilization that he could see in the same region. Around the same time Robert Hooke (1635–1703) in England argued likewise for the organic origin of many fossils, but he assumed that they represented organisms that would have been almost contemporary with early mankind. When, in a famous phrase, he suggested that it might be possible to "raise a chronology out of them," he meant that fossils might be used to supplement the coins and other artifacts used by antiquarians, thus amplifying the chronology of human history by charting the parallel history of the animate world.[8] Seventeenth-century naturalists such as Steno and Hooke envisaged only dimly, if at all, the possibility of a long *prehuman* history of the earth.

It is easy in retrospect to pour scorn on the tradition of chronology in biblical scholarship and the related work of naturalists such as Steno and Hooke. But its literalism was a simple consequence of the precritical approach to biblical texts, and there was little nonbiblical evidence that threw any serious doubt on the short time scale that it proposed. Chronologists were not responsible for establishing the belief in a short cosmic history; their work merely codified and lent an air of precision to an already-taken-for-granted view of the natural world. What is most significant about that view is not its short timescale by modern standards, but rather its almost unexamined assumption that the history of the cosmos must be virtually coextensive with the history of mankind—indeed, the history of civilized literate mankind. In this way the culturally dominant Christian cosmology, although in principle centered on God, was in practice centered more on mankind. The history of the earth was seen only as the stage for the drama of human history, the drama of the creation, fall, and redemption of a unique set of rational beings.

NEW MODELS OF EARTH HISTORY

Within the traditional temporal image of the cosmos, the origin of the earth could only be regarded as an integral aspect of the creation of the cosmos as a whole. What was religiously important about that primal creation was the assertion, or alternatively the denial, that it had indeed been the work of a transcendent Creator. The origin of the earth could not become an object of inquiry on any other level until the earth could be seen as a part of the cosmos that was not wholly unique. But even the slow acceptance in the seventeenth century of Copernicus's heliostatic system of the cosmos did not have that effect, since it merely altered the position of the earth within a bounded cosmos similar in structure to the earlier geostatic system. A far more radical alternative, however, had continued a submerged existence from earlier centuries. This was the image of a cosmos without center or boundaries, a cosmos that was infinite, eternal, and possibly even uncreated. Giordano Bruno (1548–1600) described such a cosmos in the late sixteenth century, and in the mid-seventeenth century René Descartes (1596–1650) tried to rehabilitate that suspect image. In view of the recent condemnation of Galileo's work, the French philosopher felt obliged to keep a "low profile" in relation to Catholic authority and its localized secular power, and so he framed his speculations in an ambiguously hypothetical style. In *Principiae Philosophiae* (1644) he sketched a possible model of a universe of indefinite limits, within which the earth, and perhaps countless similar bodies, could have had their own origins and histories. He outlined a possible physical history of this hypothetical earth, suggesting natural explanations of its origin and of its main surface features. This account, which left aside the question of the primal creation of the cosmos, was quite compatible with the brief earth history sanctioned both by tradition and by most of the nonbiblical evidence. In this way Descartes effectively detached earth history from cosmic history (see fig. 1B). Following this new model, it became possible to conceive that the history embodied in the biblical narrative, which all agreed had been written primarily for the salvation of mankind, referred to terrestrial events directly relevant to mankind and not to the cosmos as a whole.

II

The implications of Descartes's model were first articulated by Thomas Burnet (ca. 1635–1715), a Cartesian scholar living in England, which was much freer from intellectual restrictions than absolutist Catholic states. In his *Sacred Theory of the Earth* (1680–1689) Burnet tried explicitly to discover the "true" interpretation of the biblical records of the great physical events of earth history by drawing on the explanatory principles of Cartesian natural philosophy. He was not reconciling Scripture with natural knowledge, for he saw no conflict there; rather, he was using natural knowledge to amplify and illuminate the biblical narrative, which he restated in physical terms (see fig. 1B). The present irregular and "imperfect" state of the earth was framed in the past by the Flood and in the future by the "Conflagration" (or burning of the whole earth) that was widely expected to bring the present order to an end.[9] Beyond those twin catastrophes were the matching "perfections" of Paradise and the Millennium, which in turn were bracketed by the Creation and the final End. The whole symmetrical scheme lay under the sovereignty of Christ as the "alpha and omega" of history, a traditional attribution that now referred only to the earth, not to the cosmos. Burnet conceived the brief and finite history of the earth as being flanked by vast "oceans" of past and future cosmic time. But in asserting the finitude of the earth he explicitly refuted the unnamed eternalists who claimed that the earth and mankind had not been created by the divine will at all.

A PROLIFERATION OF THEORIES

Burnet's work was the prototype of a new kind of writing that became known as the "theory of the earth."[10] This proved to be a highly flexible conceptual resource, in the sense that grand speculative theorizing about the shape of earth history could be, and was, used to support and justify highly diverse cosmologies. In the later seventeenth century, theories of the earth proliferated in great variety, particularly in England, where Burnet's Cartesianism was soon replaced by various aspects of the newer Newtonian natural philosophy. For example, in *A New Theory of the Earth* (1696) the Newtonian William Whiston (1667–1752) used comets to give what he claimed were better explanations of the great physical events of earth history. Above all, the older idea that the earth had decayed from an original perfection gave way to a new emphasis on the way that the harmonious equilibrium of the present earth had been derived from original chaos.[11] Of all such theories, perhaps the most influential was that of John Woodward (1667–1728), who made use of his substantial firsthand knowledge of strata and fossils. In his *Essay towards a Natural History of the Earth* (1695) Woodward claimed that the whole sequence of strata had settled in order of specific gravity out of a chaotic global mixture at the time of the Flood. He thought this would explain the tendency for specific fossils to be embedded in particular strata. This example makes clear the way in which, in this kind of speculation, physical interpretations of events such as the Flood generally diverged far from any literal interpretation of the biblical narratives, even *before* the impact of critical methods of biblical interpretation began to be felt. The basic historicity of the narratives was not necessarily questioned; but it was assumed that behind the conventional or "vulgar" interpretation lay a true or "philosophical" interpretation, which could be discovered by the light of the latest natural philosophy. This attitude was characteristic, for example, of the latitudinarian churchmen in England, who sought to make Christian beliefs acceptable to the enlightened intellects of their age.

It is easy to see how this conception of a privileged interpretation, dependent on scholarly or philosophical knowledge, could lead to a total inversion of the traditional task of biblical interpretation. Traditionally, nonbiblical sources, whether natural or historical, had received their true meaning by being fitted into the unitary narrative of the Bible. This relationship now began to be reversed: the biblical narrative, it was now claimed, received its true meaning by being fitted, on the authority of self-styled experts, into a framework of nonbiblical knowledge. In this way the cognitive plausibility and religious meaning of the biblical narrative could only be maintained in a form that was constrained increasingly by nonbiblical considerations.

The same inverted relationship could equally well be used, however, to promote radically anti-Christian cosmologies. If the biblical narrative were to be interpreted in the light of nonbiblical resources, it could be claimed that it had no validity whatever, except as a record of the superstitions of an unenlightened primitive people. This was the strategy characteristically adopted by the early Enlightenment *philosophes* in France. They generally used the theory of the earth as a means with which to attack Christian orthodoxy and the secular cultural power that it still wielded in many of the Catholic states. For example, Voltaire (1694–1778), in order to remove any evidential foothold for the historicity of the Flood, went so far as to deny the natural emplacement of fossils altogether. He failed to win support for that view, however, since anyone who had observed fossils in strata could see at once that it was untenable. Later theorists in this tradition therefore focused their attention on denying the validity of the supposed evidence for the Flood, claiming, for example, that it had been merely local or that, if universal, it had mysteriously failed to leave any traces and therefore could be ignored for all scientific purposes.

Another line of argument with similar cosmological goals involved reviving the older eternalistic model of the cosmos and applying it to the earth. The earth, it was claimed, had always been and would always be under the dominion of the same purely natural laws. Its history stretched indefinitely or even infinitely into past and future and involved no unique and unexplained events such as the Flood; indeed, earth history was "without vestige of a beginning, without prospect of an end." That famous phrase appeared in James Hutton's *Theory of the Earth* (1795), but the sentiment had often been foreshadowed earlier, particularly in continental Europe, in such works as the Baron d'Holbach's *Système de la nature* (1770). Writers such as Holbach and Hutton generalized the present relative stability of the earth into a permanent feature of the terrestrial system, past and future. This was often given meaning in deistic terms by being attributed to a wise and providential design directed toward the permanent well-being of mankind. Most significantly, the virtual eternalism of such theories was extended, often explicitly, to the history of mankind— for example, in George Toulmin's *The Antiquity and Duration of the World* (1780). Mankind could thus be claimed as uncreated and therefore not subject to any of the traditional moral and social constraints.[12] These theories were far from anticipating modern geology, despite their casual references—for example, in Jean-Baptiste Lamarck's *Hydrogéologie* (1802)—to "millions of centuries" for the history of the earth. Such vast spans of time were invoked primarily as an essential component of eternalistic theories that had clear and generally overt cosmological goals.

PREHUMAN EARTH HISTORY

Meanwhile, however, the bulk of empirical studies of the earth was taking a different direction. Such studies were not necessarily integrated into any high-level theory of the earth. Rather, they were often directed toward more mundane and practical goals, such as the discovery of mineral resources. In such work the implausibility of Woodward's explanation of the strata soon became clear, but his emphasis on the strata as a temporal sequence was retained and enlarged. It became commonplace among eighteenth-century naturalists to distinguish the most ancient, or "Primary," rocks, having no fossils, from the regular sequences of "Secondary" fossil-bearing strata; these were distinguished in turn from irregular and patchy superficial deposits containing the bones of exotic animals such as elephants. The record of the biblical Flood or Deluge was now identified only with these relatively recent deposits. Admittedly they did not suggest the kind of episode recorded in Genesis, and they contained no human remains; but they did seem to be the product of some exceptional catastrophe, and they were difficult to explain away in terms of ordinary observable processes. Since these superficial deposits were identified as "diluvial" (that is, dating from the Flood), the regular sequence of Secondary strata was necessarily described as "antediluvial" or pre-Flood. Since it too contained no human remains, it gradually came to be accepted as a record of *prehuman* earth history (fig. 1C).

This implicit separation of the origin of the earth from the origin of mankind had profound consequences. It created a new kind of history, a history without human documents, which required new conceptual tools. These were borrowed by naturalists from human historiography.[13] Antiquarians had already concluded that "monuments"—a term that covered artifacts such as coins as well as architectural remains—were more trustworthy as historical evidence than textual documents, and in the eighteenth century this term was appropriated by naturalists. They argued that fossils and strata should be more reliable monuments than human artifacts, since Nature could hardly be suspected of historical bias or forgery. In this way, natural monuments were transformed from being merely a source of evidence to supplement the annals of human history, as they had been for Steno, into evidence for a long history that predated mankind altogether. By the late eighteenth century, fossils and strata were routinely described as monuments or as "Nature's archives"; they helped to define "epochs" of local validity, which gave greater precision to the global categories of Primary and Secondary. The naturalists who pioneered this new kind of earth history rarely attempted to quantify its time scale. On occasion this may have been a matter of prudence, to avoid disturbing readers who still took the traditional biblical chronology for granted.[14] But a more general reason was the scarcity of evidence to go on, beyond the impression given by thick piles of strata that seemed to have been deposited slowly on a former seafloor. Nonetheless, such evidence was sufficiently compelling by the late eighteenth century for many naturalists to assume implicitly—and sometimes explicitly, if only in passing—that prehuman earth history must be reckoned at least in tens or hundreds of thousands of years.

This view of a prehuman earth history of inconceivable duration was seized upon for diverse cosmological purposes. For example, the posthumous publication of Gottfried Wilhelm von Leibniz's theory of the earth *Protogaea* (1749), which had originally been intended as a prelude to a conventional human history of his patron's territories, provided a useful model for interpretations of earth history that stressed natural causes but were not eternalistic.[15] Thus the French naturalist Georges Louis Leclerc, Comte de Buffon (1707–1788), abandoned the virtual eternalism of his earlier theorizing and worked out a directional model based on analogical experiments with cooling globes. This led him to date the successive "epochs of nature," past and future, with an impressive air of precision; in his *Epoques de la nature* (1778) he gave a total of about a hundred thousand years.[16]

Although Buffon's work could seem like a secular parody of the Creation narrative, other naturalists sought to harness recent empirical results to the cause of defending the plausibility of the traditional Christian viewpoint. For example, the Swiss naturalist Jean-André Deluc (1727–1817) conceded the vast time-scale of prehuman earth history; and, even though he regarded it as peripheral to the main issues of religious concern, he absorbed it into the biblical narrative at least to his own satisfaction. He criticized eternalistic theorists like Hutton, above all for implicitly denying the biblical account of the origin and early history of mankind. In *Lettres physiques et morales* (1779) he focused his attention on the detailed evidence of "natural chronometers" derived from the observable rates of natural processes, believing these indicated that an exceptional physical disturbance— which he identified with the biblical Flood—had affected many areas of Europe only a few thousand years ago, turning seafloors into land areas. In effect, Deluc believed that in the study of the earth the issues of concern to Christians could be narrowed down to those affecting the creation and history of mankind. Within that narrowed limit, the tradition of biblical chronology remained in his opinion valid; outside it, for the epochs before the Flood and before mankind, a more symbolic interpretation of Genesis was quite acceptable (see fig. 1C). This was no simplistic compromise between Christianity and science; rather, Deluc, in seeking to defend traditional Christian beliefs from the skeptical attacks of openly antireligious *philosophes,* was trying to define just what areas of natural philosophy were of legitimate concern in maintaining those beliefs as valid guides to the meaning and conduct of life.

The inference that earth history as a whole had been inconceivably lengthy, even if human beings were relative newcomers to the scene, was not very widespread outside the small circles of naturalists engaged in this kind of study. Such conclusions certainly remained suspect among some conservative religious groups in Western societies; but this was at least partly because high estimates of the earth's antiquity were tainted by their long association with eternalistic cosmologies. Some social groups, however, found it possible to accept the idea of a high antiquity for the earth without rejecting the religious authority of the biblical writings. This was due above all to the emergence of the new critical school of biblical scholarship centered in the German cultural area. Unlike much earlier work of biblical criticism, this was not necessarily directed toward rationalistic or antireligious goals. Its conclusions could indeed be used to argue reductively that the Bible was now more worthless than ever. But alternatively it could be claimed that the new critical perspective gave even the "legendary" parts of the Bible their truly religious meaning. They could now be regarded as a precious record of early *religious* insights into the relation between mankind, the created world, and God. In any case, since the new methods treated the biblical texts as the products of diverse periods and cultures, the early chapters of Genesis, bearing a close relation to physical and historical events for which there was non-biblical evidence, became a natural early focus of debate, as for example in J. G. Eichhorn's *Die Urgeschichte* (1779).[17]

But even if the new biblical criticism was not used in the service of antireligious cosmologies, it altered profoundly the traditional concerns of theories of the earth, because the search for physical evidence that would confirm, or undermine, the more or less literal meaning of the opening biblical narratives became irrelevant from this perspective. The Flood could be taken to have been a purely local event, and the narrative of Creation could be regarded as a prescientific story designed primarily to express in religious terms the creaturehood of the natural world and mankind. Above all, the time scale of earth history became religiously irrelevant. For those who abandoned the traditional precritical approach to the biblical texts, and with it the idea of a total earth history of only a few thousand years, it made no difference to religious belief and practice whether naturalists estimated that history in tens of thousands or hundreds of millions of years. The only religious problem was now that of finding some human meaning in those vast spans of prehuman history.

THE NEW SCIENCE OF GEOLOGY

Toward the end of the eighteenth century many naturalists engaged in the study of the earth tried to dissociate themselves from the use of their work by rival cosmological interests. In the three main cultural areas of Europe, the preferred terms for the study of the earth became *Geognosie, géographie physique,* and (somewhat later) *geology.* Such terms marked a firm rejection of large-scale theorizing and a new emphasis on the value of cumulative observation, at a time when the economic value of mineral surveying was becoming apparent. The new empiricist rhetoric was therefore not socially neutral. It expressed an attempt to establish the study of the earth as a practical pursuit that would be free of cultural pressures from *either* side: from the traditional concerns of biblical chronologists *and* from the secularizing concerns of eternalistic theorists. Empiricism was justified at the time on the grounds that earlier theories of the earth had signally failed to lead to any clear progress in knowledge. But after the revolution in France in the late eighteenth century it was strongly reinforced by widespread political suspicion of the naturalistic speculations of the *philosophes* and their antireligious cosmological goals.[18]

The new science of geology therefore emerged in the early nineteenth century as a strongly bounded field of knowledge, conspicuous for what it did *not* contain. Geologists (as they may now at last be called without anachronism) excluded as unscientific almost all that had previously made the earth rich in cosmological meaning: the origin of the earth, its ultimate fate, and, above all, the origin and early history of mankind (see fig. 1D). Within its self-imposed frontiers, however, geology developed its cognitive heartland with great practical success. Local sequences of strata were classified in terms of a programmatically global sequence of Systems (Carboniferous, Silurian, etc.), which were accepted as the records of a sequence of corresponding Periods of literally immeasurable length. Deluc's unique, recent diluvial event was generalized into a notion that such catastrophes had been a repeated feature of the earth's history; and this was used by the French zoologist Georges Cuvier (1769–1832) and others to account for the apparently abrupt changes in fossil animals and plants between adjacent sets of strata. Far from bringing the supernatural into geology, this so-called catastrophism turned even the puzzling Flood or Deluge into one of a series of natural events, as can be seen, for example, in Léonce Elie de Beaumont's *Sur les révolutions du globe* (1829–1830). Geology, with its carefully maintained cosmological neutrality, was prized as a science in which Christians of all denominations, together with freethinkers and those who were later termed agnostics, could cooperate amicably. At least in Europe, if not in America, those geologists who regarded themselves as Christians generally accepted the new biblical criticism and therefore felt the age of the earth to be irrelevant to their religious beliefs. The more sensitive issue of the origin of mankind—and, by implication, the nature of human beings as creatures in God's image— was effectively excluded by the newly drawn boundaries of geology.

It should be added, however, that this did not seem at the time to be avoiding the issue, because human remains were conspicuously absent from even the geologically recent diluvial deposits.

In the early nineteenth century, therefore, geologists opened up with increasing self-assurance an astonishing drama of vanished worlds that had been inhabited by strange, extinct organisms. They set this drama within an assumed time-scale that dwarfed even the whole history of civilization (see fig. 1D). In popularized form this new drama of earth history was eagerly absorbed by many among the wider public. But not all were prepared to accept on trust what ran so strongly counter to both tradition and common sense.

MOSAIC GEOLOGY

In conscious opposition to the self-styled philosophical (i.e., scientific) geologists, a so-called scriptural or Mosaic geology emerged, in which the cognitive validity of the biblical narrative was reasserted in terms inherited from the biblical chronology of the sixteenth and seventeenth centuries. Mosaic geology was so called because its proponents retained the precritical view that Genesis and the rest of the Pentateuch had been written under divine inspiration by Moses himself. It flourished more in the English-speaking world, including the United States, than in continental Europe, probably because critical methods of interpretation were accepted earlier and more widely among intellectual Christians on the Continent. Mosaic geologies ranged from work by erudite, though precritical, biblical scholars down to the most unsophisticated popularizations, from theories supported by at least some empirical fieldwork down to books by those who had never studied a rock or a fossil at first hand. What united all these writers was the conviction that Genesis, if rightly interpreted, embodied an authoritative narrative account of the origin and history of the earth and mankind. But the contrast that they frequently expressed between the "fanciful" theories of the geologists and their own "commonsense" conclusions shows that more was involved than simple religious or social conservatism. The geologists' startling assertions about earth history were indeed derived from increasingly esoteric inferences that the ordinary person could no longer follow easily. Mosaic geology was, therefore, in part a cultural reaction to the social and cognitive exclusion of all but self-styled experts from an area of speculation that, in the heyday of theories of the earth, had been open to all.

The scientific geologists claimed to regard Mosaic geology as marginal to science and worthy only of derision. But in fact they took very seriously its threat to the cognitive and social status of their own enterprise. This is shown particularly by their reaction to any work from within their own social circle that transgressed the tacit boundaries of the science. Early in the century, for example, Cuvier's extension of Deluc's arguments for the historicity of a recent Flood event was widely criticized on just these grounds. Cuvier argued for the low antiquity of civilization—and, by implication, of mankind—by comparing the records of all ancient civilizations. Although he was probably no orthodox Christian, and although he used the biblical records with all the scholarly impartiality of the German biblical critics, his excursion into chronology seemed to others to imperil the cosmological neutrality of geology.[19] This was particularly the case because, when imported into the English-speaking world in Robert Jameson's edition of the *Theory of the Earth* (1813), Cuvier's work was used openly to reassert the literal authority of the Bible. Cuvier's English follower William Buckland (1784–1856) met with a similar critical response, when he argued in *Reliquiae Diluvianae* (1823) for the historicity of a recent, though nonmiraculous, Flood event, basing his case in large part on a detailed study of a supposedly antediluvial hyena den. Such episodes seem, however, to reflect highly specific social circumstances. Buckland's diluvialism, for example, was evoked by a local situation at Oxford, in which he felt he had to defend the new science of geology from the old charge that such speculations encouraged religious skepticism. Once that social threat receded, he quietly dropped his claim that geology gave evidence of a universal Flood event, and later became one of the first geologists to support Louis Agassiz's (1807–1873) theory of a recent Ice Age as a better explanation of the puzzling diluvial phenomena.[20]

The famous remark by the British geologist Charles Lyell (1797–1875) that he was determined to "free the science from Moses" therefore needs to be interpreted in its proper context. Other geologists, whether Christians or freethinkers, agreed with Lyell that the cognitive boundaries of geology needed to be maintained in order to exclude Mosaic geology and its practitioners. They also agreed that scientific geologists such as Buckland, who—for whatever reason—transgressed those boundaries, did the science a disservice. But this exercise in boundary maintenance reflected a conflict not between science and religion but between one social group and another. The geologists were struggling for a cultural place in the sun, for greater social recognition of their cultural authority, in competition with older elites. In the early nineteenth century they did not represent antire-ligious interests but rather those of a pragmatic alliance between lib-eral. Christians, whom even Lyell accepted as "enlightened saints," and a varied assortment of freethinkers; both groups valued geology especially for its potential practical utility.

SCIENTIFIC NATURALISM

Within the cognitive boundaries of early-nineteenth-century geology, one cosmological gloss won widespread acceptance because it was not socially divisive. This related not to Genesis but to natural the-ology. Geology, or more particularly the analysis of extinct organisms preserved as fossils, gave a new temporal dimension to the sense of divine design in the world. The traditional static concept of design was dramatically enlarged by the understanding that divine provi-dence had underlain equally all the successive phases of earth history, even before the existence of mankind, a view expressed, for example, in Buckland's Bridgewater treatise on *Geology and Mineralogy Consid-ered with Reference to Natural Theology* (1836). In this way cosmological meaning could be attributed to the vast prehuman history that geology had opened up. Rather than being socially divisive, this perspective, because it was so broad, acted as a social cement between conflicting religious groups.[21] It was acceptable to a deistically inclined geologist like Lyell, with his uniformitarian or virtually eternalistic theory of earth history, as well as to a quite explicitly Christian geologist such as Adam Sedgwick. And any geologist privately inclined to a more materialistic viewpoint, like the young Charles Darwin (1809–1882), could simply omit the rhetoric of design, leaving his geology unim-paired in the eyes of others.

A second cosmological interpretation of geology was, however, more debatable and had more divisive implications. The general belief among geologists that the earth had cooled gradually in the course of its history seemed at first to be an adequate explanation of an observed directional change in the character of its faunas and floras. This could readily be assimilated within the new sense of a dynamic providential design. In the middle decades of the century, however, evidence began to emerge that could be made to support the view that the history of life had been not only directional but progressive, moving from "lower" to "higher" forms independently of environ-mental change.[22] Indeed, it was possible for geologists such as Agassiz and Richard Owen (1804–1892) to attribute this to the progressive unfolding of an overarching divine design, by which the living world had slowly been prepared for its culmination in the creation of mankind. But this kind of progressionist interpretation could be radically transformed into a strictly naturalistic image of earth history. This possibility first became apparent in the *Vestiges of the Natural History of Creation* (1844) by the Scottish writer Robert Chambers (1802–1871), notwithstanding his substantial use of providentialist arguments. Chambers used the nebular hypothesis of Pierre Simon Laplace (1749–1827), in a new context of progressionist geology and evolutionary biology, to provide a naturalistic explanation of the origin of the earth and the progressive evolution of life toward mankind. It is significant that such reinterpretations of geology (and biology) in terms of the cosmology of scientific naturalism first arose among generalists like Chambers, outside the social circle of the geological specialists; for of course any such theory transgressed the cognitive boundaries that had been defined by that circle. Because such theories could not simply be ignored—Chambers's work was extremely popular among the general public—they forced open the tacit boundaries of the science and obliged geologists to respond to them in kind. If those with an ax to grind in the interests of scientific naturalism claimed that the latest geological discoveries supported that cosmology, then those concerned to defend a Christian cosmology were obliged to show how the same evidence could be interpreted differently, and to do so on the same popular level; Hugh Miller's *Footprints of the Creator* (1847) exemplifies such an effort.

In the middle decades of the nineteenth century, therefore, the frontiers of geology were reopened, as it were, at both ends of the time scale (see fig. 1D). The question of the origin of the earth was forced back into scientific discourse, though it is significant that this opening was first exploited not by geologists themselves but by physicists such as William Thomson, later Lord Kelvin (1824–1907). At the other end of their temporal territory, geologists at last accepted the evidence that human beings must have coexisted with extinct mammals in a geologically recent (but humanly remote) period, because stone implements were found with those animals' bones in circumstances that could no longer be doubted. These discoveries opened up a new conceptual space of prehistory, between the relatively well-established history of civilized literate mankind and the strictly geological history of prehuman periods (see fig. 1D). The question of the origin and antiquity of mankind could thus no longer be tacitly cordoned off from scientific geology. Yet that question, like the origin of the earth, was debated in the later nineteenth century not in terms of conflict between science and religion but between rival cosmologies. The naturalistic interpretation of human origins was neither

II

neutral nor disinterested: it was generally used in the service of the often strident cosmology of scientific naturalism. And scientific naturalism was itself the cosmology of specific social groups, including the self-consciously professionalizing scientists, who used it as a means of wresting cultural power from the hands of older social elites, particularly, of course, the clergy.[23]

These conflicts in the later nineteenth century were more directly concerned with the relation of mankind to the rest of the animal world than with the earth as such. At this point, therefore, they pass beyond the strict limits of this essay. The specific question of the age of the earth did arise once more in the mid-nineteenth century, but only in this biological context was it of cosmological significance. Ever since the rise of geology with its self-imposed cognitive limitations, there had been a tacit embargo on quantitative estimates of geological time. As the Prussian naturalist Alexander von Humboldt (1769–1859) had put it in his *Essai géognostique* (1823), such estimates belonged to "géologues hebraizans," that is, to Mosaic geologists. Geologists got on very well without quantitative estimates; and by the middle of the century Lyell's explanatory use of unquantified but virtually limitless time had been assimilated into much routine geological practice. The quantitative magnitude of the time scale reemerged into scientific discourse only after Darwin in *The Origin of Species* (1859) hitched his concept of natural selection to a Lyellian concept of geological time and used natural selection in the service of a far-reaching naturalistic theory of species change and, implicitly, of the origin of mankind. Darwin rashly committed himself in print to a "guesstimate" of geological time that even modern geologists would find extravagant.[24] This gave his contemporaries the opportunity to attack his theory at a weak point on impeccably scientific grounds, though in reality their reasons for doing so were no more disinterested cosmologically than his. On the basis of observed rates of sedimentation, for example, geologists such as John Phillips (1800–1874) estimated the total age of all preserved strata to be on the order of a hundred million years. They found this estimate gratifyingly compatible with those proposed by Kelvin for the total age of the earth, using the quite independent evidence of the thermodynamic history of the sun.[25] In the face of such a consensus on the relatively limited time scale of earth history, natural selection seemed to be eliminated from being a plausible motor for evolutionary change. But the dynamic behind this criticism lay in opposition not to evolutionary theories in general but to the particular form of Darwin's theory, in which "blind chance" seemed to play such a crucial role and providential design was apparently excluded from organic nature. Still less was the scientists' opposition to Darwin's geological time scale rooted in any desire to reinstate the time scale of precritical biblical chronology. Such an attempt did indeed continue throughout the nineteenth century, but only in the tradition of Mosaic geology, not among those whose work was accepted within the circle of scientists. And even Mosaic geology was pushed inexorably into an increasingly marginal position, both cognitively and socially, as the intellectual spokesmen for Christian opinion, even in the English-speaking world, abandoned precritical forms of biblical interpretation and restated their beliefs in terms that took account of the newer critical methods.

CONCLUSION

Soon after 1900 Kelvin's ever more restrictive estimates of geological time were burst open by the discovery of radioactivity in rocks. But by then the question had lost all its earlier associations with rival religious and secularist cosmologies. The great expansion of estimates of the age of the earth during the twentieth century (to their present level of several billions of years) facilitated the revival of Darwinian interpretations of evolutionary theory, but it had no new religious implications. For Christians who accepted critical methods of interpretation in their understanding and practical use of the biblical documents, the religious meaning of texts such as the Creation narrative remained undisturbed by changing estimates of the quantitative magnitude of earth history or the history of mankind. Some individual scientists did assert that those histories bore a deeper cosmological meaning; the Catholic priest and paleontologist Pierre Teilhard de Chardin (1881–1955), for example, argued for a religious interpretation, while the French biologist Jacques Monod (1910–1976) favored an antireligious one. But neither assertion found general support among scientists; to most of them, either conclusion seemed evidently imposed on the scientific evidence rather than derived from it. More clearly than ever, any such cosmological gloss could be seen to represent an individual decision to attribute a certain set of values to a scientific story of earth history that in itself was geared to far more limited cognitive goals.

Yet this modest and perhaps tame conclusion continues to be challenged from two opposed directions. On the one hand there are the successors of the strident nineteenth-century proponents of scientific naturalism, such as Monod, who insist that the scientific story does carry an intrinsic cosmological implication, namely an atheistic one. And on the other hand there are the successors of the equally strident

II

nineteenth-century proponents of biblical literalism, who insist that the scientific story is radically false because it is incompatible with their own "scientific" evidence and with a precritical method of interpreting the Bible. Such Christian fundamentalism, including its component strand of creationism, has become in recent years a powerful cultural force in some Western societies—for example, in the United States and the Netherlands (see chapter 16). But it has been evoked, at least in part, by the way that a few scientists have made well-publicized and often arrogant claims to a privileged monopoly in the attribution of human meaning to the natural world.

Late-twentieth-century Christians who reject the precritical assumptions of the fundamentalists find it necessary to steer a difficult middle course between this Scylla and Charybdis. But so do late-twentieth-century agnostics who reject the arrogant scientistic pretensions of the new scientific naturalists. Being in the same boat, those two groups can probably agree about the present outcome of the earlier debates on the origin and history of the earth and of mankind's place in it, namely that Christian beliefs about the meaning and conduct of human lives have no legitimate point of contact of any significance with the modern scientific story of earth history. This is not because mainstream Christian theology has compromised with atheistic secularism, as the fundamentalists claim. Nor is it because orthodox Christianity has been defeated in a conflict with science, as the new scientific naturalists claim. The true reason, I suggest, is twofold. First, earth scientists as a social group have *collectively* chosen the historical option of abandoning any cosmological ambitions, as the most effective route to the achievement of more limited cognitive and technical goals. And second, mainstream Christian theologians have recognized that the religious meaning of biblical texts is to be found in terms of whatever input from a "God-labeled" source (that is, in traditional language, revelation) may be embodied in the religious insight of the ancient cultures that produced those texts.[26] Of course the Christian and the agnostic are likely to differ profoundly in their estimate of the cognitive and practical value of that insight for the construction of individual and social lives in the modern world. But that is another story.

NOTES

I am under an obligation to state that the work for this paper was begun while I was still Professor of the History and Social Aspects of Science at the Vrije Universiteit, Amsterdam, the Netherlands. Preliminary versions of the

The Shape and Meaning of Earth History

paper were given to colloquia at the Science Studies Unit, Edinburgh University, and the Department of History and Philosophy of Science, Cambridge University; I am grateful to those who made helpful comments and suggestions on these earlier occasions, as well as at the conference at Madison. Neal Gillespie and Jim Moore made valuable written comments on the draft discussed at the conference.

1. James R. Moore, *The Post-Darwinian Controversies: A Study of the Protestant Struggle to Come to Terms with Darwin in Great Britain and America, 1870–1900* (Cambridge: Cambridge Univ. Press, 1979), part 1, gives a valuable account of the polemical contemporary context of these works.

2. For further discussion of this point of view see Martin Rudwick, "Senses of the Natural World and Senses of God: Another Look at the Historical Relation of Science and Religion," in *The Sciences and Theology in the Twentieth Century,* ed. A. R. Peacocke (London: Routledge & Kegan Paul, 1981), pp. 241–261.

3. This point is well made by Mary Hesse, "Criteria of Truth in Science and Theology," in her *Revolutions and Reconstructions in the Philosophy of Science* (Hassocks: Harvester Press, 1980), chap. 10.

4. For work that explores the extension of the term *cosmology* beyond its original anthropological setting see, for example, Mary Douglas, *Implicit Meanings* (London: Routledge & Kegan Paul, 1975). The same notion is used in the context of the history of science in Barry Barnes and Steven Shapin, eds., *Natural Order: Historical Studies of Scientific Culture* (Beverly Hills, Calif.: Sage Publications, 1979). I deliberately avoid the contentious term *ideology* in this context.

5. See Anthony T. Grafton, "Joseph Scaliger and Historical Chronology: The Rise and Fall of a Discipline," *History & Theory* 14 (1975): 156–185; and J. D. North, "Chronology and the Age of the World," in *Cosmology, History, and Theology,* ed. Wolfgang Yourgrau and Allen D. Breck (New York: Plenum Press, 1977), pp. 307–333.

6. See Don Cameron Allen, *The Legend of Noah: Renaissance Rationalism in Art, Science and Letters* (Urbana: Univ. of Illinois Press, 1949); Arnold Williams, *The Common Expositor: An Account of the Commentaries on Genesis, 1527–1633* (Chapel Hill: Univ. of North Carolina Press, 1948); and Janet Browne, *The Secular Ark: Studies in the History of Biogeography* (New Haven: Yale Univ. Press, 1983).

7. See Martin Rudwick, *The Meaning of Fossils: Episodes in the History of Palaeontology,* 2d ed. (New York: Science History Publications, 1976), chaps. 1 and 2.

8. On the close relation between naturalists and antiquarians see Cecil J. Schneer, "The Rise of Historical Geology in the Seventeenth Century," *Isis* 45 (1954): 256–268; and Paolo Rossi, *The Dark Abyss of Time: The History of the Earth and the History of Nations from Hooke to Vico* (Chicago: Univ. of Chicago Press, 1984).

9. The political context of Burnet's chiliastic expectation of an imminent millennium is explored in Margaret C. Jacob and Wilfrid A. Lockwood, "Po-

litical Millenarianism and Burnet's *Sacred Theory*," *Science Studies* 2 (1972): 265–279. Burnet's scheme is elegantly summarized in visual form in the frontispiece of his book, which is reproduced in Rudwick, *Meaning of Fossils*, p. 79.

10. See the important interpretative synthesis in Jacques Roger, "La theorie de la terre au XVIIe siècle," *Revue d'histoire des sciences* 26 (1973): 23–48. See also Roy Porter, "Creation and Credence: The Career of Theories of the Earth in Britain, 1660–1820," in *Natural Order*, ed. Barnes and Shapin, pp. 97–123; and "The Terraqueous Globe," in *The Ferment of Knowledge: Studies in the Historiography of Eighteenth-Century Science*, ed. G. S. Rousseau and Roy Porter (Cambridge: Cambridge Univ. Press, 1980), pp. 285–324.

11. See Marjorie H. Nicolson, *Mountain Gloom and Mountain Glory: The Development of the Aesthetics of the Infinite* (Ithaca, N.Y.: Cornell Univ. Press, 1959); and Gordon L. Davies, *The Earth in Decay: The History of British Geomorphology* (London: Macdonald, 1969).

12. Roy Porter, "George Hoggart Toulmin's Theory of Man and the Earth in the Light of the Development of British Geology," *Annals of Science* 35 (1978): 339–352.

13. I am grateful to Dr. Rhoda Rappaport for allowing me to draw on her unpublished research in this section.

14. On the absence of any consensual view see Rhoda Rappaport, "Geology and Orthodoxy: The Case of Noah's Flood in Eighteenth-Century Thought," *British Journal for the History of Science* 11 (1978): 1–18.

15. Bernhard Sticker, "Leibniz' Beitrag zur Theorie der Erde," *Sudhoffs Archiv* 51 (1967): 244–259.

16. Jacques Roger, ed., "Buffon: Les epoques de la nature, edition critique," *Mémoires du Muséum d'histoire naturelle*, n.s., ser. C, 10 (1962). Privately, Buffon suspected that the true time scale was even longer.

17. My interpretation of the precritical concept of biblical narrative, and of the way it was supplanted by critical methods of interpretation, is much indebted to Hans W. Frei, *The Eclipse of Biblical Narrative: A Study in Eighteenth and Nineteenth Century Hermeneutics* (New Haven: Yale Univ. Press, 1974).

18. See Roy Porter, *The Making of Geology: Earth Science in Britain, 1660–1815* (Cambridge: Cambridge Univ. Press, 1977).

19. See Dorinda Outram, *Georges Cuvier: Vocation, Science and Authority in Post-Revolutionary France* (Manchester: Manchester Univ. Press, 1984).

20. See Nicolaas A. Rupke, *The Great Chain of History: William Buckland and the English School of Geology, 1814–1849* (Oxford: Clarendon Press, 1983); Martin Rudwick, "The Glacial Theory," *History of Science* 8 (1970): 136–157.

21. See John H. Brooke, "The Natural Theology of the Geologists: Some Theological Strata," in *Images of the Earth: Essays in the History of the Environmental Sciences*, ed. L. J. Jordanova and Roy S. Porter (Chalfont St. Giles: British Society for the History of Science, 1979), pp. 39–64.

22. For the earlier interpretation of progression see Martin Rudwick, "Uniformity and Progression: Reflections on the Structure of Geological Theory in the Age of Lyell," in *Perspectives in the History of Science and Technology*, ed. Duane H. D. Roller (Norman: Univ. of Oklahoma Press, 1971), pp. 209–227; for the later or true progressionism see Peter J. Bowler, *Fossils and Progress:*

Paleontology and the Idea of Progressive Evolution in the Nineteenth Century (New York: Science History Publications, 1976).

23. See, for example, Frank M. Turner, "The Victorian Conflict between Science and Religion: A Professional Dimension," *Isis* 69 (1978): 356–376.

24. J. D. Burchfield, "Darwin and the Dilemma of Geological Time," *Isis* 65 (1974): 300–321.

25. See Joe D. Burchfield, *Lord Kelvin and the Age of the Earth* (New York: Science History Publications, 1975).

26. For a reinterpretation of revelation framed in these terms see John Bowker, *The Sense of God: Sociological, Anthropological and Psychological Approaches to the Origin of the Sense of God* (Oxford: Clarendon Press, 1973).

III

Minerals, strata and fossils

Animal, vegetable or mineral? The opening move of the traditional guessing game preserves part of what was the taken-for-granted structure of the sciences, at least until the end of the eighteenth century. 'Natural history' was at that time still a highly esteemed branch of human knowledge, and no merely amateurish pursuit. It was not an archaic synonym for what would now be called biology, for it ignored the boundary between the living and the non-living: it included mineralogy as one of three divisions or 'kingdoms' of equal importance (the others were, of course, zoology and botany). But 'mineralogy' was much wider in meaning than the modern science of the same name; it was roughly the equivalent of 'earth sciences' today. The term 'geology' had indeed been proposed, but it was a neologism that was neglected or even rejected, for reasons that will become clear later in this chapter. In fact the shift from 'mineralogy' to 'geology', as the most usual term for what would now be called the earth sciences, encapsulates the dramatic changes in the culture of inorganic natural history that occurred between the late-eighteenth and the mid-nineteenth centuries.

A science of specimens

In the late eighteenth century, throughout Europe and wherever European culture extended, mineralogy was first and foremost a matter of mineral specimens: specimens collected, sorted, named and classified. Specimens were extracted from mines and quarries, hammered out of coastal cliffs or mountain crags, picked out of stream-beds or off the surface of fields, and assembled indoors in museums or private 'cabinets'. Those who collected these specimens called themselves 'mineralogists' or, more broadly, just 'naturalists'. Some, for example the owners or managers of mines, made their collections for strictly practical reasons; but most did so as a socially acceptable part of polite culture, valuing above all the unusual and the spectacular, with motives that might be at the same time aesthetic, scientific and monetary. Rare or valued specimens were exchanged between enthusiasts, purchased when a

Figure 16.1 A highly emblematic portrait of Horace Bénédict de Saussure (1740–99), one of the most distinguished of the late eighteenth-century naturalists who studied the mineral kingdom. He is dressed to match his social status in the polite society of Geneva, but he is portrayed outdoors as if doing fieldwork. He has in his hands a miner's hammer – the badge of the mineralogist – and a rock specimen obtained by its use; by his side are mineral specimens and the bag in which he has collected them, and instruments for surveying the topography and measuring the inclination of rock strata; and he looks up – in a gaze recalling the pious poses of saints in an earlier iconography – towards the Alpine peaks in the background. By 1796, when Saint-Ours painted the portrait on which this engraving is based, the fifty-six-year-old Saussure had in fact suffered a paralysing stroke and his fieldwork career was over. From a print in the author's collection.

deceased or bankrupt naturalist's collection was put up for sale, and – not least – bought from the miners, quarrymen and peasants whose daily toil enabled them to find what these noblemen and gentlemen (and a few ladies) were prepared to pay for.

At least in the more serious collections, specimens were then compared with those of other naturalists, identified and named. As standards of comparison, collections of specimens that had been named authoritatively were particularly valued, and were exchanged between individuals or institutions. Comparisons were often made, however, not with other real specimens, but with what were in effect the *proxy* specimens pictured in publications (Figure 16.2). These were usually engravings, which were sometimes

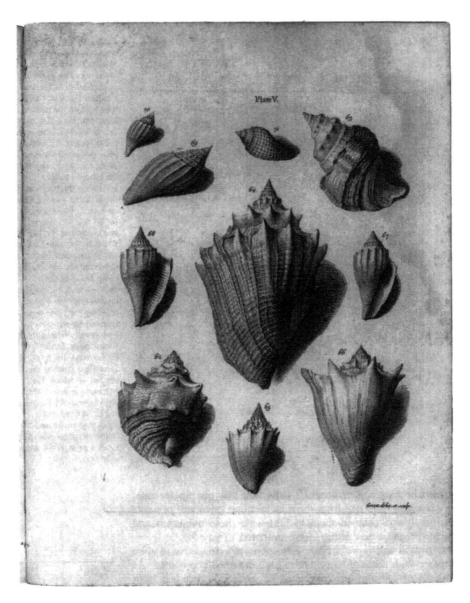

Figure 16.2 A typical set of eighteenth-century illustrations of fossils: these are of well-preserved mollusc shells from Secondary (in modern terms, Cenozoic) strata in the south of England. Such engraved representations – minerals were depicted in a similar style – formed highly effective proxies for the real specimens. From Gustav Brander's *Fossilia Hantoniensia collecta* (London, 1766), illustrating specimens preserved in the British Museum in London.

hand-coloured with astonishing *trompe l'oeil* realism. Books with illustrations of mineral specimens, often recording a celebrated collection or material from some famous locality, were in effect proxy museums, and they spread their authors' descriptions and identifications as widely as the volumes were bought and sold.[1]

In all this, mineralogy differed little from the other branches of natural history. As in botany and zoology, the fundamental scientific goal was simply to describe, name and classify the diverse riches of nature. Minerals, no less than plants and animals, were to be described in terms of their natural *species*: species such as quartz and felspar, no less than species of daisies and deer. But most mineralogists, like other naturalists, were not content merely to identify and name their specimens. They wanted to construct a classification that would assemble similar minerals into a nesting set of groups, and so reveal the hierarchical structure of the diversity of the whole mineral kingdom.

In this task of identification and classification, it was increasingly regarded as imperative to examine the interior of minerals, as it were, as much as that of plants and animals. While the botanist dissected the intimate sexual parts of the flower, and the zoologist the literally internal anatomy of the animal body, the mineralogist resorted to the laboratory, and performed chemical analysis on his specimens in order to discover their true nature. In this way mineralogy had developed some of its strongest links with chemistry. The emergence of what became known as crystallography, at the end of the eighteenth century, provided a further set of characters for the same task of constructing a truly natural classification of minerals; but it also brought to mineralogy the prestige of being geometrical and quantitative.

To ask about the *origins* of natural species, however, seemed as meaningless in mineralogy as in botany and zoology; or at least, such questions were often regarded as abandoning natural history for the speculative realm of metaphysics. Classifications were intended to reflect the diversity of the world; how its natural kinds had come into being was generally considered to lie beyond scientific investigation, simply because it belonged, in effect, outside time. However, it was in mineralogy that this static conception of the natural world first began to be undermined, as a result of the emergence of problems for which questions of origin seemed both appropriate and soluble.[2]

A science of fieldwork

One of the distinctions that was clarified in the eighteenth century was that between *minerals* and *rocks*. The former term took on a more restricted – and modern – meaning; rocks were interpreted as aggregrates of, usually, more than one kind of mineral. Thus

III

granite was understood as a rock composed of crystals of minerals
such as quartz, felspar and mica, and limestone as a rock composed
mainly of grains of calcite. Even if the origins of mineral species
were considered to be beyond the realm of natural science, the
origins of composite entities such as rocks were clearly not. Many
rocks, notably 'pudding-stones' (in modern terms, conglomerates
or consolidated gravels) and sandstones, were said to be of 'mechanical' origin, being evidently composed of the debris of pre-
existing rocks; but many others, such as granite and marble, were
composed of crystalline minerals and were considered to be of
'chemical' origin. Whether a chemical origin implied crystallization
from an aqueous solution or from a true melt was hotly debated:
the contemporary state of chemistry made the former, stressing the
chemically active role of water, seem generally the more plausible.

Questions of origin remained problematic, however, for many
rocks, particularly for fine-grained ones such as basalt.[3] Significantly, in such cases evidence had to be sought outside the laboratory, in the *field* relations of the rocks. Using fieldwork, the French
naturalist Nicolas Desmarest (1725–1815) demonstrated, for
example, that at least some basalts – including some with the
spectacular hexagonal jointing that made them look like gigantic
crystals – were connected to present or former volcanoes, and must
originally have been molten lavas.[4] But the field relations of other
basalts, found far from any volcanoes and sandwiched between
sandstones or other rocks that had clearly been sediments, later
suggested to other mineralogists that basalt was a rock of sedimentary origin: this view was propounded forcefully by Abraham
Werner (1749–1817), who taught at the great mining school at
Freiberg in Saxony. The argument that followed, peaking in the
1790s, pitched the proponents of heat against those of water, or
'Vulcanists' against 'Neptunists'. On the specific issue of the origin
of basalt, the Vulcanists eventually won the argument, mainly on
the strength of the field evidence. However, most mineralogists –
even the Vulcanists – considered that *most* rocks were probably of
aqueous origin (though they thought the water might have been
very hot and chemically active in some cases) and that volcanoes
were relatively minor agents in the earth's economy.[5]

The basalt controversy was important in the long run, less
because it settled the origin and classificatory position of one kind
of rock, than because its resolution entailed *fieldwork* as an essential
part of scientific practice. Until quite late in the eighteenth century, all three branches of natural history were still mainly indoor
sciences. Travel and fieldwork were indeed considered essential,
but they were undertaken primarily to collect specimens, which
were then gathered indoors (or at least into a botanic garden) for

the closer work that made their study truly scientific. It was in mineralogy that this predominantly indoor culture first began to be seriously challenged.

A science of mineral distributions

By the late eighteenth century, mineralogy was already far wider than the modern science of the same name, because it encompassed the geographical dimension of the science of the earth. Some of its most prominent practitioners, such as the Genevan naturalist Horace Bénédict de Saussure, insisted that fieldwork was indispensable, not just for collecting specimens – a task that had often been delegated to assistants or employees – but for seeing with one's own eyes how the various minerals and rock masses were spatially related to one another and to the physical topography of the areas in which they were found.[6] Added to that was the importance of witnessing for oneself the more spectacular features of the mineral world, such as erupting volcanoes and high mountains and their glaciers. Published descriptions of travels could convey only a pale intimation of the grandeur of these phenomena. Even the pictures that increasingly accompanied such texts were no more

Figure 16.3 A view of an extinct volcano in central France, with a solidified lava flow revealed at the river's edge to be basalt with prismatic or columnar jointing. The carriage indicates not only the scale, but also the means by which some gentlemanly naturalists did much of their fieldwork. From an engraving in Barthélemy Faujas de Saint-Fond, *Recherches sur les volcans éteints du Vivarais et du Velay* (Grenoble, 1778).

Figure 16.4 A view by Pietro Fabris of the 1767 eruption of Vesuvius, from a hand-coloured engraving in *Campi Phlegraei* (Naples, 1776–9), Sir William Hamilton's monograph on the volcanic region around Naples. Landscapes such as this were effective proxies for the first-hand experience of the more spectacular features of the mineral world. Hamilton (now perhaps best known as the husband of Admiral Nelson's mistress Emma) was the British ambassador to the court in Naples, and used his residence there to become an outstanding expert not only on volcanoes but also on the antiquities of the region.

than proxies for the first-hand visual experience of remote or distant places; at their expensive best, however, the proxies could be remarkably vivid (Figure 16.4).

'Physical geography' or 'mineral geography' therefore became for many mineralogists the preferred name for their scientific activity. Topographical maps became indispensable tools, with which the distributions of minerals and rocks could be plotted and their spatial regularities perceived.[7] Topographical maps drew attention to river patterns and drainage basins, the location and direction of mountain ranges, the form of coastlines and the distribution of more striking features such as volcanoes; they enabled generalizations about the form of the earth's surface to be perceived and expressed. The occurrence of distinctive or useful rocks and minerals could then be plotted on a map, using conventions adapted from standard cartographic practice: either as scattered symbols, denoting outcrops or quarries, or more boldly – by extrapolation – as a patchwork of colour washes (Figure 16.5).

No mineral geographer, however, could be blind to the third dimension that – at least potentially – converted distributions at the earth's surface into structures in the earth's interior. The relative abundance of rock outcrops and other natural sections in hilly and mountainous regions, and the concentration of useful mineral resources there, focused mineralogists' attention on the hard rocks they termed 'Primary', in preference to the generally softer 'Secondary' rocks of the lower-lying regions.[8] 'Primary' and 'Secondary' denoted the relative structural position of rocks, and only sub-

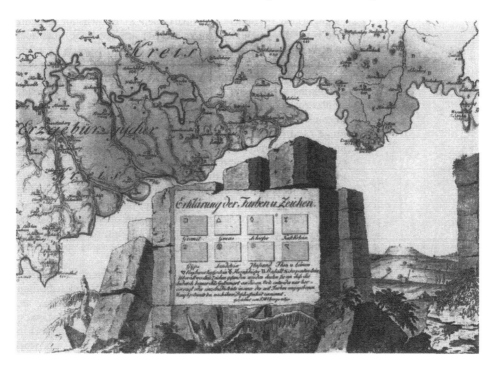

ordinately their presumed relative age (Primary rocks were sometimes termed 'Primitive'). The hard rocks of upland regions were 'Primary' because they appeared to constitute the foundations of the earth's crust; the softer rocks of lowland regions were 'Secondary' because they manifestly overlay the others, and were at least partly composed of their debris (although often lower in topographical position, Secondary rocks could be seen to overlie or lap against Primary ones, wherever the junction was exposed).

The distinction between Primary and Secondary was taken for granted in the eighteenth century, just for practical convenience of description. Volcanic rocks of any age were generally treated as another category on the same level; and 'Alluvial' was used for superficial deposits of sand and gravel (not rocks at all, in the everyday sense).

A science of rock formations

These four broad categories – Primary, Secondary, Volcanic, Alluvial – were much too general to do justice to the diversity of rocks found in many regions. On the other hand, the individual layers, beds or 'strata' (for example, specific coal seams) that were

Figure 16.5 The key to a late eighteenth-century mineral map: the distributions of eight kinds of rock (granite, limestone, sandstone, etc.) are represented both by spot symbols and (in the original) by colour washes; but they are not arranged in any particular order, and the map represents a pattern of areal distribution rather than a three-dimensional structure. From the *Mineralogische Geographie* (Leipzig, 1778) of part of Saxony, by Johann Charpentier.

distinguished by miners and quarrymen were often not recognizable beyond a single mine or quarry, or at most some small local area. What came into use in the late eighteenth century, as a category of intermediate generality, was the *formation*.[9] The formation was a concept of immense practical value, despite the impossibility of defining it precisely. A formation was an assemblage of broadly similar rocks, separated more or less sharply from the adjacent formations; the equivalent German term *Gebirge* (literally, 'mountain range') and French term *terrain* both indicate its geographical connotations. A formation might, for example, be termed a sandstone, even if it included some intercalated strata of limestone or shale, provided it had some distinctive overall character and was clearly separated from (say) a limestone formation on one side, or above, and a shale formation on the other side, or beneath. Formations, unlike most of their constituent strata, could often be traced across country throughout some wide region, varying perhaps in thickness and detailed composition, but retaining the same position relative to other formations.

The use of the formation concept made it apparent that minerals required two distinct and complementary kinds of classification: one appropriate to specimens as analysed in the laboratory and stored in the museum, the other to the larger spatial relations of rocks observed in the field.[10] The basic and continuing work of defining, naming and classifying minerals and rocks was work centred on the examination of specimens, and it aimed to display and order the diversity of mineral 'species' and of the rocks that were their aggregates. In contrast, a classification centred on fieldwork included such categories as bed or stratum, Primary and Secondary, and – now – formation; it aimed to display the three-dimensional spatial relations of 'mineral bodies' or rock masses.

The branch of mineralogy that dealt with the classification of rock masses and their spatial relations became known as *geognosy* (literally, knowledge of the earth). The formation concept was central to its practice. Its usual form of publication was a sequential description of the formations found in some specific region. This was often accompanied by a map showing the areal extent of their 'outcrops' at the surface, and one or more sections showing their inferred relations below the surface: together, these allowed the reader to imagine the structure of the area in three dimensions (see Figures 16.6, 16.7). But 'geognosts' (as they called themselves) aimed to define and describe formations that would be recognizable beyond a single region, and ideally even on a global scale. That required a corresponding concept of *correlation*, by which a given formation was identified with its equivalents in other regions or even on other continents, even if it did not have exactly the same character everywhere.

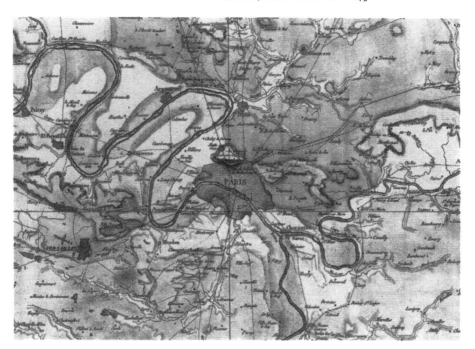

In that task of recognizing formations in different regions, and thereby making the classification as widely applicable as possible, many different criteria were tried out empirically. The kinds of rock were always basic, but many of the same rock types – for example, sandstone or limestone – characterized more than one distinct formation. That criterion was therefore supplemented by others: for example, the altitude at which a formation was usually found, or the degree to which its constituent layers were usually tilted out of the horizontal. However, those criteria proved fallible in practice; what seemed to be the same formation might be found high on mountains in one region and at low elevations in another, or highly tilted in one region and almost horizontal in another. The criterion of 'superposition' proved more reliable: true formations, whatever their altitude or degree of tilt or folding, seemed always to retain the same relative position in the three-dimensional stack of rocks revealed in natural or man-made sections.

Geognosy embodied a primarily structural conception of mineral science. Formations were typically described as 'above' or 'below' others; it was their structural order, as three-dimensional rock masses, that seemed to be reliably invariable, even when in a given region certain formations were missing. The Prussian geognost Leopold von Buch (1774–1853), in a public lecture in 1809,

Figure 16.6 Part of the engraved 'mineralogical map' (hand-coloured in the original) illustrating the monograph by Georges Cuvier and Alexandre Brongniart on the *Géographie minéralogique des environs de Paris* (Paris, 1811). This was based on fieldwork in which the standard procedures of geognosy were supplemented by study of the abundant invertebrate fossils in some of the formations; it allowed relative ages to be assigned to Cuvier's much rarer but spectacular vertebrate fossils. The continuous lines radiating from the centre of Paris indicate the positions of a series of sections; the combination of map and sections enabled the region to be envisaged as a three-dimensional structure.

III

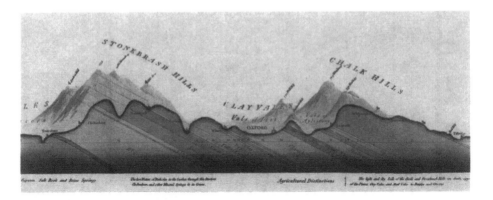

Figure 16.7 Part of William Smith's *Geological Section from London to Snowdon* (London, 1817), showing the succession of formations (in modern terms, mostly of Jurassic and Cretaceous age) in southern England. The section was intended to be 'read' in conjunction with Smith's great geological map of England and Wales (1815), to give a sense of the three-dimensional structure of the country (the 'Stonebrash' and 'Chalk' hills are, respectively, the Cotswolds and Chilterns). The vertical scale – and hence also the 'dip' of the strata – is exaggerated, in order to clarify the relations between the formations. The boundaries between them are drawn boldly with ruled lines, in a style reflecting Smith's work as a civil engineer, although this entailed major extrapolation from the evidence observable at the surface in outcrops and quarries.

explained the concept of formations by using the homely analogy of a row of houses, in which the identity and relative positions would remain unaltered even if some houses were demolished; his fellow Prussian, Alexander von Humboldt (1769–1859), later proposed an elaborate algebraic notation ('pasigraphy') to express the physical place of any formation in a putatively universal order (*Essai géognostique*, 1823). All geognosts were well aware that this structural order of position also represented a temporal order of origin, since it was axiomatic that a structurally lower formation must have preceded in origin any formations that lay above it. But this temporal element was always subordinate to the structural; geognosy was essentially a spatial science, a three-dimensional extension of mineral geography.

A science of characteristic fossils

Around the end of the eighteenth century, yet another criterion – fossils – began to be added to the practice of correlation in geognosy. The mineral specimens that eighteenth-century naturalists collected and classified included many that they considered to be of plant or animal origin. Seventeenth-century debates about the nature of distinctive mineral objects ('fossils' in the original sense) had long been resolved by settling the criteria by which those that were truly the remains of once-living beings could be distinguished from those of inorganic origin. Phrases such as 'extraneous fossils' denoted those of organic origin, but the adjectives were slowly dropping out, leaving the noun with its modern meaning.

Fossils were collected assiduously from Secondary strata, but their perceived significance was limited. The conception of them as 'extraneous' to the rocks in which they were found subtly discouraged any use of them as potential criteria for defining or identifying formations: on the ancient philosophical distinction, they seemed merely 'accidental' characters, not 'essential' ones. Fossil shells were recognized as analogous to those of living marine molluscs (see Figure 16.2); but that simply confirmed that most Secondary rocks had been deposited in the sea and that the sea must formerly have covered the present continents, which was no news to any naturalist. Above all, however, fossils were neglected because scientific attention was focused mainly on the Primary rocks, with their valuable mineral veins and spectacular mineral specimens, and they, by definition, lacked any trace of fossils.

This relative neglect of fossils in geognostic practice ended dramatically in the early nineteenth century when, from two different directions, a new attention was given to the soft and richly fossil-bearing Secondary rocks of some low-lying areas. The English mineral surveyor William Smith (1769–1839) found empirically that fossils were a highly effective means of distinguishing between otherwise similar formations, across wide tracts of the English countryside, where there were only scattered rock outcrops or quarries: specific fossils, he claimed, were 'characteristic' of specific formations. At about the same time, the French comparative anatomist Georges Cuvier (1769–1832), having been attracted to the study of *fossil* bones, realized that their relative ages could be clarified by following geognostic procedures. He and his mineralogist colleague Alexandre Brongniart (1770–1847) augmented that practice, however, by giving close attention to the fossil shells found abundantly in some of the formations around Paris: in their work the fossils became 'essential' features of the formations. Maps and sections, and lists or pictures of the relevant fossils, were published both by Cuvier and Brongniart (1808–11) and by Smith (1815–19).

The priority dispute that ensued had nationalistic overtones – not surprisingly, since France and Britain were at war until 1815 – but the end result was simply to equip geognostic practice internationally with a powerful new tool for correlation. Fossils proved to be generally – though not invariably – reliable indicators of equivalent formations, not only within a given region but also internationally and even globally. The 1820s and 1830s saw the widespread application of the new fossil-based methods to Secondary rocks in many parts of the world; by about 1840 geognosy had been transformed by the empirical success of the fossil criterion.

A standard sequence of formations, now assembled into still larger groupings or 'systems' (for example, 'Carboniferous' and 'Cretaceous'), had been accepted consensually as being valid throughout Western Europe, the most thoroughly explored part of the earth's surface. Its limits were being extended to the Russian empire and to North America, and tested in still more remote

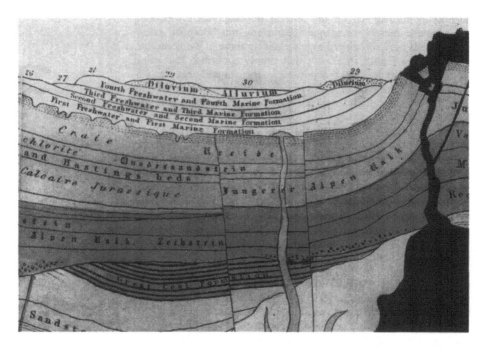

Figure 16.8 A portion of the large generalized or 'ideal section' that illustrated the popular *Geology and Mineralogy* (London, 1836) by the English geologist William Buckland, showing part of what was by then an internationally accepted sequence of major formations of sedimentary rocks (note the French names in italics and German ones in Gothic); igneous or volcanic rocks erupt from the depths. The 'Diluvium' was the peculiar 'boulder clay' or 'till' that was generally attributed to the most recent 'catastrophe'; the 'Alluvium' was the still more recent material (e.g. river gravels) from the human epoch; note the minor role of both in the whole sequence of formations, and hence implicitly the vast scale of *pre-human* earth history. Far thicker formations (not shown in this part of the section) underlay the 'Great Coal Formation' and represented even earlier periods of the earth's history.

regions throughout the world. Even the rocks that Werner had called 'Transition', in which fossils were usually rare or poorly preserved, were yielding to the same treatment (giving rise, for example, to 'Silurian' and 'Devonian'); this pushed the sequence of systems down towards the Primary rocks.

Such descriptive work – later termed 'stratigraphy' – became the foundational practice of what was now almost universally called 'geology'. It was carried out both by gentlemanly 'amateurs' – whose work was anything but amateurish – and by a new and growing breed of professional geologists. The latter were now to be found not only in the management and administration of mines, but also in the new 'geological surveys' instituted and financed by governments in many parts of the world. The first state-supported survey was in France: a team of three geologists began work, significantly, by visiting England in 1823 to study the methods that by then were standard among the members of the Geological Society of London.[11] By 1834 their geological map of France was virtually complete. In the British political climate, less hospitable to state intervention of any kind, an analogous survey started in a precarious and *ad hoc* manner in 1832, but was not established on a permanent basis until the end of the decade. By that time some of the states of the USA had also founded surveys, spreading the model beyond Europe.

Stratigraphical geology remained as structural in orientation as the geognosy from which it had developed. Of course, it provided a basis for a historical understanding of the earth and of life at its surface, but it was not itself primarily historical. Formations continued to be described as 'above' or 'below' one another far more often than they were said to be 'younger' or 'older', and the focus continued to be on their three-dimensional relations as rock masses. Likewise, the study of formations remained as thoroughly descriptive in character as the natural history that had been its origin; it provided materials for causal inferences about the earth and its life; but it was not itself primarily a causal science.

A 'theory of the earth'

Historical and causal analyses of the earth belonged to a different intellectual tradition, which in the late eighteenth century was regarded as distinct, even by those who aimed to contribute to both; only gradually, in the early nineteenth century, did it merge with the descriptive tradition. Mineralogy, mineral geography and geognosy were all regarded as branches of descriptive natural history; theorizing about the history of the earth and its causes, on the other hand, belonged to 'natural philosophy' or, in the old broad sense of the word, to 'physics', the science of natural causes.

Ever since the seventeenth century, causal and historical interpretations of terrestrial phenomena had been termed 'theory of the earth', and in the eighteenth century many important works bore that title. The phrase denoted not so much any particular theory, but rather a genre in which a set of initial conditions (for example, the earth as a molten globe) was coupled with a set of physical principles (for example, the laws of cooling bodies), and used to generate a hypothetical sequence of events or stages through which the earth might have passed in reaching its present state.[12] The *Theory of the Earth* by the Scottish 'natural philosopher' James Hutton (first published in 1788, enlarged into book

form in 1795) was a late example of the genre; by the time it
appeared, the sheer proliferation of such theories, with each author
proudly expounding his own and emphasizing its originality, was
leading to a reaction against such unconstrained causal speculation.
Saussure, for example, was prominent among those who argued
that the variety of these theories proved that all were premature,
because they were too little constrained by observational evidence.
So when in 1779 Jean-André Deluc (1727–1817), a Genevan
resident in England, proposed the word 'geology' (in a mere footnote!) as the terrestrial counterpart of cosmology, it was correctly
taken to be a synonym for 'theory of the earth', and was therefore
treated with caution or even rejected outright.

Only in the early nineteenth century did the word 'geology'
begin to lose its speculative connotations, as 'geologists' – as they
then began to call themselves – recognized that more restricted
kinds of causal interpretation might be legitimate. The changing
status of the word is signalled by its adoption in 1807 by the
first scientific society specifically for the study of the earth (the
Geological Society of London), notwithstanding its founders'
explicit rejection of 'theory of the earth' and strong emphasis on
the value of collecting 'facts'. By 1830, when the similar Geological Society of France was founded in Paris, the word had completely lost its earlier dubious reputation, and was used in its
modern sense.[13]

One of the earlier overarching theories, however, gave the science
a conceptual legacy that transcended its genre. In Les Époques de
la nature (1778), the great French naturalist Georges Leclerc, comte
de Buffon (1707–88), postulated an initially molten globe that had
gradually cooled to its present state. The theory itself was based on
little fieldwork, and was widely regarded as outmoded even when
published; but the elegant text embodied metaphors that were
powerfully influential, although not original to Buffon. His hypothetical story was divided into the six 'epochs' of his title, and he
referred to features such as extinct volcanoes and the marine fossils
found high above sea-level as the 'archives' or 'monuments' of
nature, because they could be regarded as relics surviving from
some former state of things. The language of 'epochs' was quickly
adopted by others such as Desmarest, but in the service of more
modest and local interpretations.[14] Likewise, natural features were
used increasingly as evidence for reconstructing an earth history
which – because it was ultimately contingent – could not be predicted in advance on the basis of any overarching theory. It is no
coincidence that this matched the contemporary use of human
archives and monuments by historians and antiquarians, in the service of a new historicism, in place of an earlier and deductive style
of 'conjectural history'. As with the human world, the diversity of

the natural world began to be historicized: a static 'natural history' of the earth began to turn into a truly temporal *history* of the earth.

A science of the history of nature

This newly historical element was evident (as in Desmarest's work) even before geognosy was transformed by the addition of the fossil criterion; but the new attention given to fossils from around the turn of the century greatly accelerated the change. Large bones, apparently those of elephants and other tropical mammals, had long been found in relatively cool climates and high latitudes in both Old and New Worlds. But only in the 1790s did Cuvier's careful study of these bones decisively confirm earlier suspicions that they belonged to species that were distinct from any living mammals and probably extinct. That in turn made it seem more plausible that many fossil shells – far more common than any bones – also belonged to truly extinct species, and not, as had until then seemed possible or even probable, to species still lurking alive in the unexplored depths of the ocean.

With that growing belief in the generality of extinction, it made sense to treat the 'characteristic' fossils of the various formations as being indeed 'essential' to them; for they now became indices of unrepeatable *historical* change, as well as of repeatable environmental conditions. Cuvier's and Brongniart's joint study of the formations around Paris (1808–11) was accepted immediately as a decisive exemplar of a new practice that combined the older geognostic framework with a newly historical and causal dimension. Unlike Smith's rival work on the English formations, with its purely empirical use of the fossil criterion, the French study treated both formations and fossils as evidence for a truly historical interpretation of the Paris region: in terms both of a changing environment – an alternation of marine and freshwater conditions – and of a unique and irreversible history of life.

At much the same time, Cuvier transformed the concept of the earth's 'revolutions' – until then simply a vague notion of major changes in the past – into a much more concrete argument: that *some* (not all) such changes had been 'catastrophes', or sudden changes in environment, which could have caused the extinctions he claimed. Like most of his contemporaries, Cuvier rejected the older style of 'theory of the earth' as premature, and prudently abstained from suggesting what might have caused the sudden catastrophes. But he also rejected the widespread view that the 'actual causes' currently at work in the world were adequate to have produced all the effects observed.[15] The subsequent controversy led geologists in the 1820s to examine much more closely – either by direct field observation or by analysing

III

Figure 16.9 The frontispiece to Charles Lyell's *Principles of Geology* (London, 1830). This engraving of the surviving columns of the 'Temple of Serapis' near Naples deliberately symbolized Lyell's claim that all the natural 'monuments' of past events can be attributed to the action of processes no more catastrophic than those that are directly observable or recorded in human history. Although the temple (probably in fact a market) was built on dry land in Roman times, a later submergence was recorded by the borings of marine molluscs part way up each column; yet at a still later time the ruin had been elevated back to its modern position at sea-level. This epitomized Lyell's theory of non-directional or steady-state earth history; it suggested how a long succession of similar small-scale changes could have produced even the elevation of mountain ranges and the subsidence of continents, given enough time; and it neatly integrated geological with human history, and past with present, by using a human monument as a witness to geological change.

historical records – just what those agents *were* capable of effecting.

Developing that tradition, the London geologist Charles Lyell (1797–1875) argued persuasively in his *Principles of Geology* (1830–3) that the power of 'actual causes' had indeed been underestimated, and that much geological change had been very gradual, or at least, no more violent than the natural events and processes recorded in human history (Figure 16.9). 'Catastrophists' and 'uniformitarians', as they were called in the 1830s, were never sharply separated parties, and by around 1840 they had in practice reached a kind of compromise: most geological features were agreed to be the work of agencies still observable in the present world, such as sedimentation, erosion, vulcanism and crustal elevation; but it was widely believed that these processes might have slowly declined in their intensity, and might have been much more powerful in the distant past.

A compromise seemed inescapable, because some phenomena continued to resist Lyell's kind of explanation. The peculiar features (erratic blocks, till or boulder clay, etc.) that had been attributed to the most recent catastrophe were particularly puzzling, because no 'actual cause' seemed adequate to account for them. Cuvier had equated the most recent catastrophe with the flood recorded obscurely in the ancient records of many human cultures. His English follower William Buckland (*Reliquiae diluvianae*, 'Relics of the Deluge', 1823) accentuated its identification with one such record – the 'Deluge' of Genesis – and termed the deposits 'Diluvium' (see Figure 16.8). But in practice the 'geological deluge' was conceived as an event very different from any literal reading of Genesis. Its later transformation into an 'Ice Age', due mainly to the Swiss naturalist Louis Agassiz's *Études sur les glaciers* (1840), finally severed any such connection, for all but the scientifically marginal 'scriptural geologists' (mainly in Britain and North America). Above all, however, the acceptance of some kind of glacial period in the geologically recent past served to confirm that the earth had had an unexpectedly eventful history.[16]

A history of earth, life and man

By around 1840 most geologists conceded in practice that the course of earth history could not be predicted in advance by any grand theory – neither by Lyell's steady-state theory nor by the more generally favoured theory of a gradual cooling – but only reconstructed by detailed analysis of the organic and inorganic 'archives of nature'. That conclusion was embodied most strikingly in the first tentative attempts to represent in pictorial form what the world might have looked like in that remote pre-human past (Figure 16.10). It was embodied more formally in the proposal in 1840, by the English geologist John Phillips (1800–74; the nephew of William Smith), that the whole history of life could be summarized in three great eras, the Palaeozoic, Mesozoic and Cenozoic (the eras of ancient, intermediate and recent kinds of animals). With such terms, the historicization of the older geognostic classification of formations and systems was in principle complete.

That this immensely long and complex history was almost entirely *pre-human* was a conclusion that was transformed in these same decades from conjecture to consensus. Back in the late

III

Figure 16.10 'A more ancient Dorset': a view of life at a remote pre-human period (in modern terms, Jurassic), as drawn by the English geologist Henry De la Beche and lithographed by the artist George Scharf (1830). This was one of the first true pictorial reconstructions of extinct animals, based on a detailed analysis of well-preserved fossils and showing them in their inferred habitat. Large-jawed ichthyosaurs, thin-necked plesiosaurs and flying pterodactyls (in modern terms, pterosaurs) are the most prominent animals. Such imaginary landscapes were quickly adopted for popular books, and served to make the new earth history vividly real to a wide public.

eighteenth century, Buffon's seventh and human epoch (added only shortly before publication) made explicit a speculation already widespread among naturalists: that the whole of human history was but a brief final chapter in a far longer story, recorded in the thick sequences of fossil-bearing Secondary rocks. Buffon's total time-scale of tens of thousands of years – based on scaling up the results of experiments with cooling model globes – was modest by later standards; but it was quite vast enough to be, in a literal sense, almost unimaginable. In the subsequent decades, quantitative estimates of geological time were rarely made explicit, simply because there was little concrete evidence to base them on, and they were widely regarded as merely speculative. But the practice of geologists leaves no doubt that by around 1840 they had surmounted the imaginative hurdle of thinking about vast expanses of time, as successfully as astronomers had become accustomed to thinking about the vast expanses of space.

Geologists' estimates of time were, of course, far too large to be compatible with the traditional chronologies derived from a literal reading of Genesis. However, such literalism had already been discarded by most savants or 'men of science', whether they were personally religious or not, chiefly as a result of the development of scholarly biblical criticism. After the mid-eighteenth century, no major naturalists were seriously hampered by ecclesiastical criticism, still less by persecution, on account of the time-scales they proposed. Buffon, for example, received only the most perfunctory criticism in 1778 for his *Époques*, in contrast to that which had greeted his earlier theory of the earth in 1749. After the turn of the century the relation between geology and Genesis became a marginal issue for geologists, except for the public relations of their science, and even then only locally (mainly in Britain and North America).

What was more problematic was the relation between the complex history of the earth and its life, as it was reconstructed with increasing confidence and precision by geologists in the early nineteenth century, and the far shorter span of human history, as it was extended backwards by analysis of ancient documents (of which Genesis was only one). Both natural and textual sources continued to be deployed in conjunction with one another, as in Cuvier's and Buckland's work, because both seemed relevant to what was now perceived as a single *history*. Until after mid-century, however, the two sources proved extremely difficult to integrate: the fossil traces of early human life remained sparse and highly problematic, and the concept of a long human 'pre-history' preceding any literate civilization was slow to gain acceptance.[17]

Conclusion

The practice of mineralogy first began to diverge from that of botany and zoology when its problems demanded a geographical, distributional or spatial dimension, and therefore a heightened emphasis on fieldwork. That practice then became increasingly three-dimensional or structural in character, and developed into 'geognosy', the study of rock masses or 'formations' and their world-wide correlation. Geognosy in turn was transformed (into what was later termed stratigraphy) by the striking empirical success of the new criterion of fossils. Initially distinct from all such descriptive practices was the causal project of 'theory of the earth', which aimed to model the likely course and causes of the earth's temporal development. This did not become truly historical, however, until its deductive style was abandoned: its concepts of nature's 'epochs', and of specific features as nature's 'archives', were absorbed into a more inductive and contingent style, by being combined with the newer fossil-based geognosy. By about 1840, a long and complex earth history, dwarfing the whole of subsequent human history, had become a consensual feature of the scientific view of the world.

Further Reading

Ellenberger, François, *Histoire de la géologie*, vol. II, *La Grande éclosion et ses prémices, 1660–1810* (Paris, 1994).

Gohau, Gabriel, *A History of Geology* (New Brunswick, NJ, 1990).

Les Sciences de la terre aux XVIIe et XVIIIe siècles: Naissance de la géologie (Paris, 1990).

Gould, Stephen Jay, *Time's Arrow, Time's Cycle: Myth and Metaphor in the Discovery of Geological Time* (Cambridge, MA, 1987).

Grayson, Donald, *The Establishment of Human Antiquity* (New York, 1983).

Greene, Mott, *Geology in the Nineteenth Century: Changing Views of a Changing World* (Ithaca and London, 1982).

Laudan, Rachel, *From Mineralogy to Geology: The Foundations of a Science, 1650–1830* (Chicago, 1987).

Porter, Roy, *The Making of Geology: Earth Science in Britain, 1660–1815* (Cambridge, 1977).

Rudwick, Martin, *The Meaning of Fossils: Episodes in the History of Palaeontology* (London and New York, 1972; Chicago, 1984).

The Great Devonian Controversy: The Shaping of Scientific Knowledge among Gentlemanly Specialists (Chicago and London, 1985).

Scenes from Deep Time: Early Pictorial Representations of the Prehistoric World (Chicago and London, 1992).

Rupke, Nicolaas, *The Great Chain of History: William Buckland and the English School of Geology, 1814–1849* (Oxford, 1983).

Secord, James, *Controversy in Victorian Geology: The Cambrian–Silurian Dispute* (Princeton, 1986).

Minerals, strata and fossils

I am grateful to Jane Camerini, Rhoda Rappaport and Kenneth Taylor for helpful criticism of a draft of this essay. I have also gratefully borrowed the notion of 'proxies' from Mark Hineline, 'The visual culture of the earth sciences, 1863–1970' (Ph.D. dissertation, University of California, San Diego, 1993). The research for this essay was supported by the National Science Foundation (grant no. DIR–9021695).

1 Many such volumes were costly, and could only be owned by the rich or by institutions; hence it was important to most naturalists either to belong to a scientific institution with a good library, or to have access to the library of an affluent patron with scientific interests.

III

2 There were some speculations about the origins of animal, plant and mineral species, but rarely as more than marginal discussions of possible common origins of similar varieties, or – at most – of related species within a genus; it would be anachronistic to regard such theories as truly evolutionary.

3 Thin-sectioning techniques and polarized light microscopy were not developed until the mid-nineteenth century; before that time it was difficult to distinguish specimens of basalt from those of greywacke, for example, since little could be discerned with a hand-lens, and the results of gross chemical analysis were enigmatic.

4 Desmarest's study of the extinct volcanoes of Auvergne in central France (first published in 1774) proved literally decisive: a steady stream of other naturalists, going there in the subsequent decades to see for themselves, came away convinced of the volcanic origin of at least *these* basalts.

5 On Werner and the Neptunist–Vulcanist controversy, see Martin Guntau's chapter in this volume.

6 Saussure's four-volume *Voyages dans les Alpes* (1779–96), was a particularly influential model in this respect. By the early 1800s, demonstrable fieldwork experience had become an essential qualification for any author wishing to be taken seriously in this kind of science.

7 This kind of science therefore flourished first in countries, notably France, where accurate maps had already been made for political or economic reasons of state. See, for example, Josef Konvitz, *Cartography in France 1660–1848: Science, Engineering and Statecraft* (Chicago, 1987).

8 The distinction corresponds more closely to that between 'hard-rock' and 'soft-rock' terrains, in the colloquial usage of modern geologists, than to any modern stratigraphical or age-related terms. 'Primary' rocks included, in modern terms, igneous and metamorphic rocks of almost any age; 'Secondary' rocks included many sediments now regarded as Tertiary or Cenozoic in age, but also some that would be assigned to the Palaeozoic or even the Precambrian.

9 Werner's little book entitled *Kurze Klassifikation* ('Short Classification', Dresden, 1787), and the students from many countries who attended his lectures at Freiberg, spread the idea widely: see Guntau's chapter in this volume. Werner later interpolated a category of 'Transition' rocks between Primary and Secondary.

10 For animal and plant species, diagnostic features described from specimens were sometimes supplemented by those of habitat and geography; but there was no separate classification based *primarily* on fieldwork, until the development of biogeographical categories in the nineteenth century.

11 The geological map of England and Wales published in 1820 by the Society's first president, George Greenough, had become in effect the cartographical model; it had largely superseded the similar map by William Smith (1815).

12 Such 'theories' thus corresponded closely to the modern concept of scientific models. The genre was founded, in effect, by Thomas Burnet's *Sacred Theory of the Earth* (London, 1681–9), which ingeniously combined Cartesian with biblical motifs.

13 The nineteenth-century meaning of 'geology' was even wider than the modern, because it usually embraced the older 'mineralogy', and also such specialties as palaeontology and geophysics (not named as such until later): in effect, it was close to the modern phrase 'earth sciences'.

14 Desmarest's study of the extinct volcanoes of Auvergne (1779) was an influential example of this kind of geohistorical analysis. Specifically, he argued that basalts now capping some of the plateaux in Auvergne represented lavas that had flowed down valleys at a remote epoch in the past; but that subsequent erosion had left them high and dry above the present valleys, some of which were occupied by lavas of a far more recent epoch.

15 The introduction to Cuvier's great four-volume *Recherches sur les ossemens fossiles* (Paris, 1812) was later reprinted as a short popular book, which made his argument known throughout the Western world. Cuvier regarded his 'catastrophist' view as the minority position at the time, and the actualist interpretation (as it has since been termed) as the more usual opinion. An influential example of the latter, far

more widely read than Hutton's book, was the *Illustrations of the Huttonian Theory* (Edinburgh, 1802) by the Scottish mathematician John Playfair.

16 Agassiz's theory of an Ice Age was so extreme in form (he argued that most of the earth's surface had been covered in ice, and all forms of life wiped out) that it was met with great scepticism. On the other hand, geologists soon accepted a more moderate version, with extensive glaciers and drift ice, conceding that the climate in the geologically recent past had been much colder than at present, contrary to what had been assumed from the model of a steadily cooling earth.

17 An excellent account of the British side of that later development is in Bowdoin Van Riper, *Men among the Mammoths: Victorian Science and the Discovery of Human Prehistory* (Chicago and London, 1993).

IV

THE EMERGENCE OF A NEW SCIENCE

From Mineralogy to Geology: The Foundations of a Science, 1650-1830, by Rachel Laudan (Chicago and London: University of Chicago Press, 1987), xii + 278 pp., £21.95/$32.95.

I

AT FIRST GLANCE, the title of Dr Rachel Laudan's book might seem to echo the "From A to Z" formula much adhered to by an earlier generation of historians of science. In this case, however, the "From ... To ... " points to a problem of great importance and continuing interest in the history of science, namely the question of the origins of any new or reconstituted discipline or specialised area of scientific study. One of the effects of the general renunciation of positivistic certainties in the history of science has been to make such origins problematic, since the appearance of a distinctively new science can no longer be attributed simply to the natural differentiation of a cumulatively growing body of knowledge. In recent years, historians have increasingly recognised the complex character of the developments that have led to fundamental reconstitutions of the map of knowledge. Major areas of science are now seen as having emerged, or been shaped, or invented, at specific times and places and under highly specific intellectual and social circumstances. It is not necessary to accept Foucault's view that such reconstitutions are necessarily revolutionary or abrupt in character. Indeed, the evidence for gradual and evolutionary origins seems at present to be much stronger than evidence to the contrary. It has certainly become highly problematic to talk of "biology" or "physics", for example, before the nineteenth century, without careful qualifications.

Another major science that "emerged" in the decades around 1800—or at least, that began to have its modern label attached with a recognisably modern meaning—is the science of geology. Dr Laudan's book examines, in the words of her subtitle, "the foundations of a science". The book's title states succinctly her central claim, namely that the historical foundations of geology are to be found in the pre-existing science of mineralogy. That claim might seem unsurprising to the point of being banal: after all, both take the solid constituents of the earth's crust as part of their subject-matter. But in fact Dr Laudan's argument represents an important work of "historical revisionism". The origins of geological science have generally been seen in the tradition of reconstructing the history of the earth, or in that of theorising about the causal explanation of its features, rather than in that of classifying its constituent materials.

II

Dr Laudan begins her interpretative survey of the origins of geology by reviewing the history of what became known as mineralogy, from Agricola in the mid-sixteenth century to Becher in the late seventeenth. Much of this material is familiar to students of the history of chemistry, but it is valuable to have it reassessed in what was, for many of these writers, its primary context of mineralogy. The main intellectual problem was that of making sense and order

IV

out of the bewildering variety of inanimate solid objects and materials that are found naturally occurring on or in the earth. The very word "mineral" had a much wider connotation than in modern usage, which is itself a reflection of the problem. As Dr Laudan rightly emphasises, the shifting categories of "elements" and "principles", of "earths", "metals", "salts" and so on, all betray a profoundly unstable taxonomy. (She contrasts this with a relative stability in zoological and botanical taxonomy at the same period; but to later taxonomists such as Ray and Linnaeus, all the three branches of natural history seemed equally in need of their reforming attention.)

Dr Laudan sees this kind of taxonomic mineralogy turning towards what could later become geology, in the "chemical cosmogony" of Becher's *Physica Subterranea* (1668) and the tradition to which it gave rise in the following century. What distinguishes this from straightforward mineralogy is its attempt to postulate a chemical or mineralogical route for the development of the earth, from an assumed beginning as a fluid chaos to its present condition as a structured solid. For such a cosmogony, the focal concept, in modern terms, is that of a chemical differentiation; its central analogy for understanding this process is also chemical, namely, what goes on in the then newly recognised space of the chemical laboratory. The main reference point in the Book of Genesis is the primal separation of water and earth.

What makes this an interesting suggestion, and one that the author could well have developed and justified in much more detail, is the contrast between this chemical tradition and that of mechanical cosmogonies, and the claim that the chemical one was a much more important component in the later emergence of geology. This reverses the more common historical judgement, in which the theories of the "mechanical philosophers", and above all that of Descartes, have been seen as the primary foundation for a science of the earth. In these theories the focal concept, in modern terms, is that of mega-tectonic change, for example, crustal collapse; the central explanatory analogy is that of Galilean experimental mechanics; and the main point of reference in Genesis is the Flood. In fact this Cartesian tradition gets rather short shrift in Dr Laudan's account. There are brief summaries of Descartes's ideas, and those of Steno, Whiston and Woodward; but Thomas Burnet is not mentioned at all, although it was his "theory of the earth" that other writers in the later seventeenth century felt most obliged to discuss, if only to refute it.[1]

More important is the fact that Dr Laudan describes the conceptual structure of these "mechanical" cosmogonies in terms that could in fact be applied equally to the chemical ones. They all postulated a set of initial conditions at the origin of the earth, assumed the operation of some general natural laws, and thence derived a "likely story" of the historical development of the earth from its beginning to the present. In all of them, this would then be checked empirically against the relevant sources of "testimony": of Genesis, which was taken almost universally to be the only reliable historical record of the earliest periods; and of Nature, for which the scanty known details of rocks, physical geography, etc., provided a rather undemanding constraint. The only major difference between mechanical and chemical cosmogonies—and it was certainly not unimportant—was of course that the initial conditions and the general laws were regarded as having been, respectively, either mechanical or chemical in character. But at this

[1] Roger, Jacques, "The Cartesian Model and its Role in Eighteenth-century 'Theories of the Earth' ", in Lennon, Thomas M. *et al.* (eds), *Problems of Cartesianism* (Kingston and Montreal: McGill-Queen's University Press, 1982), pp. 95–125; Gould, S.J., *Time's Arrow, Time's Cycle* (Cambridge, Mass.: Harvard University Press, 1987), Ch. 2.

IV

stage in the argument the sceptical reader, with the benefit of hindsight into what later became geology, might reasonably suggest that perhaps both cosmogonical traditions were significant factors in that emergence.

Dr Laudan's case for the greater importance of the chemical approach is given some support, however, by her interpretation of eighteenth-century work. She rightly emphasises the mineralogical context of much in the "Becher–Stahl School" that is usually treated simply as part of the history of chemistry. In a rare gesture towards the institutional history of science, she points to the economic basis for the growing state support for mineral science, particularly in the foundation of schools of mines in several parts of Europe in the later eighteenth century. The first and most notable of these was the Bergakademie Freiberg established in 1765, which later provided Werner with an institutional base. In this context of a growing concern with mines and minerals, "rocks" gradually came to be distinguished from "minerals"—thus confining the latter term to roughly its modern meaning—and began to be studied and classified in their own right. The consolidation of rocks became a major set of problems; and laboratory work on the bewildering variety of "earths" led towards a consensus in favour of the "wet way" of solubility and precipitation, rather than the "dry way" of fusion and crystallisation, as the more plausible explanation of how rocks come to be rock-like.

At the same time, this work was put to the service of chemical cosmogonies that enlarged and extended Becher's example. Becher's root idea of a chemical differentiation from a chaotic fluid was articulated early in the century into the notion of a primitive ocean that had been a thick chemical soup, from which different kinds of rock had been precipitated successively in the course of time. By the middle of the century there was a general recognition of two major categories of rocks, which in practice were defined as much by their ostensible relative age—indicated by their relative position—as by their mineralogical character. Of the many names proposed for these categories, perhaps the most widely adopted were "primary" and "secondary". Though Dr Laudan does not note the point, these two terms could be interpreted in either a chronological or a formal sense. They roughly correspond to Rouelle's *terre ancienne* and *nouvelle*, which are clearly temporal in meaning, but also to Lehmann's *Ganggeburge* and *Flotzgeburge*, which are clearly mineralogical. The point is not a pedantic one. The evaluation of Dr Laudan's central contention about the origins of geology hinges on this recognition of the essential ambiguity of such terms as "primary" and "secondary", which in turn are important because they were the forerunners of a much wider range of names such as Jurassic, Carboniferous, etc., which survive into modern geology.

What is not in doubt, however, is that the distinction between primary and secondary was interpreted in terms of a sequence of chemical processes. These processes were not only chemical: some of the "secondary" rocks—for example, conglomerates, which are consolidated gravels—seemed evidently derived from the erosion of pre-existing rocks, and were accordingly given a mechanical explanation. This suggests that Dr Laudan's distinction between chemical and physical cosmogonies may be more sharply drawn than the facts of history warrant, at least for the eighteenth century. She treats the physical cosmogonies of the eighteenth century only briefly, on the grounds that such theories had little impact on mineralogy. Yet although, for example, Buffon's *Époques de la Nature*, published in 1778, was built around the postulate of a cooling globe, his explanatory repertoire also included chemical processes, albeit with a then unfashionable focus on the agency of heat. There is much evidence, not least in the proliferating use of the language of "epochs" after 1778, to suggest that Buffon's work was highly important for the growth of temporal conceptions of the earth in the later eighteenth century; but there is no reason to regard this as antithetical to the importance of those whom Dr Laudan terms chemical cosmogonists. The historical problem is to discover how they came to be related. That "discovery of time"—the author uses the phrase that was given currency by Professor Stephen Toulmin and Dr June Goodfield's pioneering survey[2]—was impeded, in her opinion, not by religious prejudice but by the deep-rooted assumption of a temporally static Nature. Specifically, the static world-view was embodied in, and epitomised by, the practice of taxonomy. Dr Laudan rightly rejects the arrogant philosophical assumption that a focus on taxonomy is a mark of scientific immaturity, and she recognises the central importance and high achievements of taxonomy in eighteenth-century science. But she argues that the adoption of a thoroughgoing temporal conception of the natural world was impeded by the attempt to extend Linnaeus's hierarchical taxonomy of static natural kinds to the mineral kingdom. Linnaeus himself had little success in this direction, and the attempts by French mineralogists such as Romé de l'Isle and Haüy to apply the Linnaean methods were, Dr Laudan claims, tangential to the science of the earth until much later—Dolomieu, her third "key figure" in this category, can surely not be dismissed as merely marginal to geology. She argues that, in contrast to France, the "botanical model" for mineralogy was decisively rejected by the "majority tradition" which she locates in Scandinavia and the German states, in the work of mineralogists such as Cronstedt, Bergman and Werner. This may well be so; but the contrast between the French focus on external crystal form as the key to mineral taxonomy, and the German–Scandinavian insistence on the primacy of chemical composition, is not self-evidently parallel with any contrast between static and developmental categories. Neither seems intrinsically better adapted than the other to form a fundamental conceptual scheme that could generate a more historical approach to the study of the earth.

III

It is at this point in Dr Laudan's argument that her main hero, Werner, is brought on stage. Following the lead of Alexander Ospovat in the United States, and of East German historians with a long-standing interest in what went on at the Freiberg school of mines, she assigns to Werner a truly heroic role, and thereby completes the task of rescuing him from his more traditional position of cardboard villain in the history of geology.

Although not himself primarily a chemist, Werner was steeped in chemical mineralogy, and his cosmogony—never embodied in any major publication—stood squarely in that tradition. But what Dr Laudan sees as his outstandingly important innovation was to define the category of "formation" in primarily temporal terms. Each *Gebirgsformation* was a defined mass of rock intermediate in size, scope and generality between the well-established handful of major units (primary and secondary, together with volcanic and alluvial), and the practical miners' recognition of a vast number of very small and purely local units, such as individual coal seams and their adjacent strata. Dr Laudan argues persuasively that this intermediate level of gemerality was crucial to the practical value of Werner's concept of "formations". Werner claimed that in principle his 20 or 30 formations were "universal", that is, of global validity, and he fitted his classification causally into a fairly standard cosmogonic scheme of a gradual

[2] Toulmin, Stephen and Goodfield, Jane, *The Discovery of Time* (London: Hutchinson, 1965; Chicago: University of Chicago Press, 1977).

IV

Book Reviews

chemical differentiation from a primitive ocean. However, empirical details led him to modify this simple model in important ways. Like others earlier in the century, he invoked mechanical as well as chemical agencies to explain the composition and structure of the *Flötzgebirge* or "secondary formations". The evidence of repeated rock-types led him to postulate a certain degree of fluctuation in the putative sea-level in the course of the earth's history, rather than its simple and uninterrupted retreat, as the primitively universal ocean was transformed into the present, more localised areas of sea.

This much is uncontroversial, and it is good to have Werner rehabilitated once more, and rescued from the old canards that he was a catastrophist, or a biblical literalist, or an adherent to a very short time-scale, and so on. Furthermore, Dr Laudan suggests that Werner's use of formations as the primary entities in the study of the earth helped shift the locus of practice away from the laboratory and out into the field. This is an intriguing suggestion that deserves further study, though there were certainly many other factors involved in the shift towards fieldwork; Dr Laudan does not mention the powerful example of Saussure's *Voyages dans les Alpes* (1779-96), and his much more explicit advocacy of field-work.

What is more problematic is the author's claim that, in Werner's concept of formations, "time is of the essence" (p. 94). The claim is important, because if valid it would locate a crucial component of "the discovery of time" in the work of Werner and his famous and influential "school". According to Dr Laudan, Werner defined formations as unique historical entities rather than as atemporal natural kinds; he thereby gave mineralogy—or rather, *avant la lettre*, geology—an intrinsically historical classification that was incompatible with a static world-view. In the longer run, according to Dr Laudan, this led geology to draw away from mineralogy—in the modern sense—chemistry and natural philosophy, to become the first fully historical natural science.

The problem with this reading of Werner is that his slender publications, supplemented—thanks to the Werner scholars—with some scrappy manuscript notes and students' synopses of his lectures, do not seem to give the historical element the primacy that Dr Laudan's interpretation requires. Werner's adoption of the term "*Geognosie*" for his main field of teaching and research does not, in itself, express any markedly temporal approach. As Dr Ospovat justly para-phrased it, "geognosy", like the slightly later usage of "geology", simply denoted "the abstract systematic knowledge of the solid earth".[3] Werner's brief but important booklet setting out his *Kurze Klassifikation* (1786) was simply a proposed taxonomy for *Gebirgsarten*, or kinds of formation, defined primarily in terms of their mineral composition. Above all, when Werner outlined his work on formations—in fact he generally used the completely non-temporal term *Gebirge*—he described it as "the search for their essential differences, based on their mode and time of formation; and on the classification or characterization of these differences according to the nature of these rock masses".[4] Of this description, Dr Laudan quotes only the phrase "mode and time of formation"; and even this, which forms the nub of her "historicist" reading of Werner's project, suggests that the causal origin of formations was to be as important a criterion as their relative ages.

The fundamental problem with this "historicist" reading of geognosy, as Dr David Oldroyd recognised,[5] is that it places back into Werner's time a temporal

---

[3] Werner, A.G., *Short Classification and Description of the Various Rocks*, ed. Ospovat, Alexander M., (New York: Hafner, 1971), p. 102.
[4] *Ibid.*, p. 19.
[5] Oldroyd, David R., "Historicism and the Rise of Historical Geology", *History of Science* (September 1979), pp. 191–213; (December 1979), pp. 227–257.

element that did not become significant in this tradition until later. Dr Laudan is undoubtedly correct to emphasise the importance of the criterion of "superposition" in the classification of formations, both for Werner and for his followers. But the problem arises from our natural inclination to read back into their work the temporal sense of terms like superposition, sequence, succession and series, when at the time their meaning was much more strongly that of structural order.

Examples cited by Dr Laudan herself (pp. 139-142) indicate clearly that this is how the followers of Werner used the notion. In a lecture in 1809, von Buch—no mean follower for anyone to have—used the revealing analogy of a numbered sequence of houses along a street, to explain how in identifying formations what mattered was their relative positions, rather than the materials of which they were made: the formations, like the houses, form a sequence only in a structural sense. Just as von Buch clearly thought the dates of the buildings irrelevant to his analogy, so we can infer that the ages of the formations was not a primary component in their definition. Likewise, when Fitton in 1818 claimed that Werner "was the first to draw the attention of geologists, explicitly, to the *order of succession* which the various natural families of rocks are found in general to present" (quoted on p. 139), it is all too easy to overlook the ambiguity of terms like "order" and "succession", and to assume that they carried a temporal rather than a primarily structural meaning. That Fitton meant them to be understood structurally seems clear from the way he summarised Werner's principle: "in the series [of formations] A, B, C, D it may happen that B or C, or both, may be occasionally wanting, and consequently D be found immediately above A; but the succession is never violated, nor the order inverted, by the discovery of A above the formation B, or C, or D . . ." (quoted on p. 140). This is the language of "above" and "below", not of "before" and "after"; the language of structural relationships between three-dimensional rock-masses in the present, not of a chronological or historical sequence of events or processes that formed those rock-masses in the past. A third example confirms this reading. In his profoundly influential *Essai géognostique* (1822), Alexander von Humboldt struggled to clarify just what made a formation "independent"; that is, on what criteria it merited a place in the series that one could hope to recognise globally or—in the words of his title—"dans les deux hemisphères". His definition could hardly be more explicitly structural: "The essential character of the identity of an independent Formation is its relative position, or the place which it occupies in the general series of formations" (quoted on p. 142). To express this, von Humboldt even developed an ingenious algebraic notation (*pasigraphie*)—not mentioned by Dr Laudan—to denote these relationships of position.

Werner and his followers were obviously well aware that the superposition of formations implied a temporal order of formation—the ambiguity is of course neatly expressed in that word. But it does not follow that the idea of a history of the earth was the dominant goal of their intellectual endeavour. The general understanding of the relation between the two was accurately expressed by von Humboldt, when he wrote: "What clearly proves the *independence of a formation*, as M. de Buch has well observed, is its immediate superposition on rocks of a different nature, which, consequently, ought to be considered as more ancient" (quoted on p. 140). Superposition, a structural category, was primary; age was merely a derivative consequence.

In any case, the coherence and importance of the line of work that stems from Werner is not in doubt. Dr Laudan proposes the term "radiation" to express the way in which the example of Werner's work both spread and diverged in the course of time, being ultimately modified almost out of recognition, yet never "overthrown". The term is quite useful for a pattern that historians of science, unlike some philosophers, will find familiar. Dr Laudan argues that the Wernerian tradition split around 1810 into two branches, which she terms historical and causal, and that they developed in conceptual isolation until synthesised again by Elie de Beaumont about two decades later. This formulation comes close to defining as Wernerian all of what was by then called "geology"; and Dr Laudan can only cite the English "practical men" such as Smith, and the anomalous Hutton, as outsiders. However, this is certainly consistent with her assessment of the far-reaching—and grossly under-recognised—importance of her central figure. She reduces Dr Nicholas Rupke's "English school" of Buckland and others[6] to a mere local variant, distinctive only for its concern with diluvialism and the relation between geology and Genesis; and she rightly stresses the overtly Wernerian character of the early Geological Society of London, the first such body to be devoted to that science. But her apparent historical antipathy to British geology in this period gets the better of her when she dismisses it as amateurish, on the grounds that its practitioners lacked technical training in mineralogy; her own interpretation has shown how Werner helped shift the locus of work from the laboratory to the field, which made that expertise much less indispensable than it had been before. On the other hand she is right to emphasise the way the Wernerian approach was widely disseminated by text-books, journals, and the control of institutional positions; though this point, like others that involve a "social" dimension, gets too little discussion or analysis in her book.[7]

As hinted above, Dr Laudan follows Dr Roy Porter and others in her revisionist interpretation of Hutton as a figure who, far from being "the founder of modern geology", was of rather marginal significance. His famous *Theory of the Earth* (1795) was highly anomalous in its day for proposing a cyclic rather than developmental pattern for earth history; for using Boerhaavian heat theory to explain how this endless round of change might be generated; and for being founded in the metaphysics of deism, not of Christianity. Dr Laudan's discussion of Hutton's work focuses, however, on the empirical problems that his contemporaries found in his work, rather than on its metaphysical dimension. She avoids the mistake of dismissing Hutton as a "mere" theorist, an attribution that has been difficult to sustain since Dr Gordon Craig published the superb drawings that Hutton got Clerk to make for him in the course of their fieldwork together.[8] But she rightly points out that in Hutton's method the function of fieldwork could only be to find illustrative evidence to support a theory already conceived, rather than to generate new theoretical insights. The chronology of his work confirms that this was indeed his procedure. Dr Laudan points out that the cool reception of Hutton's work is not surprising, even if one discounts the political suspicions of his deism at the height of the Revolution in France. Quite apart from the basic implausibility—in the eyes of his contemporaries—of a quasi-eternalistic cyclic system, there were serious problems on a much lower level with crucial parts of his theory. In particular, the central role that he ascribed to heat as the only agent of consolidation encountered serious difficulties in the case of limestones, where every experiment—not to mention the lime-kilns that dotted the countryside at that time—seemed to belie his case. Hutton's few followers, such as Playfair and Hall, had to minimise important parts of his theory or modify it radically to make

[6] Rupke, Nicolaas A., *The Great Chain of History: William Buckland and the English School of Geology (1814-1849)* (Oxford: Clarendon Press, 1983).

[7] Porter, Roy, *The Making of Geology: Earth Science in Britain, 1660-1815* (Cambridge and London: Cambridge University Press, 1977); Bailey, Edward Battersby, *James Hutton: the Founder of Modern Geology* (Amsterdam, London and New York: Elsevier, 1967).

[8] Craig, Gordon Y., (ed.), *James Hutton's Theory of the Earth: The Lost Drawings* (Edinburgh: Scottish Academic Press, 1978).

it palatable even to themselves. Dr Laudan concludes, I think correctly, that Hutton, whatever his iconic significance for modern geologists, was of only marginal importance in the establishment of geology as a science, and then mainly through the very indirect survival of his ideas, via Playfair, in the work of Lyell.

IV

Having relegated "the father of modern geology" to the sidelines of history, Dr Laudan next traces the fortunes of the Wernerian tradition that she terms "historical". As I have indicated, that adjective is in my opinion subtly anachronistic, in that it attributes to the Wernerians a temporal sense that was at best a subordinate part of their approach to the study of the earth. On the other hand, there is no doubt that their sophisticated analysis of the architecture of formations in the earth's crust formed potentially the raw material for a full-blooded history of the globe.

Among the lasting achievements of the Wernerians, as Dr Laudan shows, was their gradual clarification of the distinction between a rock and a formation, a *roche* and a *terrain*, a *Gestein* and a *Gebirgsart*. As already mentioned, this then entailed working out in practice just which sets of strata in a given region displayed the "independence" that merited the status of a "formation". A major problem then became the correlation of formations between different and even distant regions. This was only resolved gradually, and not until the 1820s, by a process of practical evaluation of diverse criteria. The "mineral character" of the rocks comprising a formation—in modern terms, their lithology—was the preferred criterion at first. This is not surprising, since chemical composition had been the starting point of Werner's classification, and the foundation of his cosmogonical scheme of a gradual chemical differentiation. The altitude at which strata were found, and the angle at which they were tilted or dipping, were also considered relevant for a time, because both were related to relative antiquity on the Wernerian cosmogony: if the primitive universal ocean had sunk gradually in level, precipitating the various rocks on to the irregular and even mountainous surface of the primitive earth, the oldest rocks would be those preserved at highest altitudes in mountain ranges, and those most strongly tilted. But these criteria soon became problematic, when relatively young secondary formations were found high up on Alpine mountain-tops and, elsewhere, tilted spectacularly into a vertical position.

In the end, as Laudan rightly claims, another criterion came to be given the greatest weight, because in practice its use led to the most consistent conclusions. Fossils were far from having been neglected by earlier writers, but generally they had been invoked only as evidence for major changes in the environment: the shells of marine molluscs on hilltops far from the sea proved that the oceans had once been quite different in extent; the bones and teeth of elephants in northern Europe proved that those latitudes had once enjoyed a much warmer climate. Dr Laudan makes the important suggestion that the crucial new factor in the use of fossils in geology arose from the recognition of extinction. Fossils had earlier been regarded as unquestionably "accidental" characters, which could not properly be used to define or identify a formation. But Cuvier and Brongniart's celebrated monograph on the Paris basin, published in 1811, made the startling assertion that "What *essentially characterizes* this formation is the fossils it contains . . ." (quoted on p. 146). Dr Laudan claims that, until the acceptance of extinction, there was no reason for anyone to anticipate that fossils might indicate the age of a formation rather than just the conditions under which it was formed; a particular environment, and the organisms it supported, might well have recurred many times, whereas once a certain set of species or genera had become extinct they could never recur in a later formation.

IV

This is an intriguing suggestion, but it will need more thorough documentation than the author provides. She argues that it is supported by the "significant overlap" between those geologists who stressed the value of fossils and those who believed in the reality of extinction; but this hardly amounts to much, since it is difficult to find any geologist who did not accept extinction, within a few years of Cuvier's classic work on fossil mammals. Dr Laudan attributes the new interest in fossils not to Cuvier but to his intellectual "grandfather", Blumenbach; but as evidence for this she cites only a publication of Blumenbach's dating from 1806, a full decade after Cuvier's first and sensational claims for mammalian extinction.

Dr Laudan's claim for the link with extinction also depends on relegating William Smith to a marginal position. Smith is a figure as iconic as Hutton for modern geologists. In an earlier article she argued quite persuasively that Smith, when tracing formations across England to make his famous geological map (1815), had used fossils only where other criteria failed, and that this had led him into some revealing major errors.[9] In her present book, she develops this interpretation of Smith; she argues that even in Britain geologists learnt the geological use of fossils from the Continent, not from Smith, and concludes that "it is hard to imagine that the development of geology would have been much different if he had never published" (p. 168). This may well be true, but it needs closer documentation to establish it. The point is important beyond the trivial issue of assigning credit for priority. As Dr Laudan rightly stresses, Smith worked in virtual isolation from the Wernerian tradition, and, as a self-taught "practical man", came moreover from a quite distinct social class. His elimination from the mainstream story therefore entails an important redefinition of the social history of geology.

Still, Dr Laudan has made a good case for concluding that, one way or another, the newly detailed attention given to fossils from the turn of the century was largely responsible for a huge improvement in the description and correlation of formations throughout Western Europe and even beyond. By about 1830 much of the geological "succession" was consensually established in a form that survives recognisably into modern geology. However, bearing in mind the essentially structural emphasis of this work in "stratigraphy"—as it came to be termed—its interpretation in terms of earth history is another matter. Much more evidence is needed to show just how, when and by whom a pile of formations with characteristic fossils came to be seen as a record of a complex series of historical events. Dr Laudan claims that what I termed the "directionalist synthesis" in the geology of the 1820s was nothing other than the Wernerian tradition, with only the addition of the geophysical theory of a cooling earth.[10] But that extra element was historically decisive, precisely because it suggested a way to translate the pile of formations into temporal terms, by proposing a causal explanation for the changing character of those strata and their fossils.

As already mentioned, Dr Laudan sees the tradition of causal theory in geology as a part of the Wernerian heritage that diverged from the "historical" (or, as I would prefer to term it, the stratigraphical) tradition during the early nineteenth century. But her claim that the two lines of work were largely independent, even within the work of individuals such as von Buch and von Humboldt, who contributed to both, needs more careful documentation. She argues, for example, that von Buch's influential theory for the elevation of mountain chains—which, as

[9] Laudan, Rachel, "William Smith: Stratigraphy without Palaeontology", *Centaurus*, XX, No. 3 (1976), pp. 210–226.
[10] Rudwick, Martin J.S., "Uniformity and Progression: Reflections on the Structure of Geological Theory in the Age of Lyell", in Roller, D.H.D. (ed.), *Perspectives in the History of Science and Technology* (Norman: University of Oklahoma Press, 1971), pp. 209–227.

Dr Mott Greene has shown,[11] remained a central problem for causal geology throughout the century—did not depend at all on his—and others'—simultaneous work on "historical" geology. Yet the problem of explaining mountains causally was in fact altered dramatically in just these years, as a result of stratigraphical research. It was this work, and this alone, that established that mountains had been elevated—by whatever causal means—at very different periods of earth history, including geologically recent times. Dr Laudan follows Greene and myself in emphasising the historical importance of Elie de Beaumont and his theory of successive mountain elevations (1829-30). I would regard this work as a distinguished example of a unified research tradition in which causal and stratigraphical elements had long been combined fruitfully. On Dr Laudan's interpretation, however, Elie de Beaumont occupies a culminating position, for it is in his work that she sees the coalescence of the "historical" and causal branches of the Wernerian tradition, and, at the same time, the transcendence of that tradition and hence in effect its demise.

The last chapter of Dr Laudan's historical survey is devoted to the figure who has been usually regarded as occupying this kind of pivotal position, namely Lyell. Having relegated Hutton and Smith to marginal positions, she ends her survey with yet another gesture of iconoclasm, suggesting that the historical evidence "tends to favor the position that Elie de Beaumont was more influential than Lyell" (p. 221). This contrasts strikingly with Wilson's identification of Lyell with the "revolution in geology" that set the science on its feet for the first time.[12] Dr Laudan enlarges on an earlier article[13] to argue that Lyell's primary commitment was not to the method of actualism, i.e., to the use of the present as a key to the past, but rather to the ultimately Newtonian methodology of *verae causae*, which he found in the writings of Hutton's exponent Playfair and, in his own day, in the work of John Herschel. In practice the two methods are not easy to distinguish, but Dr Laudan's argument is certainly plausible. Nevertheless, her generally excellent analysis of Lyell is left somewhat up in the air, as a result of her doubts about his longer-term significance, particularly about his influence outside the English-speaking world. But this, as Dr Laudan recognises in conclusion, is only one example of our current ignorance of the history of geology after the relatively well-investigated 1830s.

V

The final chapter of Dr Laudan's book is entitled "Postscript", and reads as if it is literally just that. Here, and in a concluding section of the introductory chapter, which also reads as if it was added at a late stage, she sets out some of her historiographical principles and summarises her conclusions.

It is here that Dr Laudan's antipathy to any kind of "social" element in the history of science comes explicitly to the surface. "What makes science worthy of study", she claims, "... are scientists' beliefs about the world" (p. 17). Only the most narrow-minded historian of, say, scientific institutions might disagree. Dr Laudan moves however from that unexceptionable statement to the far more questionable assertion that the central field for the history of science must always be the history of scientific ideas, and that the social history of science must play

---

[11] Greene, Mott, *Geology in the Nineteenth Century: Changing Views of a Changing World* (Ithaca: Cornell University Press, 1983).

[12] Wilson, Leonard G., *Charles Lyell: The Years to 1841. The Revolution in Geology* (New Haven: Yale University Press, 1972).

[13] Laudan, Rachel, "The Role of Methodology in Lyell's Geology", *Studies in the History and Philosophy of Science*, XIII, No. 3 (1982), pp. 215-250.

IV

"second fiddle" to intellectual history. This is surely to muddle matters unnecessarily. Current work in the history of science gives no warrant for drawing such a sharp distinction between the social and intellectual dimensions. As I have mentioned in this review, Dr Laudan herself has made several valuable suggestions about the ways in which they might have been substantially related, in the case of the origins of geology. However, her apparent antipathy to any social analysis of the history of science has prevented her from exploring these insights as thoroughly as she might have done.

Dr Laudan also criticises "most microstudies in the history of geology" on the grounds that they assume that British geology in this period was typical of the science as a whole (p. 18). In fact, neither of the unnamed but evident targets of this criticism—Dr James Secord's and my own recent accounts of two early nineteenth-century British controversies[14]—made any such assumptions. Dr Laudan's emphasis on the importance of the work of German-speaking scientists can only be applauded, and she brings a new and welcome prominence to many authors who have been neglected in anglophone histories. Unfortunately, she redresses the historical balance only in part. She gives far too little prominence to, and shows too little familiarity with, the work of those who wrote in what was perhaps the most important of the three major scientific languages of the period. If her account of the origins of geology had given as much attention to authors writing in French as it does to those who wrote in German, some of her major theses might have needed substantial modification. For example, it is questionable whether her chemical and physical traditions in the eighteenth century, or her "historical" and causal traditions in the early nineteenth century, can be so sharply distinguished, if French writers are taken into account; or that the study of landscape—in modern terms, geomorphology—and of the structural features of the earth's crust were as marginal to the science of the earth as she implies. Dr Laudan has taken an important step towards transcending the bias of anglophone histories of geology in favour of persons who wrote in English; but there is much more to be done historically before our interpretations will come near to reflecting the internationality that characterised the scientific work itself.

The author contrasts her own general interpretation with the "received view" of the history of geology (pp. 222-223), but this unspecified adversary turns out to be a rather superannuated Aunt Sally. Few historians of geology would quarrel with any of the three claims of her own which she emphasises: that the "historical" tradition—which, for reasons I have set out above, is better termed stratigraphical—did not displace the causal tradition in the early nineteenth century, but rather complemented it; that the relation of theories to empirical data was no new problem at that time, and that fieldwork was not self-evidently the best or only way to resolve it; and that geologists' attitudes to the "Genesis-and-geology" issue were far from monolithic. As I have just indicated, her claim that geology became a science not in Britain but on the Continent can also be accepted, if it is modified to include the French as well as the Germans, and if it allows some role for lesser scientific breeds such as the British. Above all, her central claim that geology had important antecedents in the eighteenth-century science of mineralogy also deserves an enthusiastic welcome, provided it is not made exclusive. Some place needs to be found both for the francophone tradition of causal analysis, and for a way of tracing its fusion with the mineralogical tradition that flourished further east.

[14] Secord, James A., *Controversy in Victorian Geology: The Cambrian-Silurian Dispute* (Princeton: Princeton University Press, 1986); Rudwick, Martin J.S., *The Great Devonian Controversy: The Shaping of Scientific Knowledge among Gentlemanly Specialists* (Chicago and London: University of Chicago Press, 1985).

I have dealt here only with a few of the issues that are raised in this important "revisionist" survey of the emergence of geology as a new science. It has been worth discussing at length, because whatever its shortcomings they are those of a pioneer attempt to tackle a difficult problem in the history of science. Dr Laudan has made a notable contribution to the historical understanding of the origins of scientific disciplines. It is in fact a compliment to the book that it suggests so many areas in which further historical research is needed. Only by assembling a body of thoroughly documented case-studies, dealing with different scientific disciplines and different periods of history, shall we be in a position to begin to make broader generalisations about the intellectual and social ways in which the map of scientific knowledge has been profoundly reconstituted in the course of history.

V

THE EMERGENCE OF A VISUAL LANGUAGE FOR
GEOLOGICAL SCIENCE 1760 – 1840[1]

"We ought to talk less and draw more. I personally should like to renounce speech altogether and, like organic nature, communicate everything I have to say in sketches." *Goethe*[2]

I. INTRODUCTION

There is a striking discrepancy between the way in which modern geologists communicate their ideas to each other, and the way the antecedents of these ideas are analyzed by historians. Whether in talks at conferences or in published scientific papers and books, modern geologists make extensive use of visual materials—maps, sections, colour slides, diagrams of all kinds. They do this so much as a matter of course that to call these materials 'visual aids' seems both inadequate and pretentious. In this respect, modern geologists are following a tradition which reaches back to the formative period of their science in the early nineteenth century. But if the geological books, articles, lecture-notes, field notebooks and correspondence of that period are compared with recent published work on the *history* of geology, a contrast is at once apparent. In modern historical analysis the strong visual component of the original source-materials is generally either missing altogether or else reduced to a virtually decorative role. Few of the original illustrations are reproduced, and even when they are, they are rarely integrated with the text by being used substantively for conceptual exposition. The quantity of illustration may be limited by publisher's fiat.[3] But a more fundamental reason for the meagre use of illustrations in the history of science lies in the lack of any strong intellectual tradition in which visual modes of communication are accepted as essential for the historical analysis and understanding of scientific knowledge.

This intellectual situation is in turn related to the social history of the history of science itself. At least until recent years the profession of the history of science has been recruited mainly from those whose initial training was in the physical sciences; and these sciences are deeply rooted in an educational tradition of 'numeracy' which is strongly mathematical in emphasis and non-visual in outlook. Likewise the modern historical profession, with which the history of science is now becoming more closely linked, is deeply rooted in an educational tradition of 'literacy' which is verbal and stylistic in emphasis. And the philosophy of science, with which the history of science has had institutional links of great distinction, represents as it were an amalgam of both 'numerate' and 'literate' traditions. These factors have together produced a situation in which acceptance of the conceptual importance of *visual* modes of discourse will require a rather fundamental change of intellectual values within the history of science. In the hierarchy of our educational institutions, visual thinking is simply not valued as highly as verbal or mathematical dexterity; and an academic discipline such as the history of science has inevitably been infected by this bias of educational values.[4]

Within the history of science, the two most obvious exceptions to these remarks actually confirm the generalization. The history of medicine and the history of technology have been distinguished for their awareness of the centrality and importance of visual communication; and it is surely no coincidence that they have been developed primarily by those whose initial education was in the highly visual *practices* of medicine and technology. The present article is intended as a small contribution towards an analogous development in the history of geology. A study of the conceptual uses of visual images in an early nineteenth century science may help in a small way to counter the common but intellectually arrogant assumption that visual modes of communication are either a sop to the less intelligent or a way of pandering to a generation soaked in television.

If we survey a broad range of late eighteenth century books and journals on topics relevant to the future science of 'geology', one of their most striking features (with of course a few exceptions) is the scarcity and poor quality of their illustrations. A similar survey of material published in the 1830s shows a remarkable change : the texts are now complemented by a wide range of maps, sections, landscapes and diagrams of other kinds. During the period in which 'geology' emerged as a self-conscious new discipline with clearly defined intellectual goals and well established institutional forms,[5] there was thus a comparable emergence of what I shall call a *visual language* for the science, which is reflected not only in a broadening range of kinds of illustration but also in a great increase in their sheer quantity.

In his important book on *Prints and visual communication,* William Ivins pointed out that what art historians tend to regard as a 'minor' art form has in fact been a major means of communication in Western cultural history. Whereas verbal texts were indefinitely repeatable without serious loss of meaning—notwithstanding copyists' errors and changes in linguistic context—visual images could not be repeated without serious distortion until the invention of technical means of mass-reproduction. On this view the successive development of woodcuts, copper engravings

V

and etchings, all within the fifteenth century, was as important culturally as the development of printing at the same period; and Ivins documented the rapid utilization of prints as a means of conveying invariant visual information about a widening range of subjects—machinery, paintings, plants and animals, travel scenes and so on. It seems plausible to suggest that the second major phase of technical innovation in this field—the development of aquatints, wood engravings, steel engravings and lithographs in the decades around 1800—played a comparable role in the natural-history sciences in general and geology in particular.[6]

More was involved, however, than a broadening range of techniques for visual expression. As Ivins emphasized, prints are a means of *communication,* and as such the new modes of representation required new modes of perception by those who looked at them. Furthermore, the relation between the object depicted and its visual representation was never straightforward, however 'realistic' the intentions of the illustrator: artistic representation is always a visual *language,* which has to be learned and which changes over time.[7]

The same point can be made from within geology. For example, modern geologists tend to treat their everyday use of geological maps as essentially natural and unproblematical, perhaps forgetting their own initial difficulties in learning to read such maps and the difficulties of teaching students to do the same. But the metaphor of 'map-reading' emphasises that such a task is indeed a kind of linguistic skill that has to be learned. In other words, a geological map—or any other visual diagram in geology—is a document presented in a visual language; and like any ordinary verbal language this embodies a complex set of tacit rules and conventions that have to be learned by practice. Again, like an ordinary language, these visual means of communication necessarily imply the existence of a social community which tacitly accepts these rules and shares an understanding of these conventions. It is therefore worth studying the historical development of the visual language of geology not only for the way in which it gradually enabled the concepts of a new science to be more adequately expressed, but also as a reflection of the growth of a self-conscious community of geological scientists.

The development of the visual language of geology makes a particularly instructive case-study in the development of visual communication in science generally. It shared with other natural-history sciences a concern with configurations that could not be adequately conveyed by words or mathematical symbols alone. But it involved the visual representation of a far wider range of different kinds of phenomena than other sciences: not only specimens that could be stored in a 'cabinet' or museum, but also the configuration of topography, and the penetration of that topography to form a three-dimensional picture of the structure of the earth's crust. As the science emerged, even that structural goal became inadequate, and the causal and temporal interpretation of the observed structural configurations required—and was perhaps made possible by—the development of ever more abstract, formalized and theory-laden modes of representation. By about 1840, these forms of visual communication in geology no longer functioned as supplements to verbal description and verbal concepts; still less were they merely decorative in function. They had become an essential part of an integrated visual-and-verbal mode of communication.

In this article I want to explore the ways in which the visual element of this integrated discourse came into being. I shall first briefly survey the range of documentary materials that are available for the historical analysis of the visual language of geology, and I shall comment on the technical and economic constraints that may have influenced its development. I shall then analyze the various different forms of expression of this visual language;[8] I shall suggest their possible sources and interconnections; and I shall try to show how they were developed in the course of time into more specialized vehicles for the communication of an increasingly theoretical content. As is proper for an article in this journal, my treatment of the subject is exploratory rather than definitive: I shall survey the range of relevant material and suggest an interpretation of it, in the hope that others will supplement my material and improve on my interpretation.[9]

II. MATERIALS AND TECHNIQUES

The importance of illustrations in published books and articles, as visual statements of a 'public' kind, needs no emphasis. Other kinds of material should be mentioned briefly, however, if only to point out the range available for research. Pre-eminent among such materials, and already well appreciated by historians of the earth sciences, are the many maps which were issued in sheet form and not in atlases. The bibliographical problems of sheet maps are well known; but the historian of geology faces the same problems with other sheet materials. Single plates of memorable events, striking localities and unusual specimens; sample plates to illustrate and promote new techniques of reproduction or to advertize works to be published by subscription; special plates or diagrams issued singly without accompanying text; special charts designed to illustrate particular books but issued separately and in some cases later: examples of these and other kinds of illustrative material are only slowly coming to light, because they are often inadequately catalogued in libraries and archives, either preserved as loose sheets or bound in miscellaneous albums. Then there are the semi-private materials which were accessible to a limited social group, and which were sometimes duplicated in small numbers but not formally published : in this section the plans and sections of mine-workings, which often had 'classified' status for reasons of industrial secrecy, are of special importance; while from a different social context there are enlarged drawings made as visual aids for university lectures on geology, and the various lithographed caricatures and sketches which form such revealing evidence of the distinctive 'clubbish' ethos of the early Geological Society of London.   Finally, in a private category but often forming the rough drafts for visual statements that were later made public, are the sketches, maps, sections and other diagrams (and occasional caricatures) which are scattered profusely through the field notebooks and scientific letters of most geologists, at least by the early nineteenth century.

All these materials need to be taken into account, before the development of a distinctive visual language for geology can be adequately assessed. Needless to say, such a balanced assessment is not yet possible; but at least the range of visual materials allows us to see some of the historical questions that would be worth asking.   For example, it would be worth discovering whether the field notebooks of eighteenth century 'geologists' included as many sketches and diagrams as those of most of their early nineteenth century successors.   The ability of most of the latter to draw field sketches with reasonable competence must surely be a factor of some importance in the development of the science, and it is probably related to the fashion for amateur drawing and watercolour painting, which was widespread in the social classes from which many of these geologists came, particularly in England.   A few were artists of actually or potentially professional calibre, such as Thomas Webster, the official 'draughtsman' of the Geological Society of London, and John MacCulloch, one of its early Presidents; several such as Henry De la Beche were accomplished amateurs; while most others were at least competent enough for their field sketches to be re-drawn to a publishable standard by an artist such as Webster, and some had wives whose considerable artistic skills were used to aid their husbands' work (for example, Mary Buckland and Charlotte Murchison).   This widespread artistic skill points to the historical signi- ficance of a very general cultural trend at this period towards a greater degree of visual awareness, particularly of 'romantic' subjects such as mountain scenery and intrinsically curious subjects such as fossils, among the leisured social classes.[10]

Another question that can only be answered by a more detailed study of the source-materials is whether the general paucity of illustrations in late eighteenth century books and articles on 'geology' was compensated to any appreciable extent by widely circulated visual materials of other kinds.   I suspect that it was not, but that is at present little more than a conjecture.   If my conjecture is correct, and the illustrations contained in books and periodicals are a fair sample of *all* public visual statements in the science, then the striking increase in their quantity, quality and diversity over the period which I am surveying cannot be dismissed as an artefact of differential preservation. On the other hand, if it is accepted as a real historical trend, it is tempting to interpret it in purely economic terms, as a by-product of technical innovations which made published illustrations cheaper to produce and more effective for their purpose. This is an important argument as far as it goes, but I suggest that it is not adequate to explain the emergence of a visual language for geology.

In the late eighteenth century the main medium available for the illustration of works in the natural-history sciences was still the copper engraving. For the size of print-run that was usual for scholarly books at that period, the copper engraving provided an excellent means of reproducing certain kinds of illustration, but the whole process required much highly skilled labour, and was therefore very expensive. Consequently, illustration was generally reserved for the few subjects that the author deemed most important or the publisher most marketable. In many books the only illustration was a single frontispiece, which for this reason always deserves especially close historical scrutiny, since it functioned as a visual summary of what the author and/or the publisher considered most important about the book.[11]

That considerations of cost were partly responsible for the usual paucity of illustration is suggested by the occasional exceptions. An outstanding example is Sir William Hamilton's *Campi Phlegraei* (1776-79), which illustrated the volcanic activity of Vesuvius and its environs with a lavish series of superb hand-coloured engravings. The sumptuous format, bilingual text and advertized price of this work make it clear that it was produced for the luxury end of the market;[12] but a comparison with other moderately expensive works indicates the inadequacy of the economic explanation. For example, few 'geological' works of this period were as highly admired, or as frequently and extensively quoted, as de Saussure's great four-volume *Voyages dans les Alpes* (1779-96).[13] While less sumptuous than Hamilton's book, it was a handsomely produced work, and dealt with subjects equally appropriate for visual illustration; yet in fact it contains, in proportion to its lengthy verbal text, surprisingly few illustrations, and those rather mediocre in quality. Still more striking are the many multi-volume works of topographical travel (including much 'geological' material) which contain few if any illustrations whatever. In all these works the extra cost of more generous illustration could easily have been saved by a substantial shortening of the highly verbose texts. That this was not in fact done surely suggests that these texts were not perceived by their authors as being inappropriately verbose, and that more generous illustration was likewise not perceived as either useful or necessary.

I suggest, therefore, that while these late eighteenth century 'traveller-naturalists' had a well developed visual awareness of the topographical phenomena that they studied, they generally thought of communicating what they had seen primarily in verbal terms, and only subordinately, if at all, in visual terms.[14] A man like Hamilton can then be interpreted as exceptional not so much as an observer, but as one who perceived the value and necessity of visual communication. Yet even he was dependent not only on his ability to finance a lavishly illustrated book, but also on being able to employ an exceptional artist, Pietro Fabris, who could exploit the engraving medium to the best advantage. This is vividly demonstrated by a comparison between the illustrations in *Campi Phlegraei* and the crude engravings that accompanied Hamilton's earlier report to the Royal Society of London (see Figures 1 and 2).[15]

For the medium itself had serious limitations as a vehicle for visual communication. Apart from their cost, copper engravings were excellent for line drawings and maps requiring fine detail. But they were far less satisfactory for subjects such as natural-history specimens and landscapes. Between the naturalist and his reader/viewer stood not only the artist and the printer, but also the engraver. The artist could at least work in close partnership with the naturalist, as Fabris did with Hamilton; but his expression of subtleties of texture and shading had to be translated by the engraver on to the copper plate by means of visual conventions of hatching and cross-hatching, which, in Ivins's vivid phrase, had become an increasingly rigid 'network of rationality' superimposed on the artist's work. Like the coarse grainy texture of modern newspaper photographs, there was thus an implicit limit of 'resolution' to the detail that an engraving could convey (see Figure 2).

A further disadvantage of engravings, as a potential medium for visual communication in science, was that they had to be printed separately from the text and generally on a different kind of paper; they were bound into the book either at the end or dispersed within the text at points that suited the convenience of the binder rather than the needs of the author. This meant that the author had little encouragement or stimulus to try to write a text that would be closely integrated with his illustrations.[16]

This disadvantage persisted with most of the new techniques that were developed at the end of the eighteenth century and the beginning of the nineteenth. The aquatint process, for example, was well adapted to the needs of landscape topographical art; but like engraving it was a highly skilled and expensive process, still more so if it was supplemented by hand colouring. In the early nineteenth century, hand-coloured aquatints became for a time the supreme medium for topographical and natural-history colour-plate books. Such works were occasionally aimed at a more specifically geological audience, as when William Daniell extracted a few of the plates from his great *Voyage round Great Britain* (1814-20) and sold them separately to illustrate the volcanic geology of the Inner Hebrides (see Figure 16). Other forms of etching, such as the mezzotint, were occasionally used for the same purpose. But such techniques had few practical advantages over the better established conventions of engravings, and were not widely utilized.[17]

The invention of lithography, on the other hand, had a far-reaching impact on the natural-history sciences, for it cut out the engraver as middle-man between artist and printer, and it also enabled finely-graded shading and textural detail to be expressed with far more precision and subtlety. It was slow to be adopted, however, partly because initially the results were rather crude in quality—like drawings reproduced with an ordinary duplicating machine today. This strengthened an artistic prejudice which scorned lithography precisely because it was so cheap in relation to engraving; the engraving trade's fear of technological redundancy may also have been a factor.[18] In any case, lithography was not exploited widely for scientific purposes until the 1820s. It then owed much to the energetic promotion by Charles Hullmandel (1789-1850) of his greatly improved 'lithotint' process, and to his association with the topographical artist George Scharf (1788-1860). As early as 1818 Hullmandel began to invade the topographical picture-book market (hitherto dominated by engravings and aquatints) with lithographic illustrations, and he specifically mentioned and illustrated the value of lithography for works on natural history and geology.[19]

Hullmandel's technique was adopted in one case that was crucial for the development of geology. The last volume (1821) of the original series of the *Transactions of the Geological Society of London* was illustrated exclusively with engravings; but when the Society took over publication of the first volume (1824) of a new series, it adopted the cheaper method: almost all the specimens, landscapes and sections, and most of the maps, were lithographs printed by Hullmandel from drawings by Scharf. This arrangement continued in subsequent volumes, and enabled the Society to publish illustrations of high quality at about one-third the price of engravings.[20] I call this development 'crucial' for geology, because it enabled the Geological Society to continue to offer authors the lavish allow-ance of plates that had characterized the early volumes of its *Transactions,* but at a much reduced cost; and furthermore, as I have already empha-sized, lithographs were actually superior to engravings for many subjects of importance to geology. By contrast, other periodicals which published geological research, such as the *Philosophical transactions* of the Royal Societies of London and Edinburgh, the *Annales des mines* and the *Mémoires du Muséum d'Histoire Naturelle* in Paris, and even the in-expensive scientific monthlies such as the *Philosophical magazine,* continued

V

to include only engraved illustrations, and those rather sparse in quantity by comparison with the Geological Society's publications.

The substitution of steel for copper in the technique of engraving effected no comparable saving in absolute cost; but it did have some importance in that it made the special advantages of engraving—particularly the fineness with which detail could be rendered—available for much longer print-runs. As the size of editions increased in the early nineteenth century, particularly with the development of machine presses and cheaper paper, the steel engraving became a convenient way (*i.e.*, cheap in cost per copy) of providing a few illustrations of good quality for inexpensive books for the (relatively) mass market. Therefore it tended to be used for the customary frontispiece in books that were being printed in a run of a few thousand copies rather than a few hundred. The frontispieces of the three volumes of Lyell's *Principles of geology* (1830-33) are a well-known geological example.[21]

The last technical innovation of this period (I exclude photographic techniques as too late for my survey) was the development of wood engraving. In the hands of a Thomas Bewick this could reach the status of a genuine if minor art form, but its importance for our present topic lies less in its aesthetic value than its cheapness and its relation to the technology of printing. It provided an excellent and cheap medium for the reproduction of small line-drawings; and even small drawings of landscapes and specimens could be rendered adequately for didactic purposes, although with far less precision and subtlety than lithographs. But above all, no special paper, printing or binding was required; the blocks used for wood engravings could be set up within an ordinary galley of type and printed by the same press, so that the illustrations could appear on the same pages as the text.

Yet the potentialities of wood engraving, like those of lithography, were slow to be exploited. In the 1820s, most inexpensive works in geology and other natural-history sciences continued to have no more than an engraved frontispiece and perhaps one or two other (copper) engraved illustrations at the back, and few if any *wood* engravings within the text. Conybeare and Phillips's standard *Outlines of the geology of England and Wales* (1822), with only a few small wood engravings of geological sections within a dense text, is a characteristic example.[22] Unlike the paucity of copper engravings in late eighteenth century works, this cannot be attributed to the authors' failure to perceive the importance of visual communication, since, for example, Conybeare's articles in the Geological Society's *Transactions* are copiously illustrated. Even a decade later the situation was virtually unchanged, as can be seen, for example, in the first volume (1830) of Lyell's *Principles of geology* and the *Geological manual* (1831) of his rival De la Beche. Yet both of these books dealt with subjects suitable for generous illustration; both authors compiled field note-books that were full of sketches; and De la Beche had even published a whole volume of lithographed *Sections and views, illustrative of geological phaenomena* (1830).[23] It seems likely, therefore, that the fuller exploitation of wood engraving was inhibited not by the attitude of geological authors but by that of their publishers. Possibly the commercial enterprise which triggered a change in the situation was Charles Knight's successful *Penny magazine* (from 1832), which with its lavish use of wood engravings first brought illustrative material on to the mass market.[24] When Knight published De la Beche's *Geology* (1835) in his *How to observe* series, its 138 'wood cuts' (*sic*) were specifically mentioned as a selling point; and likewise John Murray allowed Lyell far more wood engravings in the last volume (1833) of the *Principles* than in the earlier volumes.[25] From this date onwards, a more generous use of wood engravings became normal in inexpensive English books on geology, and the practice later spread to Continental publishers. Although, as I have suggested, the authors of such books were already well aware of the value of visual communication, this new opportunity to include many small illustrations within the text must surely have spurred them towards a closer and more conscious integration of verbal and visual communication, and must also have greatly aided the reception of that integrated communication by the ordinary reader.

Looking back, however, over the period that I have surveyed, one of the most striking changes of all is that signalled by the appearance of the first periodical to be devoted specifically to geology. It is not just that the Geological Society of London (founded 1807) was the first society of its kind anywhere, but that its *Transactions* (first published 1811) were from the beginning much more generously illustrated than the ordinary run of 'geological' books and articles published in any country, either before that date or for many years afterwards. I have mentioned that the *Transactions* exploited the cheaper and better medium of lithography earlier than other periodicals; but even before that time the Geological Society had allowed its authors generous illustration, albeit in the relatively expensive medium of copper engraving. This suggests that at least some of the leading members of the Society recognized the crucial importance of visual communication in the science that they were trying to establish, in a way that sets them apart from all but a few earlier 'geologists'. No doubt the relative affluence of their members was a necessary condition for the practical fulfilment of this conception; but that contingent circumstance does not lessen the historical importance of the early work of the Geological Society, as the point at which the necessity of visual communication in geology first found institutional expression as part of the normal practice of the science.

V

III. GEOLOGICAL MAPS

A modern geological map is an attempt to depict on a two-dimensional surface what cannot in reality be seen at all except in isolated exposures, namely the outcrops of rocks which are generally concealed by soil and vegetation. The resultant distributional pattern of outcrops, when combined with the information about surface topography which is conveyed by the base-map on which that pattern is superimposed, is then intended to yield a 'deeper' meaning (both literally and metaphorically) about the underlying three-dimensional structure of the rocks. The task of 'reading' the map successfully in this way is commonly assisted by the provision of sections and other diagrams which provide 'clues' to its decipherment. Finally, the three-dimensional structure and its interaction with the present surface topography are intended to act as a starting-point for, and evidence towards, an interpretation that is still deeper in conceptual level by being causal and temporal in its goal.

Such a description of the nature and aim of a geological map will seem to adepts of geological map-reading to be either inadequate or pretentious. My reason for attempting it is to emphasize what adepts easily overlook, namely that a geological map is a highly complex, abstract and formalized kind of representation.[26] Historians of geology have been right to stress the importance of the development of the geological map as a tool for geology; but I hope that the present article will help to show that such maps should be studied historically in the context of other aspects of the visual language of geology. An earlier generation of historically-minded geologists tended to expend their energies on championing some particular individual as the true inventor of the geological map, and in rejecting rival claimants on the grounds that they did not construct 'real' geological maps. In this part of my article I want to replace this 'catastrophist' historiography with something more uniformitarian, by suggesting briefly how the geological map emerged in a more gradual way, like other aspects of the visual language of geology.[27]

The background to this story must be the point that is rightly stressed by historians of general cartography, namely the poor quality of the topographical information embodied in early maps of any kind.[28] Until the early nineteenth century the surveys on which maps of even the culturally central areas of Western Europe were based were poor or mediocre, and no geological map could be better that the topographical map that served as its base. Furthermore, even when these surveys improved in quality, as they did first in France and later in Britain and other countries, the available cartographic conventions were ill-adapted for the accurate portrayal of the physical topography that was so important for geological interpretation. De Saussure was only able to illustrate the Alpine topography of the Mont Blanc region by a map on which the mountains are represented by the ancient and crude cartographic convention of 'mole hills'.[29] Even the best surveys, despite detailed information about the positions of towns and villages and the courses of streams and rivers, were only able to indicate the configuration of the hills by a crude form of hachuring, which gives a generally deceptive impression of the topography. It makes the valleys look like steep-sided trenches cut into a flat plateau (see Figure 3), and it makes higher hills look as though they are built up in a series of terraces. It is difficult to over-emphasize the extent to which these cartographic limitations impeded the use of maps as a medium for the communication of complex and abstract forms of geological information.

The poor quality of contemporary maps may account in part for the surprisingly limited use which most eighteenth century 'traveller-naturalists' made of maps of any kind in their published work. But like the paucity of illustrations in general, it seems also to reflect a deeper attitude, in which the potential of maps as a form of visual communication was scarcely appreciated. Thus de Saussure does not seem to have seen how even purely topographical maps might have enabled a greater significance to be extracted from his detailed verbal descriptions of geological phenomena at specific localities; and Hamilton, despite the lavish scale of his illustrations of other kinds, only published one map.[30]

It is probably significant that all the earliest examples of maps on which specifically geological information is depicted come from a different social context, namely that of mining and associated surveys with utilitarian goals.[31] If we are concerned to recover the practical and cognitive *tradition* that led towards the modern geological map, rather than with celebrating historically isolated 'precursors', then our starting point should be the *Atlas et description minéralogique de la France* (1780). For whatever the problems that delayed its progress, and despite the extremely fragmentary state in which it was eventually published, its breadth of conception and overall uniformity of purpose were without parallel in eighteenth century map-making.[32] Furthermore, however great the divergences of opinion between Guettard, Lavoisier and Monnet may have been, concerning the theoretical aims and practical construction of 'mineralogical' maps, the final result showed an important degree of iconographical consistency. The sheets of the *Atlas* adopted a well-established cartographic convention, namely the use of 'spot-symbols' to represent towns of different degrees of administrative importance, country houses and so forth. But a new and far wider range of symbols was employed in the *Atlas* to denote the distribution of points of 'mineralogical' significance. These included mines and quarries of all kinds, distinguished according to the nature of their product; natural exposures of various kinds of rocks; and localities where various distinctive minerals and fossils were to be found (see Figure 3). The extensive use of these 'mineralogical' spot-symbols resulted in maps that were above all *distributional* in character. There was virtually no attempt to suggest how a localized clustering of a particular symbol might reflect a (literally) underlying uniformity in the bedrock. The symbols denoted scattered *points* about which information was offered; they did not connect those points into any kind of pattern.

Other late eighteenth century 'mineralogical' maps show a superficially greater resemblance to later geological maps, in their use of a series of colour-washes. But in fact they are better regarded as standing in the same cartographical tradition as the *Atlas*. For example, the 'Petrographische Karte' that accompanied Charpentier's *Mineralogische Geographie der Chursächsischen Lande* (1778) used colour-washes to denote in broad outline the major rock-types that characterize the different areas.[33] This convention was probably derived from the well-established cartographical use of such colours to denote administrative and political divisions. But on Charpentier's map these colours merely supplement a scatter of equivalent spot-symbols, which pinpoint the major exposures of the rocks with greater precision; and further spot-symbols are used without equivalent colours to indicate localized occurrences of various minerals (see Figure 4). The colours are an important iconographical innovation, however, because they implicitly represent a distributional *extrapolation* from the evidence of major exposures across intervening areas of the map. In other words, the colours denote an implicit belief that the relevant rock-type would be found in the intervening areas, under the concealing cover of soil and vegetation. Probably in practice this belief was strengthened (as we know it was in the comparable but later work of William Smith in England) by the use of many other clues to the underlying bedrock, such as the character of the topography, soil and vegetation. A related iconographical device was the provision of a *key* to the colours, in the form of a series of 'boxes' with the relevant spot-symbol in each and a verbal explanation of their meaning (Figure 4). The use of separate boxes is a further indication of the essentially distributional meaning of the colours : there is no indication that they were intended to give a clue to the underlying geological structure of the area. The contrast between Charpentier's map and the French *Atlas* should therefore not be over-emphasized : both used spot-symbols as their main iconographical convention, and these are only supplemented to a limited extent by the colours in Charpentier's map. By the end of the eighteenth century it is possible to speak of an established tradition of such maps, although their overall number was small.[34]

Cuvier's and Brongniart's 'Carte géognostique', illustrating their memoir on the strata around Paris (1811), should be seen as the further development of this tradition : significantly they called their work 'géographie minéralogique'. Spot-symbols were abandoned, however, and only colours were used. The key to the colours was still in the form of a series of boxes, but the boxes were placed closely together, in a vertical column corresponding to the real order of the strata as described in the text.[35] This *structural* implication of the colours on the map was reinforced by detailed geological sections (which I shall describe later); and the integration of the sections and the map was encouraged by the way that, with these almost horizontal strata, the pattern of outcrops could easily be related to the hachured indications of topography. In other words, the colours on the map represented not only an extrapolation of surface outcrops between the actual scattered exposures of the rocks; they also indicated implicitly the underground extrapolation of those outcrops into a three-dimensional structure.

The relation between this French map and William Smith's great *Delineation of the strata of England and Wales* (1815)[36] has been a matter of passionate argument and chauvinistic prejudices, from the time the maps were published up to quite recent historical writing. I do not intend to contribute to this somewhat unprofitable debate. We simply do not yet know enough about the diffusion of information about thematic maps of all kinds between the Continent and Britain during the period of the Revolutionary and Napoleonic wars. What has become clearer from recent research, however, is the variety of cartographical precedents that were available to Smith within Britain, but also the scope of his truly innovative attempts—parallel to, but probably independent of, those of Cuvier and Brongniart—to find adequate means of depicting three-dimensional structure on a two-dimensional map. Smith evidently struggled at an early stage with this fundamental problem. In the end, he settled for a convention that involved a particularly difficult and expensive form of hand-colouring, because no pattern of uncoloured engraving, and no simple series of colour-washes, was adequate to his intentions. The topographical pattern of successive scarps, and hence the *structure* of the dipping formations of strata, were to be suggested visually by lines of colour that were shaded gradually from an intense tone along the scarp-face edge of each outcrop, fading away down the dip-slope.[37]

Although Smith himself continued to use this ingenious and effective convention on his later maps, it was not adopted by other geologists; and we can now see how much his work stands aside from the mainstream of geological map-making. This is not to belittle his achievement, the magnitude of which was appreciated as much by his contemporaries as it can be by us to-day. Nor do I wish to champion instead the Geological Society's rival compilation, the more detailed map that its first President published in 1820, after delays as long as in Smith's case. George Greenough's map was conceptually old-fashioned by the time it appeared, for his extreme empiricism forced him to limit his map's effective purpose to a virtually distributional description of what he non-committally termed 'groups' of strata.[38]   Both Smith's maps and Greenough's were undoubtedly used by other geologists as source-materials for their own work; and their effects can be traced by following the spectacular improvements in the small-scale geological maps that illustrated inexpensive books on geology in the years after 1815.[39]   But the conventions that were used in such popular maps, and in the more detailed and original maps that quite suddenly burgeoned in the volumes of the Geological Society's *Transactions* from 1816 onwards, suggest that the cartographical pattern for English geological map-making had come primarily from Cuvier and Brongniart, not from Smith.

Even before the end of the wars, the French work on the Tertiary strata around Paris was well known in Britain. Webster's research showed that the same unexpected alternation of marine and freshwater strata could be found in Southern England. His first outline geological maps were published by the Geological Society in 1814, and a more detailed geological map of the Isle of Wight and part of Dorset was engraved in 1815 and finally published in 1816.[40]   These maps used the Cuvierian (and modern) convention of plain colours to indicate outcrops; in the key the 'boxes' were arranged in the *true* order of the strata; and the later map established what became a common convention in subsequent geological practice, by showing representative sections on the same plate, in order to assist in the three-dimensional structural 'reading' of the map (Figure 5). In the volume of *Transactions* published in 1816, similar geological maps were already becoming routine : MacCulloch showed the geology of the Isle of Skye in this way, for example, and Conybeare that of North-East Ireland.[41]   Thus the originally French conventions for a geological map were quickly adopted in Britain, and became an internationally understood part of the standard visual language of geology.

It remains only to mention the way in which maps could be used to communicate geological information and ideas beyond that of the three-dimensional structural configuration of strata and other rocks. The maps that Nicholas Desmarest used to illustrate his analysis of the volcanic rocks of Auvergne (1779) are a possibly unique example that transcended the 'distributional' intentions of most eighteenth century maps. Despite the topographical limitations of hachuring, Desmarest used a remarkably subtle range of engraving techniques to depict the volcanic cones, lava flows, and isolated outliers of older basalts. It is probably no accident that this exceptional piece of cartography was equally unusual in the causal orientation of the theorizing that underlay it : Desmarest was not concerned with the two-dimensional distribution of the rocks for its own sake, but only as a means of displaying the evidence for a temporal reconstruction of the successive 'époques' of volcanic activity that the rocks recorded.[42]

In the early nineteenth century, the new structural cognitive goals that geological maps and sections made possible began in turn to be transcended by more causal and temporal explanatory ambitions; and thematic maps like Desmarest's began to appear in greater numbers and variety. For example, in the Geological Society's *Transactions* published in 1817, MacCulloch used maps to illustrate not only his observations on the Parallel Roads of Glen Roy but also their possible causal explanation; and Dick Lauder published what are in effect palaeogeographical reconstructions of the same area in his rival explanation in the Edinburgh *Transactions*. Buckland illustrated the first statement of his 'diluvial' theory by a map, published in the Geological Society's *Transactions* in 1821, showing the strange distribution of 'erratic' material. A decade later, Lyell published a special kind of palaeo-geographical map in the *Principles*, in order to show the wide areas of Europe that at some time during the Tertiary period had been part of the sea-bed. Like other such maps, this served a theoretical purpose, namely to illustrate his theory of a continual and gradual flux in the distribution of land and sea during this period. Similar examples could be mentioned in other published works of the 1820s and 1830s, but these are sufficient to suggest the range of uses to which maps were being put by about 1840, as part of the visual language of geology.[43]

IV. GEOLOGICAL SECTIONS

Like a geological map, a modern geological section is a highly theoretical construct. It is a kind of thought-experiment, in which a tract of country is imagined as it would appear if it were sliced vertically along some particular traverse of the topography, and opened along that slice in a kind of cutting or artificial cliff. The geological structure depicted on such a *'traverse'* section may be inferred from surface outcrops and/or borehole and other underground data; but however plentiful the evidence, the section inevitably embodies extrapolations derived from theory-based expectations. Another type of section, which I will here term a *'columnar'* section, depicts the succession of strata that are found in a particular locality or region in the form of a vertical column, with the correct order of succession and relative thicknesses of the various formations; but like a traverse section this too is a highly theoretical construct, because all post-depositional disturbances of the strata (folding and faulting etc.) are deliberately eliminated, and the strata are visually restored to the more-or-less horizontal position in which each formation was originally deposited.[44]

These rather inadequate explanations are intended, like my description

V

Fig. 1. The eruption of Vesuvius in 1767: hand-coloured copper engraving after a painting by Pietro Fabris, from William Hamilton's *Campi Phlegraei* (1776). This much reduced and uncoloured photograph gives a very inadequate impression of the striking quality of the original print.

V

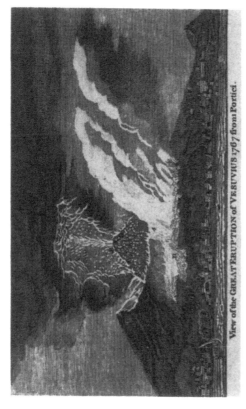

View of the GREAT ERUPTION of VESUVIUS 1767 from Portici.

Fig. 2. The eruption of Vesuvius in 1767: anonymous copper engraving from William Hamilton's *Observations on Mount Vesuvius . . .* (1772). The coarsely-hatched shading is obtrusive in this 'run-of-the-mill' engraving.

V

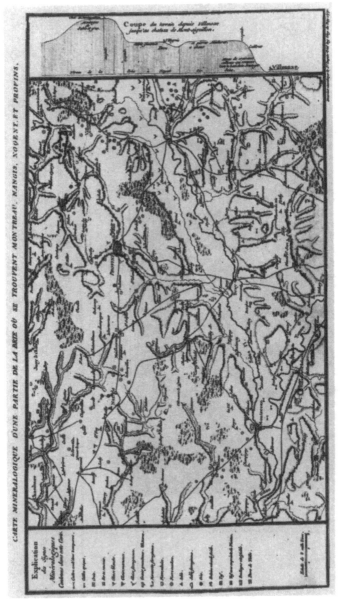

Fig. 3. A characteristic sheet of the *Carte minéralogique de la France* (1780); this particular sheet was engraved in 1770. Note the hachured topography and scattered 'mineralogical' spot-symbols on the map itself, the key to the symbols in the left-hand margin, and the measured profile (without indications of internal structure) in the right-hand margin.

FIG. 4. Part of the hand-coloured engraved 'petrographical' map in Charpentier's *Mineralogische Geographie der Chursächsischen Lande* (1778). The colour-washes of the original are supplemented by scattered spot-symbols on the map; both colours and symbols are identified in the key, which has the form of 'boxes'.

V

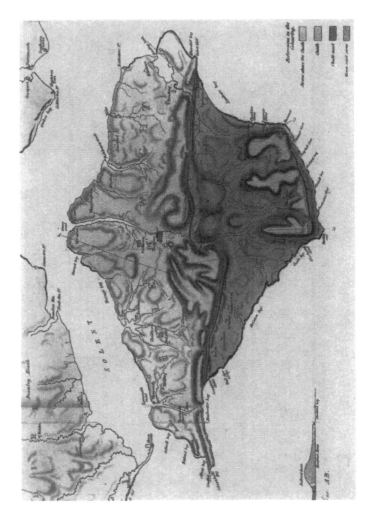

Fig. 5. Part of Thomas Webster's engraved geological map of Dorset and the Isle of Wight, published in 1816. The colour-washes of the original are identified by a key of 'boxes' arranged in the real order of the strata; a fragment of one of the associated sections is just visible, and the position of another is identified by the dotted line running north-south through the island.

Fig. 6. An engraved section of strata in Derbyshire, from John Whitehurst's *Original state and formation of the earth* (1778). The confident extrapolation from observed exposures, and the ruled lines and masonry-like style, are characteristic of the English tradition of sections drawn by men with practical engineering concerns.

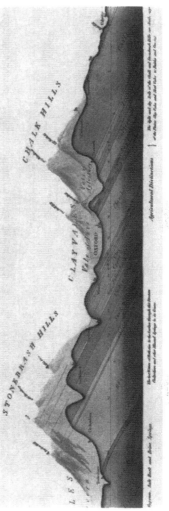

Fig. 7. Part of William Smith's coloured engraved *Geological section from London to Snowdon* (1817), showing the 'Secondary' strata in the Cotswold ('Stonebrash') and Chiltern ('Chalk') Hills. The strata are ruled with perfect regularity, and are extended faintly into a conventional landscape background.

V

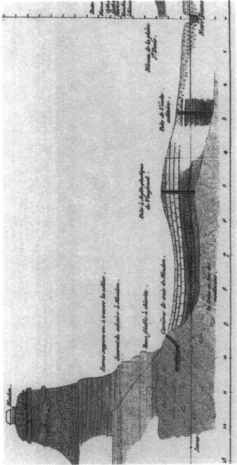

FIG. 8. Part of one of the engraved sections of the Tertiary strata around Paris, from the memoir published in 1811 by Georges Cuvier and Alexandre Brongniart. The carefully measured scales (with deliberate vertical exaggeration), free-hand style and tentative extrapolations from observed exposures are characteristic.

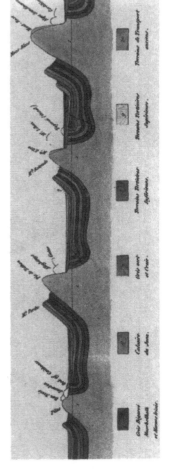

FIG. 9. Part of the coloured engraved section illustrating Léonce Elie de Beaumont's theory (1829–30) of occasional 'revolutions' of mountain-building punctuating much longer periods of tranquil deposition of sedimentary formations. The elevation of successive 'systèmes' of mountains (left to right) folded successive formations, after which new formations were deposited unconformably on their truncated edges.

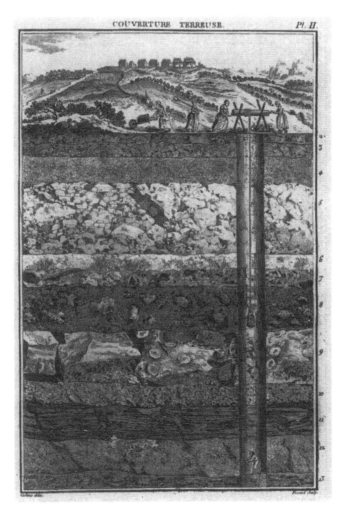

FIG. 10. An engraved section of strata, shown in relation to primitive mining techniques, from Morand's treatise on coal-mining (1768–77) in the *Description des arts et métiers*. Mining technology was the usual context of eighteenth century geological sections.

V

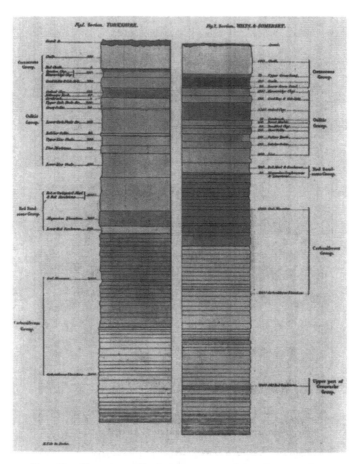

FIG. 11. Two generalized columnar sections of strata from
Henry De la Beche's *Sections and views* (1830). They show the
'Secondary' strata in north-eastern and south-western England:
the juxtaposition of two columns, each highly abstract in form,
allowed the successions in the two regions to be correlated and
compared.

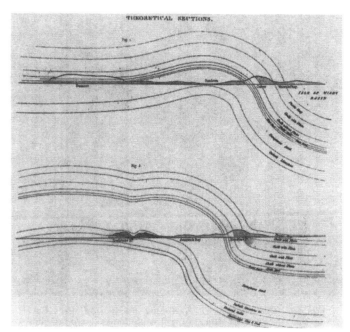

FIG. 12. Thomas Webster's two 'theoretical sections' illustrating his interpretation of the structural geology of the Isle of Wight and Dorset on the south coast of England (1816). The sections are drawn to true scale, and the folded formations are extrapolated by dotted lines both downwards and also up into the air (they imply that a vertical thickness of nearly *two miles* of strata have been removed since the strata were folded).

FIG. 13. Part of one of Thomas Webster's offshore views of the south coast of England, published in 1816. It shows Chalk strata curved from a nearly horizontal (left) to a nearly vertical (right) position, and was part of the evidence on which Webster's 'theoretical sections' (Fig. 12) were based. Such geological cliff views are stylistically similar to the coastal drawings that were often appended to navigational charts at this period.

V

FIG. 14. Engraving of the columnar basalt at Fingal's Cave on the Isle of Staffa, illustrating Joseph Banks's account, published in Thomas Pennant's *Tour of Scotland* (1774). The style is characteristic of eighteenth century 'natural-history' landscapes.

FIG. 15. Engraving of Fingal's Cave, as re-drawn to illustrate Faujas de Saint Fond's *Voyage en Angeleterre, en Ecosse et aux Iles Hébrides* (1797), showing its further distortion—probably unintentional—into an almost 'neoclassical' style.

FIG. 16. Coloured aquatint of the columnar basalt of the Isle of Staffa, by William Daniell, from his *Voyage around Great Britain* (1814–20). This fine example of early nineteenth century topographical art contrasts with eighteenth century engravings of the same locality (Figs 14, 15).

FIG. 17. Engraved view of cliffs at The Needles on the Isle of Wight, after a drawing by Thomas Webster, from Henry Englefield's *Picturesque beauties . . . of the Isle of Wight* (1816). While primarily 'picturesque' in intention, this offshore view also conveys geological information, since the lines of flints in the Chalk show how the strata have been folded into an almost vertical position (compare Figs 12, 13).

V

FIG. 18. Engraved view of the 'Parallel Roads' in Glen Roy in Scotland, from a drawing by John MacCulloch published in the Geological Society's *Transactions* (1819). MacCulloch's accurate landscapes were probably drawn with the camera lucida. Like Darwin later, he interpreted the horizontal terraces as former shorelines.

FIG. 19. Engraved view of the 'Parallel Roads' near Glen Roy, from a sketch by Thomas Dick Lauder published in the *Transactions* of the Royal Society of Edinburgh (1821). The crude style and vertical exaggeration contrast with MacCulloch's drawings of the same area (Fig. 18).

V

FIG. 20. Part of a panoramic landscape view in Auvergne, published as a coloured lithograph in George Poulett Scrope's *Memoir on the geology of central France* (1827). The conventional colouring highlights the ancient basalt on the long narrow plateau in the middle distance, and in the foreground and the further distance. Subsequent erosion of the intervening valleys has left these lava flows on hilltops.

V

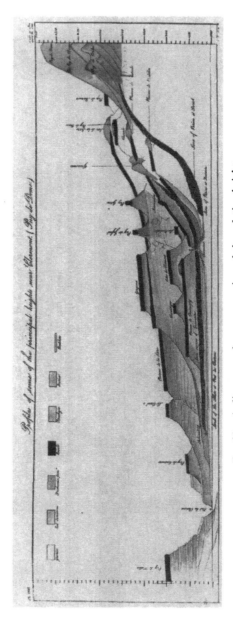

Fig. 21. A diagrammatic representation of the relative heights of lava flows in Auvergne, published as a coloured lithograph in Scrope's *Central France* (1827). This illustrated his belief that the basalts could be used as a "natural scale" for geological time.

V

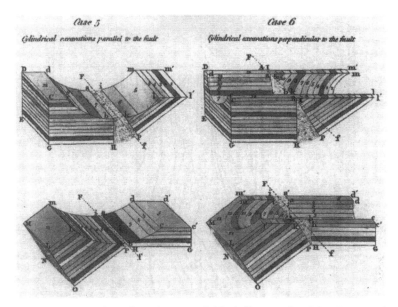

FIG. 22. Four of the coloured engraved block-diagrams by John Farey, published in 1811. They formed part of his matrix of 56 possible classes of interactions between variously faulted and dipping strata and variously eroded surface topography, showing the resultant patterns of outcrops. The 'engineering' style of the diagrams is characteristic.

FIG. 23. A typical page of coloured traverse sections from Henry De la Beche's *Sections and views* (1830), lithographed by himself. They show the visual conventions, particularly the cliff-like appearance, that had become standardized in geological sections by that date, and they illustrate various examples of folding, faulting and unconformities.

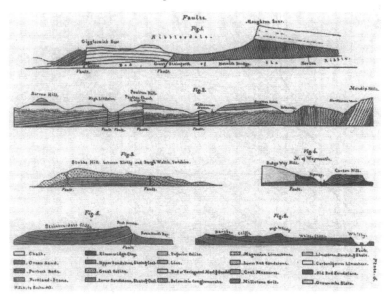

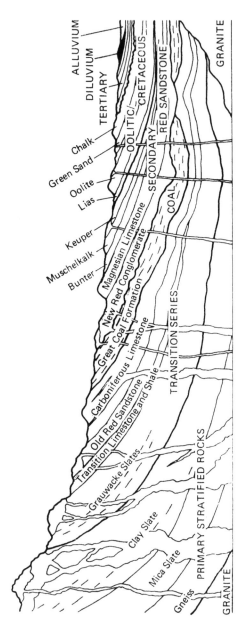

FIG. 24. Part of the coloured engraved 'ideal section' from William Buckland's Bridgewater Treatise on *Geology and mineralogy* (1836), redrawn in order to show what Buckland regarded as the average relative thicknesses of successive formations, and hence implicitly the relative time-scale of the history of the earth (simplified by the omission of igneous intrusions and volcanic rocks, and minor faults).

of a geological map, to emphasize that sections are an aspect of the visual language of geology which is far removed from straightforward observation and which embodies complex visual conventions that have to be learned by practice. Likewise, its historical development was not natural or inevitable, and it had to be constructed from other cultural resources in specific contingent historical circumstances. It is therefore not surprising to find that geological sections, like geological maps, are remarkably rare in published works before the second decade of the nineteenth century. Historians of geology have documented the few exceptions to this generalization; but it is more important to note that sections do not seem to have become a *standard* part of the visual repertoire of geology until the period when the science itself was becoming a self-conscious discipline.

Earlier sections are found in two distinct contexts. Like 'mineralogical' maps, most eighteenth century sections embodying geological information were published in a social context of 'mineral geography', *i.e.*, practical surveys and exploitation of mineral resources. For example, an impressive series of experiments in this kind of visual representation can be found in the margins of the sheets of the *Atlas minéralogique de la France*. At their simplest they are almost purely verbal descriptions of the strata exposed in some particular locality. In the next stage of visual formalization the strata are depicted as a simple columnar section, with the nature of each stratum suggested by various engraved patterns (like those which are still used in modern columnar sections) and also described in words and symbols to one side of the column. Such sections are in effect visual translations of the verbal records that would have been kept during the working of a quarry or mine. Other sections are more generalized, for judging by their captions they represent an 'average' of several exposures within the area of the map. Finally, some are true 'generalized' columnar sections, in that they synthesize the order of strata for the whole area of the sheet concerned, even indicating the range of variation in the thickness of the different formations.[45]

The simpler forms of these columnar sections do little more than amplify the information conveyed by a single spot-symbol on the adjacent map: they depict, in a kind of formalized view of a quarry face, the type of rock the occurrence of which is summarized by the symbol. To that extent, they confirm my interpretation of the *Atlas* as primarily 'distributional' in intention. It might be argued that the more generalized sections imply a 'structural' cognitive orientation, in that they suggest the sequence of strata that underlies the whole area of the sheet concerned. But this is at most a slight move in the direction of visualizing the country in three-dimensional structural terms: for even the generalized columnar sections are not accompanied by any attempt to construct a *traverse* section and thereby to integrate the vertical succession of strata with the horizontally extended features of the topographical surface. Significantly, even where a carefully measured profile traverse of the topography was included, no attempt was made to extrapolate the evidence of surface exposures downwards into any kind of structural section (Figure 3, right-hand margin).[46]

A closer approach to a fully structural orientation can be found, not surprisingly, in works that were even more directly related to mining activities. Illustrations of mining technology regularly depict strata or mineral veins in the form of imaginative sections, if only as incidental background features to the details of working methods, shafts, adits, machinery and so on.[47] Obviously mining provided a much more direct incentive to visualize in terms of three-dimensional structures than even the most extended geographical survey of surface mineral resources (see Figure 10). Most if not all of the early attempts to draw traverse sections for their own sake, i.e., to illustrate geological structures rather than mining techniques, come from a mining context and obviously drew much of their evidence from underground data. Examples include the well-known early eighteenth century sections of an English coalfield by John Strachey, and Füchsel's and Lehmann's similar sections from central Europe.[48] But it is important to emphasize how infrequent these are, relative even to the sparse illustration of eighteenth century 'geological' publications in general. By the beginning of the nineteenth century, straightforward realistic sections based on underground evidence had probably become a standard form of record-keeping in mining concerns, although few were published, probably for reasons of industrial secrecy or state security. Some of them are simple columnar sections showing the strata penetrated during the sinking of a mine-shaft; and these are sometimes extended laterally, taking into account the evidence of adits driven sideways from the shaft, so as to form a limited kind of traverse section showing the structure of the rocks within the limits of the mine-workings.[49]

A further stage in the formalization of columnar sections, whether based on evidence from mines or surface exposures, was the omission of the quasi-realistic conventions of representing the column of strata as though they were penetrated by a mine-shaft or exposed at a quarry-face, and the reduction of the section to a simple formalized column of strata drawn to some specified vertical scale. More important, however, than this stylistic change was the production of generalized columnar sections that deliberately omitted the structural complications of folding and faulting, and abstracted from an extensive tract of country a formalized succession of strata restored to their original horizontal position. Most eighteenth century generalized sections (such as the 'coupes générales' in the *Atlas minéralogique*) had been drawn from structurally simple areas of almost horizontal strata; but for structurally complex regions the abstraction of a formalized columnar section was conceptually a much greater achieve-

V

ment. One of the earliest and most ambitious sections of this kind, which was published by Westgarth Forster in 1809, synthesized evidence from mines and surface exposures across a broad tract of northern England (including what was then the world's most important coalfield) into a single measured columnar section. By 1830, the theoretical potentialities of such sections had been recognized to the point where analogous sections of equivalent strata in different regions were being juxtaposed, so as to facilitate the stratigraphical correlation of the formations and to demonstrate the lateral variation in their thickness (Figure 11).[50] This was the origin of the correlation diagrams that later became such an indispensable part of the visual language of stratigraphical geology.

Another, and initially quite separate, historical source of geological sections is the diagrams that illustrated various 'cosmogonical' theories of the formation and history of the earth. Seventeenth century examples range from the starkly geometrical style of Steno's well-known diagrams to the intricate rococo style of Kircher's. Earlier historians of geology tended to praise the first extreme and deride the second, according to anachronistic criteria of their scientific merit; but it is more historical to interpret *all* such diagrams not as reports of actual observations but as highly abstract theoretical statements in a visual language.

In the eighteenth century such 'cosmogonical' theoretical sections of the structure of the whole earth were occasionally associated with the more 'empirical' sections that I have already described. This suggests how 'cosmogonical' theorizing was not always divorced from practical fieldwork in the eighteenth century. John Strachey's work is a good example of this combination from the early part of the century; an important later example is John Whitehurst's *Original state and formation of the earth* (1778).[51] The title of this book emphasizes how it belongs to the 'cosmogonical' tradition, but its 'Appendix . . . on the strata of Derbyshire' indicates equally its empirical component, and its illustrations symbolize this combination perfectly. Whitehurst's engraved traverse sections of Derbyshire are not only printed on the same plates as some highly theoretical 'cosmogonical' diagrams, but they are themselves strikingly diagrammatic in style (Figure 6). Beneath a moderately realistic surface profile (with some exaggeration of vertical scale), the dipping strata are *ruled* in dead-straight lines with perfect parallelism. In other words, Whitehurst confidently extrapolated from the very limited evidence of surface exposures and a few mine-shafts, according to an implicit theoretical assumption that the strata extend downwards like so many tilted but perfectly parallel courses of masonry. The analogy with masonry is accentuated by the way in which he drew what we would regard as a gently curved synclinal structure as a pair of oppositely tilted rigid blocks with a clean break between them.[52] I do not think it is fanciful to see a stylistic analogy

V

between these geological diagrams and the kinds of engineering drawing with which Whitehurst and his social circle would have been familiar in their ordinary work.

The same visual style reappeared in the same social setting in most of John Farey's early sections which he circulated in manuscript, and in the one that he included in his *Report on Derbyshire* (1811) for the Board of Agriculture.[53] Below an accurate surface profile of the traverse, the evidence of dipping strata seen in surface exposures—with little or no extra evidence from mines—was again confidently extrapolated down to the base-line in the form of perfectly parallel *ruled* straight lines between the successive formations; and all faults and changes of dip were interpreted as clean breaks between adjacent rigid blocks, as though the whole section depicted a badly broken and tilted structure of masonry. The engineering conventions behind the iconography of Farey's sections are revealed still more clearly in the remarkably original block-diagrams of strata that he included in his report on Derbyshire. These were an attempt to portray, in a kind of mathematical matrix, all the possible three-dimensional "formae" that would result from the interactions between eight different types of faulting (including what we would now interpret as folding) and six types of surface erosion (Figure 22). These diagrams were unsurpassed at this period for their theoretical sophistication in the visual illustration of three-dimensional structures in geology.[54]

William Smith's published *Sections* (1817-19) were drawn in exactly the same style of ruled lines as those of his professional associate Farey. They included one new feature, however, in that like some eighteenth century mine sections a formalized landscape was drawn above and behind the line of section, so that the whole illustration looks like a lateral view of a real three-dimensional geological model of the countryside (Figure 7).[55]

This English tradition of sections in an 'engineering' style suggests how a fully structural approach to the interpretation of the complex phenomena of geology was most readily attained within a social context of practical mining and mineral surveying, by individuals whose familiarity with engineering practice 'pre-adapted' them, as it were, to the three-dimensional visualizing that structural geology required. This interpretation is strengthened by the fact that sections in a somewhat similar style, with ruled lines, were being produced at the same period in Germany, for example in the works of Johann Voigt, within a similar social context of practical mining engineering.[56]

I have already mentioned, however, that the same structural orientation also emerged in a different cultural setting, notably in the work of Cuvier and Brongniart on the geology of the Paris region. Significantly, the sections which they published to complement their map, and to make a fully structural 'reading' possible, were drawn in a style that contrasts markedly with that of the sections I have just described. Cuvier and Brongniart included a columnar 'coupe générale' that summarized the general succession of the Parisian strata; but most of their sections were traverse sections that mark a significant advance on anything in the *Atlas*. They were based on measured topographical traverses; but the evidence of surface exposures (and occasional wells) was cautiously extrapolated to form true traverse sections (Figure 8). Like the tentative reconstructions of extinct mammals that Cuvier was drawing at the same period, the sections attempt to indicate the degrees of certitude to be attached to different degrees of extrapolation from the observable evidence. Furthermore, and again like Cuvier's anatomical work, they were drawn free-hand and not ruled; and a bold vertical exaggeration was used to great advantage to make the structure and lateral variation of these thin flat-lying strata more intelligible.[57]

The visual conventions of Cuvier and Brongniart's sections were quickly adopted by the Geological Society of London. In Webster's comparison of the Parisian strata with those of southern England, published only three years after the French work, a hand-coloured traverse section of the extraordinary vertical Tertiary strata in the Isle of Wight was used to great visual effect, and subsequent volumes of the Society's *Transactions* included ever more traverse sections of this kind, hand-coloured in the same style as the geological maps that they so often accompanied and complemented.[58] It is hardly surprising that in their sections, as in their maps, the members of the Geological Society should generally have taken the French work as their pattern, rather than that of their fellow-countrymen Smith and Farey. Any chauvinistic preference they might have felt for the latter was more than outweighed by the methodological advantage of the French model: the convention of extrapolating tentatively and in a free-hand style from the observable surface evidence was much more congenial to the Geological Society's outlook than the questionable theoretical assumptions that were implicit in the ruled lines on Smith's and Farey's sections. In practice, it became acceptable to extrapolate from surface exposures as far as some base-line—generally taken at sea-level—but rarely further. In this way, standard geological sections (Figure 23) came to have a striking similarity to offshore views of cliff exposures, which I shall mention in the next part of this article. Only rarely were these conventional limitations on extrapolation exceeded; when they were, highly important theoretical implications could be expressed. For example, dashed lines were occasionally used to suggest tentatively not merely the possible underground extension of strata within a section, even below the 'base-line', but also their *former* extension above the present surface—in other words, to show where the strata might have extended before their erosion. An early but extremely bold use of this convention is found in

Webster's sections of the monoclinal fold of the strata in the Isle of Wight and Dorset, where small standard sections drawn to true scale were extrapolated far up 'into the air' and far down below the base-line of sea-level, in order to suggest the huge fold that Webster inferred from the surface dips of the strata, and the enormous extent of subsequent erosion (Figure 12).[59] Such illustrations may well have helped to confront other geologists with the causal and temporal implications of the structural configurations that their sections revealed.

By the early 1830s the conventions of traverse sections had become a well-established part of the visual language of geology everywhere. They could be used without special explanation to illustrate a wide variety of structural phenomena such as folding, faulting and unconformities (Figure 23); they were understood widely enough to be used increasingly in the form of wood engravings to illustrate popular books on geology; and they had become a powerful visual means of summarizing and synthesizing geological work on a very broad scale.[60] So effective had sections become, indeed, that De la Beche even felt it necessary to point out the theoretical dangers of one of the most common conventions that they incorporated. The great vertical exaggeration that had been so heuristic in Cuvier and Brongniart's sections could easily become misleading if applied to sections of folded strata or topography of high relief. In his book of *Sections and views* (1830), De la Beche recommended that wherever feasible traverse sections should be drawn at or near true scale (*i.e.*, with the vertical scale the same as the horizontal), in order to avoid the danger of *over*-estimating the magnitude of the phenomena which geologists needed to explain in causal terms. He followed his own advice in the lengthy sections that were appended to his *Report on the geology of Devonshire* (1839), the first-fruits of the new Geological Survey.[61] But this sober empiricism tended to make the structure revealed by such sections difficult to interpret, and more diagrammatic sections with vertical exaggeration continued to be a popular form of illustration.

A different trend is shown by the various theoretical or 'ideal' sections that were published from about 1830 onwards. Examples include the large coloured chart that was published to illustrate Brongniart's *Tableau des terrains* (1829), the long coloured 'ideal section' which forms the first and largest illustration in Buckland's Bridgewater Treatise on *Geology and mineralogy* (1836), and the small coloured frontispiece to Lyell's *Elements of geology* (1838). All these sections adopted the conventions of ordinary traverse sections, but represented an 'ideal' portion of the earth's crust in order to illustrate the spatial relations between the different classes of rocks and the interpretation of those relations in temporal and causal terms. Brongniart's and Buckland's sections display implicitly their fundamentally 'directionalist' theories of the history of the earth : immensely thick piles of strata were shown resting on a basement of 'granite', and ranging successively through Primary, Transition, Secondary and Tertiary formations to the geologically recent (and quantitatively insignificant) 'diluvium' and modern alluvium. Lyell's section was even more clearly and overtly theoretical, for it illustrated his 'uniformitarian' belief that there was no such overall directionality in the history of the earth, and that all the major classes of rock—sedimentary, volcanic, and what he called plutonic and metamorphic—had been formed in the same way at *every* period.[62]

One final example of a formalized theoretical section shows the conceptual flexibility of this means of visual expression. In his *Révolutions de la surface du globe* (1829–30), Élie de Beaumont argued that in the history of the earth long periods of tranquil deposition of the successive formations had been punctuated occasionally by sudden and violent mountain-building movements, each confined to certain linear zones of the earth's surface. In the long coloured section that illustrated his work, this theory was given visual expression in an ingenious way (Figure 9).[63] The horizontal dimension of the section now represented primarily the geological time-scale: the formations of strata deposited in each successive tranquil period were shown tilted and disturbed by each subsequent mountain-building episode. Thus although the section used the standard conventions of coloured strata, the theoretical content was quite different and highly original.

By the 1830s, therefore, both columnar and traverse sections had become a standard part of the visual language of geology, with implicit conventions that were generally accepted and widely understood not only by practising geologists but also by the wider audience for geology; and some geologists had already begun to extend these conventions still further in theoretical directions.

V. LANDSCAPES

The suggestion that geological maps and sections constitute a 'visual language' for the science of geology will probably not seem strange to historians of science. But there is one further major category of illustrations, namely landscapes, which has received far less historical attention than it deserves. I have already referred several times to the paucity of illustrations of any kind in most late eighteenth century books and articles dealing with subjects that were later to be defined as 'geological'. Among such illustrations, however, if we leave aside the separate category of pictures of specimens, the great majority are landscape views of particular localities. Yet with very few exceptions the portrayal of 'natural-history' landscapes of geological interest was decidedly crude. The engravings

V

that illustrate de Saussure's *Voyages dans les Alpes* are a case in point: the dramatic topography that de Saussure described in words might seem to have been an appropriate subject for visual illustration, but in fact the rather sparse engravings of mountain scenery were markedly unconvincing. A comparison with the more adequate engraving of the gentler environs of Geneva suggests that what was at fault was not so much a lack of skill on the part of de Saussure's artist and engraver, but rather the lack of an appropriate artistic tradition that could cope adequately with the wild irregularities of mountain landscapes.[64]

Even in Pietro Fabris's work for Hamilton's *Campi Phlegraei*—perhaps the most striking exception to my generalization—the straightforward topographical scenes and even the pictures of distant eruptions (Figure 1) are far more convincing than his attempts to represent the irregularities of a lava-flow or the interior of the crater of Vesuvius.[65]   Apart from active volcanos, perhaps the single phenomenon that most fascinated late eighteenth century authors and readers in this area of science was the extraordinary prismatic columns of the puzzling rock basalt; but here too the inadequacy of contemporary artistic traditions is reflected in the many crude and—to our eyes—highly unrealistic engravings of columnar basalt published at this period, for example those that illustrate Faujas de Saint Fond's great treatise on the *Volcans éteints du Vivarais et du Velay* (1778).   Equally awkward is the engraving of the basalt of Fingal's Cave to illustrate Joseph Banks's account of the Isle of Staffa (Figure 14), which was inserted by Thomas Pennant in his *Tour of Scotland* (1774); and when Faujas later plagiarized this to illustrate his own travel book, he exaggerated the vertical scale and managed to transform this romantic spot into a kind of natural Versailles banqueting hall (Figure 15)![66] Such unconscious distortions could be attributed in part to the fact that any artist's attempt to render such a strange phenomenon had to pass through the hands of an engraver, who would be even more unfamiliar with the original; but a more adequate explanation must take into account the stylistic limitations of contemporary artistic traditions, and the artistic unfamiliarity of the subjects that were of greatest 'geological' interest.

The development of an artistic tradition that was better able to perceive and portray subjects of this kind with what we would regard as adequate realism may owe something to developments within 'academic' art in the late eighteenth century.   The 'neoclassical' movement encouraged a clear and 'realistic' landscape style, even if the artist generally scorned to portray an identifiable scene and preferred to invent one to suit his own programmatic composition.   Likewise, the 'romantic' movement encouraged attempts to depict the wild mountain landscapes that had previously been considered unfit for serious artistic expression; but at the same time the artist generally exaggerated the vertical scale with scant regard for 'realism' in order to heighten the romantic impact of the scene, and often obscured the topography by his interest in swirling clouds and other atmospheric effects.

The influence of these contrasting movements on 'geological' subjects seems likely to have been slight, however, for its spokesmen generally regarded accurately topographical art with undisguised scorn.[67] In other words, if we are to find the sources of the kind of topographical tradition that could provide an effective part of the visual language of geology, we must look not so much at the 'academic' art of the period, but rather to a socially humbler level of documentary topographical drawing and painting. Here the artist-naturalists who accompanied most of the great eighteenth century voyages are of obvious importance; so too are the topographical artists who made a living out of painting the estates and country seats of the aristocracy with recognizable precision, or from drawing scenes to illustrate books of general descriptive topography; and finally there were those who taught the kind of accurate topographical drawing that was required as an adjunct to military surveys and marine navigational charts.[68]

The importance of this tradition of documentary drawing for geology can be seen in some of the published illustrations that accompany geological works of the early nineteenth century, which form a striking contrast to the general level of those of the late eighteenth century. Thus for example William Daniell's aquatints of Fingal's Cave and the Isle of Staffa (Figure 16) are not only more attractive and more accurate than the eighteenth century engravings of the same locality, but also far more informative geologically. They also have obvious stylistic affinities with the offshore coastal views that commonly supplemented the navigational charts of the period.[69] I have already mentioned that the Geological Society's official draughtsman, Thomas Webster, was a trained artist; and it was as much for his topographical skill as for his knowledge of geology that he was co-opted by the antiquarian and amateur geologist Sir Henry Englefield (1752–1822) to help him complete his book on *The Isle of Wight* (1816). In many ways this was a typical topographical picture-book of the period, with lavish engraved illustrations of country houses, ruined churches and picturesque coastal scenery, and Webster contributed as many drawings on these subjects as Englefield. But he was well able to add detail of geological interest to scenes that were 'picturesque' in their mood, as in his fine view of the vertical Chalk strata in the Needles (Figure 17). Many of his other offshore views, like those of Daniell, have unmistakeable stylistic links with navigational coastal views (Figure 13).[70]

When the Geological Society of London began to publish its *Transactions* in 1811, many of the illustrations were of landscapes of geological importance. One of the most prolific contributors to the early volumes

V

was John MacCulloch, whose superbly accurate landscapes suggest that he may have been using the camera lucida that William Wollaston had patented in 1807 (Figure 18). By 1820 the geological use of accurate landscapes had become commonplace in the *Transactions* of the Geological Society, and there was nothing to match this publication elsewhere, in either the quantity, quality or effectiveness of landscape as a means of visual communication (compare Figures 18 and 19).[71]

The 'realistic' character of documentary landscapes may have appealed to the 'empiricist' ideals of the leading early members of the Geological Society, since it could be claimed that landscapes convey an unbiassed factual impression of geological phenomena to those who had not had the opportunity to see them with their own eyes. Likewise, Hamilton had stated explicitly that Fabris's landscapes would provide their readers/ viewers with "accurate and faithful observations on the operations of nature, related with simplicity and truth"—in order to counter the premature construction of theoretical 'systems' by those who had never seen the relevant evidence. But in reality even the most innocuously 'documentary' landscape inevitably embodied some kind of theoretical content. Fabris's dramatic contrast between the violence of an erupting Vesuvius and a calm neo-classical foreground (Figure 1) could not help conveying an implicit message about the power of "the operations of nature" in relation to the life of man; while the 'unrealistic' (but well established) topographical convention of a quasi-aerial view was used without apology to clarify the nature of another eruption and to convey an explicit theoretical point about the time scale of earth-history.[72] Even eighteenth century engravings of columnar basalt, however crude by later standards, have an implicit theoretical content in their emphasis on the geometrical regularity of the hexagonal columns amid so much apparently 'chaotic' disorder (Figures 14, 15).

In early nineteenth century landscapes of geological interest, such implicit theoretical contents began to become more explicit. Offshore views of coastal cliffs were particularly common, simply because cliffs provide some of the most extensive and continuous natural exposures to be found anywhere. But the naturalistic indications of jointing and stratification in Daniell's and Webster's drawings (Figures 13, 16, 17), for example, were gradually formalized by the accentuation of the main structural features, such as the divisions between successive formations, and by the simplification of the 'fortuitous' features of the cliffs themselves. The colouring, which was likewise applied at first in a naturalistic manner, was gradually formalized into conventional colour-washes for the various types of rock.[73] As I have already mentioned, this convention gave these cliff-sections a striking similarity to the true traverse sections that were being produced at the same period (and in many cases by the same individuals)—even to the point that in some instances it is difficult to tell whether an illustration is a true section or a cliff view, without referring to the caption.

This stylistic similarity seems to reflect an important cognitive function that formalized cliff views had in the development of the visual language of geology. I have suggested that the over-geometrized style of Smith's and Farey's sections was unlikely to commend them to the early members of the Geological Society, whose sections generally suggest their preference for the more empirical conventions of Cuvier and Brongniart. Cliff views, on the other hand, could be drawn and coloured in a formalized manner like sections, so as to emphasize the main structural and stratigraphical features; yet at the same time they were much nearer to direct observation than the necessarily speculative extrapolations that traverse sections required. In this way, formalized cliff views seem to have acted as a kind of conceptual bridge by which the conventions of traverse sections became acceptable and—in a sense—believable as valid representations of an unseen reality. This interpretation is supported by the fact that the Geological Society's *Transactions* began to make extensive use of lengthy formalized cliff views as early as 1816, whereas true traverse sections remained comparatively infrequent until later volumes.

Some of these formalized cliff views show a further development, for they include some ordinary landscape beyond the actual cliffs, and the conventional colouring is extended into this landscape background so as to suggest the underlying bedrock. (The cruder landscape backgrounds of Smith's sections, published at about the same time, show a similar development.) Such colouring, like that on a geological map, indicates not what could actually be seen (except in isolated exposures) but what was believed to lie beneath the concealing cover of soil and vegetation.[74] From this form of illustration it was only a small further step to apply the same convention to ordinary landscapes that included no large natural exposures such as cliffs.[75] Panoramic illustrations reached their peak in the hand-coloured landscapes that accompanied George Poulett Scrope's *Memoir on the geology of central France* (1827). Here conventional colouring was used to convey a vivid impression of the extinct volcanic craters and lava flows of Auvergne, and to demonstrate the vast scale of erosion—and inferentially the vast time scale—that separated these 'modern' (yet pre-historic) eruptions from the 'ancient' basalts that capped many of the hill tops (Figure 20).[76]

In one of his illustrations Scrope developed the conventions of geological landscape still further in an abstract and theoretical direction. He drew a kind of idealized view of the many lava flows which had poured off the plateau of the Puy de Dôme in Auvergne, in order to show how the erosion of the soft Tertiary sediments in the basin of the Limagne had left these flows at many different heights above the present valley floors (Figure 21). This highly formalized diagram enabled him to complement in visual terms his verbal conclusion that the successive lava flows formed a kind of "natural scale" by which the slow but inexorable process of erosion could be measured.

Scrope's work was immediately influential on Lyell, and it is plausible to suggest that its impact derived in large part from the effective visual communication that his landscapes achieved. Lyell went to France soon after reviewing Scrope's book, to see the phenomena with his own eyes; and later he used Scrope's technique with almost equal effect for two out of three frontispieces of his *Principles of geology* (1830–33).[77] It is noticeable, however, that such formalized geological landscapes were not often used in subsequent work. This cannot be attributed to the cost of hand colouring, because coloured illustrations of other kinds continued to be used extensively. More probably it reflects the fact that the geological community and their wider audience were able by this time to 'read' the still more formalized language of maps and sections, and that landscapes, having served their purpose as a major element of that language at an earlier period, had become less important.

VI. CONCLUSION

In this article I have suggested that the striking increase in the quantity and quality of illustrations in works that we would term 'geological', during the late eighteenth and early nineteenth centuries, is partly explicable in terms of economic and technical changes consequent on the introduction of new ways of reproducing visual materials. But I have argued that it also reflects, more fundamentally, the development of a new range of kinds of visual expression, which paralleled—and indeed was an essential part of—the development of a self-conscious new science. Emerging from a synthesis of heterogeneous earlier traditions such as 'cosmogony', mineralogy, 'géographie physique', natural history, 'Geognosie' and mining practice, *'geology'* swiftly acquired its own coherent intellectual goals and corresponding institutional forms. But an essential part of this complex historical process was the construction of a *visual language* that was appropriate to the subject-matter of the science, and which could complement verbal descriptions and theories by communicating observations and ideas that could not be expressed in words.

I have argued that the diverse forms of visual expression in the new science were derived from extremely varied social and cognitive sources, and were inter-related conceptually and historically. Furthermore, each developed in the course of time towards greater abstraction and formalization, and thereby became able to bear an increasing load of theoretical meaning. I have tried to suggest these inter-relations and developments in the form of a diagram (Figure 25), which I hope will help at a visual meta-level to summarize my tentative interpretation of the complex historical origins and development of the visual language of geology. The vertical dimension of the diagram represents progressively greater degrees of abstraction and formalization, and the incorporation of greater theoretical loading into the forms of illustration. In social terms, therefore, it also represents the development of an increasingly esoteric language, which had to be learned by recruits to the newly differentiating group that came to call themselves 'geologists'. The vertical dimension is also intended to suggest how this increasing formalization of the visual language of geology involved—and reflects—the development of successive sets of cognitive goals. Thus the first zone represents *topographical* visual products, derived from social activities which were not in themselves even 'proto-geological' in their intentions, but which provided precedents or traditions for visual products that were more specifically oriented towards the understanding of the earth. That specific orientation defines the second cognitive zone, which I have termed *distributional.* This includes the forms of illustration that embody what I have argued was the characteristic cognitive goal of most late eighteenth century 'proto-geological' activities. In the early nineteenth century a new group of visual products became common; and their greater degree of formalization enabled

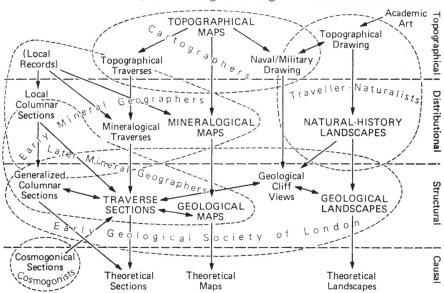

FIG. 25. A highly diagrammatic representation of the emergence and historical development of the visual language of geology. For explanation see text.

V

them to be the vehicles for what I have termed the *structural* cognitive goals of the self-styled science of 'geology'. Finally, from about 1820 onwards these 'structural' forms of illustration began to be developed still further in theoretical directions that enabled them to become a visual language for *causal* cognitive goals.[78]

The horizontal dimension of the diagram is an attempt to classify the main forms of illustration broadly according to their relation to the three-dimensional reality of the earth's surface and its internal structure. Thus columnar sections convey information or ideas about the crust of the earth at some real or notional point on its surface; profiles and traverse sections do the same for some linear tract, and maps for whole areas; while land-scapes attempt more directly to convey a reality that is experienced primarily as three-dimensional.

Within the two-dimensional space that the diagram provides, I have tried to suggest the appropriate positions and inter-relations of the forms of illustration that I have described in the body of this article. Thus the topographical zone includes the (non-visual) local records of rocks encountered in quarries, wells and mines, or observed in natural exposures. Linear traverses, topographical profiles and strip-maps form a 'linear' sub-class of the central category of topographical maps (and their marine equivalent, navigational charts). These maps were often supplemented by accurate documentary drawing, which was related socially and stylistically to topographical art; and this in turn was a kind of 'poor relation' of academic landscape art.

I have argued that most late eighteenth century 'geological' activities, and their corresponding standard forms of illustration, fall into the distributional cognitive zone. Local records were made more accessible visually by being drawn to scale as local columnar sections of strata in particular quarries or mines. These amplified the information that was generally summarized by means of symbols or verbal notes scattered over the surface of a 'mineralogical' map (or occasionally along the surface of a topographical profile, which then also became 'mineralogical' in character). The distributional pattern of major types of rock might be generalized from these symbols by means of broad colour-washes; these indicated an inferred extrapolation of surface distributions between scattered exposures of the rocks, but did not attempt to suggest a three-dimensional structure, except perhaps marginally in cases where surface distributions were complemented with columnar sections generalized for some extensive area. At the same time, the superficial appearance of particular localities or phenomena was directly portrayed, though often crudely, by means of 'natural-history' landscapes.

A shift into a cognitive zone with fully structural intentions is marked by the development of new forms of illustration in the early nineteenth century. Generalized columnar sections of a more abstract kind showed the succession of strata that were the 'raw material' for the structure of a given region, deliberately abstracted from the complications of post-depositional disturbance. Observed exposures of rocks at the surface or in mines and wells were plotted on topographical profiles and extrapolated inwards to form traverse sections that displayed the inferred structure of the region along a given traverse. The surface outcrops of rocks were plotted on maps that were truly geological in that they were designed to suggest the underground structure of the region—often by being closely associated with traverse sections. At the same time landscapes were made more explicitly geological, first by the accentuation of theoretically significant features at the expense of the more fortuitous, then by applying the deliberately conventional colouring that was being used for sections and maps, so as to suggest the three-dimensional structure of the area portrayed.

Finally, after about 1820 these truly geological forms of illustration were modified still further to enable them to become the vehicles of cognitive goals that reached beyond the analysis of three-dimensional structures to the interpretation of those structures in causal terms. Theoretical sections were used to convey interpretations of the history of the earth in all its aspects—the history of sedimentation, of igneous activity, of mountain-building and of life; theoretical maps were used to convey causal interpretation of spatial patterns and spatial events in the earth's history; and even theoretical landscapes were occasionally used for the same purposes.

The 'envelopes' that I have drawn around these varied forms of illustration are intended to summarize my interpretation of the contingent socio-historical circumstances in which these varied forms of the visual language of geology came into existence. There seems to be a broad division in the late eighteenth century between the practical enterprise of 'mineral geographers' and the work of those I have termed 'traveller-naturalists'. The former, of whom the compilers of the *Atlas minéralogique* and the German mineral surveyors such as Charpentier were characteristic, developed topographical maps in a specifically 'mineralogical' direction, supplementing them to a limited extent with sections; and they drew their data from regional surveys of mineral resources and particularly from mining and related proto-industrial activities. The traveller-naturalists, by contrast, such as Hamilton, de Saussure and Faujas, were more concerned with the topographical description of particular localities or phenomena of 'natural-history' interest, and illustrated their work primarily with topographical drawings, with only marginal use of purely topographical maps. In other words, the two areas of activity, while both relying necessarily on topographical cartography, had divergent cognitive

V

aims and social purposes, and hence also developed different standard forms of visual expression.

Around the turn of the nineteenth century, the visual language of the mineral geographers shifted into the structural cognitive zone, but did so in two separate social and geographical contexts. In France, Cuvier and Brongniart developed the tradition of the *Atlas minéralogique* into a fully structural 'géographie minéralogique', by applying the Wernerian 'geognostic' concept of strata to the Paris region, and extrapolating the evidence of surface exposures into a fully three-dimensional combination of a geological map and geological traverse sections. At the same period in England, the mineral surveyors Farey and Smith developed geological maps and sections with a similar meaning but a contrasted style; and a parallel movement in Germany was likewise related to the social world of civil and mining engineering.

These diverse antecedent traditions were first integrated into a unified and standardized visual language in the work of the first specialist society to be devoted to geology. The 'gentlemanly' social character of the Geological Society of London ensured that its members appreciated the topographical tradition of the earlier 'traveller-naturalists', and they developed a more formalized style that enabled landscape to bear a greater weight of structural meaning. At the same time, their interest in utilitarian geology and their social links with the world of mining—a concern that has often been under-estimated by historians—made them willing to adopt and develop the overtly structural maps and sections that were emerging from the work of mineral surveyors and mining engineers. In the illustrations to their *Transactions* can be seen for the first time a full integration of these divergent visual traditions; and within a few years this first specialist periodical for geology had developed a standardized visual language combining maps, sections and landscapes. This was the visual language that was subsequently adopted by the whole international community of geologists.[79]

Towards the end of the period I have surveyed, there were a few but important attempts to extend this primarily structural visual language into new forms that could more directly express theories of the causal and temporal development of the inferred structures of the earth's crust. By the end of my period, these causal forms of visual expression were still experimental in character. Although they were the antecedents of forms that have become highly important in modern geology, they seem to have been submerged later in the nineteenth century by the standard maps and sections that were needed for the geological surveying which had come to dominate the fully institutionalized science.

I have deliberately refrained in this article from attempting to analyze the relation between visual and verbal forms of communication in this (or any) science. The historical study of visual communication in science is still such a neglected area of research that I have thought it better to confine this article to an outline description and tentative interpretation of the visual language of geology, considered in its own terms. I have merely noted in passing that this visual communication was (and is) broadly *complementary* to verbal communication. It would, I think, be a characteristically sterile piece of reductionist philosophizing to debate whether the visual modes of communication that I have analyzed are in principle reducible to verbal form or not. For the historian it is surely sufficient to note that the development of a distinctive visual language was a striking feature of the emergence of geology as a new science, and that it has continued to be a prominent feature of the discourse of geologists ever since. As I suggested at the beginning of this article, it is only the non-visual traditions of professional history of science that have tended to obscure its importance.

REFERENCES

1. A short illustrated lecture based on an earlier draft of this article was given at King's College, London, on 1 September 1975 as part of the Charles Lyell Centenary Symposium. I am very grateful to the many members of that conference who made helpful suggestions and criticisms, or who told me about examples—and counter-examples—of illustrations which I had not previously seen. I am particularly indebted, as usual, to Dr Roy Porter, for his constructive criticisms and encouragement in this work.

2. Quoted in J. W. Goethe, *Italian journey (1786-1788)*, ed. by W. H. Auden and E. Mayer (London, 1962), Introduction. The volumes of the great *Corpus der Goethezeichnungen*, ed. by Gerhard Femmel (Leipzig, 1958-   ) show how far Goethe practised his own maxim, particularly if the anachronistic editorial separation of 'scientific' from 'artistic' drawings is ignored. The 'geological' drawings are in Bd Vb, *Die naturwissenschaftlichen Zeichnungen* (1967), but should be compared with, for example, those in Bd IVa, *Nachitalienische Landschaften* (1966).

3. Unfortunately this applies to my own work in this article, which would be illustrated far more fully than it is, if economic constraints could be ignored. My selection is deliberately weighted towards exemplifying those aspects of the visual language of geology which have been relatively neglected; but wherever possible I refer in the notes to other relevant published reproductions in accessible modern works. Historical books aimed at the 'general readers' market may be illustrated lavishly, in contradiction to the point I am making, but in most cases there is little integration between text and pictures, and the latter often seem to have been chosen by the publisher rather than by the author!

4. The idea that thinking can be intrinsically visual is apt to seem strange and even incomprehensible to those whose mental processes are primarily verbal or mathematical: on this point, see Rudolf Arnheim, *Visual thinking* (London, 1970). Educationalists have had to coin the ugly

V

neologism 'graphicacy' to express the belief that visual and graphical skills should be given as much weight in education as those of 'literacy' and 'numeracy'. The need for such a reform is indicated by the remarkable incompetence of many 'literate' historians and 'numerate' scientists when faced with simple visual tasks such as drawing a clear diagram or reading a map.

5. Although it covers only one country, the most thorough analysis of this development—integrating cognitive and institutional changes—is Roy S. Porter, "The making of the science of geology in Britain, 1660-1815" (Ph.D. dissertation, University of Cambridge, 1974), which should be published by the Cambridge University Press in 1977.

6. William M. Ivins, Jr, *Prints and visual communications* (London, 1953); the whole book is full of insights of potential value to historians of science, but see ch. 8 for a summary of the argument.

7. For a masterly treatment of this point in a general context, see E. H. Gombrich, *Art and illusion. A study in the psychology of pictorial representation* (London, 1960). The lectures on which the book was based were entitled "The visible world and the *language* of art".

8. Apart from this brief note, I shall omit any separate discussion of one major kind of illustration, namely that of *specimens*. Although illustrations of rocks, minerals and fossils are quantitatively one of the most important components of the visual language of geology, they have so much in common with earlier visual traditions of natural history that they deserve fuller discussion in that context. But it is worth noting that they share one feature which I shall argue is characteristic of all geological illustrations, namely the tendency to become more formalized and theory-laden in the course of time. Late eighteenth century illustrated books on rocks, minerals and fossils are often in effect 'cabinets of rarities' converted into book form: the specimens are illustrated with an extreme and—to modern eyes—indiscriminate realism (depicting, for example, the fortuitous colours of fossils and the shapes of the slabs of rock on which they are preserved), sometimes rising to virtuoso heights of *trompe d'oeil* effects. In the early nineteenth century there was a clear trend, only partly explicable in economic terms, towards the omission of colour from pictures of fossils, and the selective 'playing down' of other fortuitous preservational features of particular specimens. The same trend can be seen in the illustration of minerals (for many of which colour is equally fortuitous), where simplified line-drawings of crystal forms were used increasingly to highlight what were believed on theoretical grounds to be the more fundamental features. The technical change from copper engraving to lithography, which I discuss in the next section of this article, was particularly important for the illustration of fossils. Another aspect of the visual treatment of fossils which deserves fuller analysis is the generally hesitant and tentative manner in which possible reconstructions of extinct organisms began to be suggested visually in published form during the early nineteenth century. For reproductions of some illustrations which exemplify these points, see my book on *The meaning of fossils: episodes in the history of palaeontology* (London and New York, 1972).

9. In order to keep the text and notes of this article within reasonable bounds, I shall use a few works as representatives for much larger classes

of material. Otherwise my references would quickly become a summary of sources for the history of geology in general during this period. Likewise I shall not attempt to note all the relevant secondary literature, particularly for those aspects of the subject—notably geological maps—which have been relatively fully studied by historians of geology.

10. It would be worth making international comparisons on these points. Examples of work by some of the individuals I have mentioned will be cited later in this article. Webster's name appears on many illustrations in the *Transactions of the Geological Society of London*, in cases where he must have worked from rougher sketches submitted by the author of the article concerned. Many field sketches by Lyell are excellently reproduced in Leonard G. Wilson, *Charles Lyell. The years to 1841: the revolution in geology* (New Haven and London, 1972). Perhaps the most striking exception to my generalization about the artistic competence of early nineteenth century English geologists is Charles Darwin, whose notebooks and published works are curiously poor in illustrations of any kind: Darwin seems to have been exceptionally 'non-visual' as far as the *communication* of his observations and theories was concerned.

11. For an earlier historical period, when such frontispieces were commonly symbolic or allegorical in style, there is a well-established art-historical tradition of iconographical interpretation, which has been applied to some more specifically 'scientific' frontispieces. But there is scope for much more research on the frontispieces of scientific books of the late eighteenth century onwards, when they became less overtly symbolic in style but arguably no less rich in meaning.

12. Sir William Hamilton, *Campi Phlegraei. Observations on the volcanos of the Two Sicilies, as they have been communicated to the Royal Society of London* (Naples, 1776), with a *Supplement* (Naples, 1779). *Campi Phlegraei* has a text in English and French; it was originally sold at sixty Neapolitan ducats (see p. 90), and Hamilton spent £1300 of his own money on financing it: see B. Fothergill, *Sir William Hamilton, envoy extraordinary* (London, 1969). Hamilton also presented the Royal Society of London with an 'eidophysicon' with the aid of which back-lighted paintings on glass could give an even more vivid impression of the eruptions he had witnessed: see Bernard Smith, *European vision and the south Pacific* (London, 1960), ch. 2.

13. Horace-Bénedict de Saussure, *Voyages dan les Alpes, précédés d'un essai sur l'histoire naturelle des environs de Genève* (Neuchatel, 1779-96).

14. For example, the largely topographical works of Jean-André de Luc, such as his early five-volume *Lettres physiques et morales sur l'histoire de la terre et de l'homme* (La Haye and Paris, 1779) and his late two-volume *Geological travels in some parts of France, Switzerland and Germany* (London, 1813). The topographical works of Jean Louis Giraud-Soulavie and even Leopold von Buch are comparable in the extreme scarcity or total absence of illustrations. Likewise James Hutton, whose importance as a traveller-naturalist is now in danger of being obscured by recent historical enthusiasm for his natural philosophy, included only seven engraved plates (two re-drawn from de Saussure) to accompany the 1187 pages of text in the published volumes of his *Theory of the earth* (Edinburgh, 1795; reprinted Weinheim and Codicote, 1959). The same verbosity of text and paucity (or absence) of illustrations characterize

most of the early 'textbooks' for geology, suggesting that the value of visual communication for *didactic* purposes was likewise not generally appreciated, even into the early nineteenth century: examples include Johann C. W. Voigt's *Praktische Gebirgskunde* (2nd ed., Weimar, 1797), De Luc's *Traité élémentaire de géologie* (Paris, 1809), Scipio Breislak's *Introduzione alla geologia* (Milano, 1811), Robert Bakewell's *An introduction to geology* (London, 1813), J.-C. de la Métherie's three-volume *Leçons de géologie* (Paris, 1816) and J. F. d'Aubuisson de Voisins's two-volume *Traité de géognosie* (Strasbourg, 1819).

15. My Fig. 1 is reproduced from pl. vi of *Campi Phlegraei* (ref. 12), and Fig. 2 from pl. i of Sir William Hamilton, *Observations on Mount Vesuvius, Mount Etna, and other volcanos: in a series of letters addressed to the Royal Society* (London, 1772).

16. Sometimes this disadvantage was reduced by the fact that the engraved plates were bound in a separate volume from the text, so that the reader could have both text and illustrations before him simultaneously: a significant example is the early *Transactions of the Geological Society of London*, as originally issued. Of course where expense was no object —as in Hamilton's *Campi Phlegraei*—the plates could be interleaved with the explanatory text at the correct points.

17. Some of the Hebridean plates from Richard Ayton and William Daniell, *A voyage round Great Britain undertaken in the year 1813 . . .* (London, 1814-20) were issued separately in 1818 with a cover label which indicates that they were intended to appeal to geologists. An example of a geological mezzotint is the view of the volcanic topography of Auvergne which forms the frontispiece to Leopold von Buch, *Geognostische Beobachtungen auf Reisen durch Deutschland und Italien*, ii (Berlin, 1809).

18. Senefelder devised the process in 1797 originally for the cheap reproduction of sheet music. For its history, see Ivins, *op. cit.* (ref. 6), and for example Michael Twyman, *Lithography 1800-1850: the techniques of drawing on stone in England and France and their application in works of topography* (London, 1970).

19. In the introduction to his *Art of drawing on stone* (London, 1824), he defended lithographs against the slur of being mere "greasy daubs", urged their social value in allowing artistic knowledge to be disseminated more widely, and illustrated their scientific value with, for example, a topographical map (pl. 19) that rivals an engraving in its fine detail. He also produced a very fine specimen plate (undated, but probably about 1820) with drawings by Scharf—mostly of fossils, but also one geological cliff view—to demonstrate the value of his 'lithotint' process more directly for the geological market.

20. See the estimates submitted in 1822, printed in H. B. Woodward, *History of the Geological Society of London* (London, 1907), 63. Some of the leading members of the Society also took advantage of the Scharf/Hullmandel partnership for their other publications: see for example the drawings of caves in William Buckland, *Reliquiae diluvianae* (London, 1823).

21. Charles Lyell, *Principles of geology* (London, 1830-33; reprinted New York and London, 1969). The frontispiece to the first volume—the famous so-called 'Temple of Serapis' at Pozzuoli—is reproduced in my article on

"The strategy of Lyell's *Principles of geology*", *Isis*, lxi (1970), 4-33; and all three are reproduced in Wilson, *op. cit.* (ref. 10), Figs 35, 36, 43.

22. William Daniel Conybeare and William Phillips, *Outlines of the geology of England and Wales, with an introductory compendium of the general principles of that science, and comparative views of the structure of foreign countries* (London, 1822; reprinted Farnborough, 1969).

23. Lyell, *op. cit.* (ref. 21); Henry T. De la Beche, *A geological manual* (London, 1831) and *Sections and views, illustrative of geological phaenomena* (London, 1830). The latter book was entirely illustrated with De la Beche's own lithographs, apart from one engraving included where very fine detail was necessary.

24. See Ivins, *op. cit.* (ref. 6), 97, 107, for the specialization and division of labour within the wood-engraving trade, and for its social impact in making illustrations available to the mass market. In allowing illustrations to be printed on the same page as text, the development of wood engraving was in a sense a return to a much earlier situation in the history of visual communication, when woodcuts, although much cruder, had had the same advantage.

25. Henry T. De la Beche, *How to observe. Geology* (London, 1835); Lyell, *op. cit.* (ref. 21). A typical sample of wood engravings from Lyell's third volume, illustrating a single restricted topic, is reproduced in my article on "Lyell on Etna, and the antiquity of the earth", in Cecil J. Schneer (ed.), *Toward a history of geology* (Cambridge, Mass., 1969), 288-304.

26. The theory-laden and 'non-natural' character of geological maps is well expressed by J. M. Harrison, "Nature and significance of geological maps", in C. C. Albritton (ed.), *The fabric of geology* (Stanford, 1963), 225-32.

27. The geological analogy for this historiographical point is borrowed from Roy Porter, "Charles Lyell and the principles of the history of geology", *British journal for the history of science*, ix (1976), 91-103.

28. For a brief introduction for the relevant period, see R. A. Skelton, "Cartography", in *A history of technology*, ed. C. Singer *et al.*, iv (Oxford, 1958), 596-628. R. V. Tooley and Charles Bricker, *A history of cartography. 2500 years of maps and mapmakers* (London, 1969), is profusely illustrated with good reproductions, but unfortunately is arranged according to the areas covered by the maps described, and not chronologically.

29. De Saussure, *Voyages* (ref. 13), ii (1786), frontispiece.

30. *Campi Phlegraei* (ref. 12), before Pl. 1. The map was coloured to match the style of the other plates, but significantly the *meaning* of the colours was nowhere explained, and does not seem to be related to the 'geological' character of the area.

31. This point is well made by V. A. Eyles, "Mineralogical maps as forerunners of modern geological maps", *The cartographic journal*, ix (1972), 133-5. More specifically, see also John G. C. M. Fuller, "The industrial basis of stratigraphy: John Strachey, 1671-1743, and William Smith, 1769-1839", *Bulletin of the American Association of Petroleum Geologists*, liii (1969), 2256-73.

32. For a description of the history of the *Atlas* and an analysis of the theoretical intentions of its compilers, see Rhoda Rappaport, "The geological atlas of Guettard, Lavoisier and Monnet: conflicting views of the nature of geology", in Schneer, *op. cit.* (ref. 25), 272-87. I am indebted to Dr

Rappaport for sending me a photograph of feuille 56 (1770), from which
my Fig. 3 is reproduced. For my own study of the *Atlas* I used the
almost complete copy in the Geological Society of London (originally in
the collection of its first President, G. B. Greenough).

33. Johann Friedrich Wilhelm Charpentier, *Mineralogische Geographie der
Chursächsischen Lande* (Leipzig, 1778): "Petrographische Karte des
Churfürstentums Sachsen und der incorporirten Lande, in welche durch
Farben und Zeichen die Gesteinarten . . . angegeben worden sind" (see
detail reproduced here as Fig. 4).

34. See for example the maps in J. C. W. Voigt, *Mineralogische Beschreibung
des Hochstifts Fuld und einiger merkwürdigen Gegenden am Rhein und
Mayn* (Dessau und Leipzig, 1783), and in Matthias Flurl, *Beschreibung
der Gebirge von Baiern und der oberen Pfalz* (München, 1792). Other
German examples are mentioned briefly by Eyles, *op. cit.* (ref. 31).
Guettard's small-scale map of France (1784) was similar: it is reproduced,
unfortunately at a very small size, by Rappaport, *op. cit.* (ref. 32), Fig. 1.

35. Cuvier & Brongniart, "Essai sur la géographie minéralogique des environs
de Paris", *Mémoires de l'Institut imperial de France*, an 1810 (1811),
1-278, 2 pls and map; reprinted in Cuvier, *Recherches sur les ossemens
fossiles* (Paris, 1812), i. Since the formations were described in their true
temporal order of deposition, the corresponding 'boxes' in the key to
the map are *inverted* from their true spatial order of superposition;
nevertheless, the fact that they are in *some* real order is a significant
change from the purely arbitrary arangement of the 'boxes' on *e.g.*,
Charpentier's map.

36. *A delineation of the strata of England and Wales, with part of Scotland
. . .* (London, 1815). Excellent colour facsimiles are now available, both
of a representative portion of the 1815 map (National Museum of Wales,
Cardiff, 1975) and of some of Smith's slightly later County maps (British
Museum (Natural History), London, 1974). Joan M. Eyles, "William
Smith (1769-1839): a bibliography of his published writings, maps and
geological sections, printed and lithographed", *Journal of the Society for
the Bibliography of Natural History*, v (1969), 87-109, gives an invaluable
guide to this complex subject, and reproduces Smith's treatment of the
Isle of Wight in three successive variants of his 1815 map.

37. For a thorough analysis of the mapping activities of Smith and his contemporaries, which avoids the hagiographical tendencies of some earlier
work on Smith, see Rachel Bush (Rachel Laudan), "The development
of geological mapping in Britain from 1795 to 1825" (Ph.D. dissertation,
University of London, 1974). Dr Laudan argues that Smith's idiosyncratic convention of colouring was directly related to his—far from
'modern'—causal *theory* of the origin of the formations in a series of
sediment-bearing 'floods': the shaded limits of the formations on Smith's
maps would thus indicate their *original* distribution and dip, only slightly
modified by subsequent erosion. In the terms of my article, this interpretation would imply that Smith's maps had cognitive aims that were
not merely structural but also causal. In any case, it is noteworthy that
Smith's system of shaded colouring was clearly intended to show the
areal limit of each formation, rather than its stratigraphical base, for in
the keys the shading is often shown on the *upper* side of each box, and
not on the lower side as a more 'modernist' interpretation of his inten-

tions would lead one to expect. Smith still used the convention of spot-symbols, but only for features such as collieries. Examples of 'distributional' maps that Smith may have known include the frontispiece, a land-use map with colour-washes, to John Billingsley's *General view of the agriculture of the county of Somerset* (London, 1794), and the 'mineralogical map' (reproduced excellently in N. E. Butcher's exhibition catalogue *The history and development of geological cartography* (Reading, 1967)) to William George Maton's *Observations . . . of the western counties of England* (Salisbury, 1797), in which varieties of engraved patterns are used to suggest the chemical relationships of the different rock-types. It is well known that Smith's choice of colours for his maps was based on a somewhat similar concept of 'realistic' representation, with each colour approximating to the real colour of the rock concerned. This 'realism' was taken to its limit in sections constructed out of the rocks themselves: see Trevor D. Ford, "White Watson (1760-1835) and his geological sections", *Proceedings of the Geologists' Association*, lxxi (1960), 349-63.

38. G. B. Greenough, *A geological map of England and Wales* (London, 1819), actually not published until 1820. The key had no fewer than thirty-seven 'boxes'. For his extreme methodological scepticism, see *A critical examination of the first principles of geology* (London, 1819).

39. Compare for example the small maps in Robert Bakewell, *op. cit.* (ref. 14), William Phillips, *An outline of mineralogy and geology* (London, 1815), and Conybeare & Phillips, *op. cit.* (ref. 22).

40. Thomas Webster, "On the freshwater formations in the Isle of Wight, with some observations on the strata over the Chalk in the southeast part of England", *Transactions of the Geological Society of London*, ii (1814), 161-254, pls 9-11; his general map of the London, Hampshire and Paris 'basins' is reproduced in Wilson (ref. 10), Fig. 14. Webster's full account comprises half of the large volume by Sir H. C. Englefield, *A description of the principal picturesque beauties, antiquities, and geological phaenomena, of the Isle of Wight* (London, 1816): my Fig. 5 is reproduced from part of pl. 50.

41. John MacCulloch, "A sketch of the mineralogy of Skye", *Transactions of the Geological Society of London*, iii (1816), 1-111, pls 1-4: the map is Pl. 1; W. D. Conybeare (ed.), "On the geological features of the north-eastern counties of Ireland", *ibid.*, 121-222, pls 8-11: the map is pl. 8. The *only* coloured map in the two earlier volumes did not denote geological but topographical regions: Henry Holland, "A sketch of the natural history of the Cheshire Rock-salt district", *ibid.*, i (1811), 38-61, pl. 1. After the third volume geological maps became common.

42. Nicholas Desmarest, "Extrait d'un mémoire sur la détermination de quelques époques de la nature par les produits de volcans", *Observations sur la physique, l'histoire naturelle et les arts*, xiii (1779), 115-26. See K. L. Taylor, "Nicholas Desmarest and geology in the eighteenth century", in Schneer, *op. cit.* (ref. 25), 339-56. Desmarest's complete map of Auvergne was published posthumously in 1823.

43. MacCulloch's and Dick Lauder's work on the Glen Roy enigma is analyzed in detail in my "Darwin and Glen Roy: a 'great failure' in scientific method?", *Studies in the history and philosophy of science*, v (1974), 97-185. Buckland's map (*Transactions of the Geological Society of*

London, v (1821), pl. 37) was republished in his *Reliquiae diluvianae* (London, 1823), pl. 27. Lyell's map is in *Principles of geology*, ii (1832), opposite p. 1, and its theoretical intention is analyzed in my article on that work (ref. 21); see also his highly theoretical maps of the world, to illustrate the putative climatic effects of a clustering of continental masses in low and high latitudes, first published in the third edition of the *Principles* (London, 1834), i, opposite p. 80.

44. What I here term 'traverse' and 'columnar' sections are more commonly called 'horizontal' and 'vertical' sections respectively. The latter terms are likely to be confusing for non-geological readers, however, because a 'horizontal' section represents a *vertical* plane, albeit a plane the position of which can be marked on a map as a horizontal line across country. Modern geologists also use a variety of other more abstract forms of section, the conceptual antecedents of which are the 'theoretical sections' that I mention briefly later in this article.

45. For example, feuille 61 (1769) of the *Atlas* has almost purely verbal sections of particular quarries; feuille 25 (1766) has more formalized columns of strata, drawn to scale and cross-referenced to the symbols used on the adjacent map, and ranging in specificity from a "coupe d'une carrière" to an "ordre et coupe des bancs des montagnes du Vexin considérées généralement". Feuille 5 (undated) has a measured columnar section summarising the strata in coal-mines near Valenciennes. See also the sheets of the Paris region, reproduced (though unfortunately at a much reduced scale) as Figs 2 and 3 in Rappaport (ref. 32), and her discussion of the purpose of the sections. The range of sections that I describe is an iconographical, not a chronological spectrum.

46. There are similar profiles, with surface 'mineralogical' detail added, on feuilles 42 and 57 (1770). Measured profiles of surface topography were already a well established convention: see for example the one associated with the triangulation for the French meridian traverse of Peru in 1735-45, reproduced in Lloyd A. Brown, *The story of maps* (London, 1949), 262.

47. For example [J.F.C.] Morand, "L'art d'exploiter les mines de charbon de terre", in *Description des arts et métiers*, ii-iv (Paris, 1768-77): my Fig.10 is reproduced from ii, pl. 2, and is chosen for its relatively clear illustration of strata, although the mining methods shown are exceptionally primitive. See also the plates from the *Encyclopédie*, reproduced in Charles Coulston Gillispie (ed.), *A Diderot pictorial encyclopedia of trades and industry* (New York, 1959), pls 127-34; also the engravings in German-language treatises such as Charpentier, *op. cit.* (ref. 33).

48. Strachey's sections (1719, 1725) are reproduced for example in Kirtley F. Mather and Shirley L. Mason, *A source book in geology* (New York and London, 1939; reprinted 1964), 54, and by Fuller (ref. 31); those of Lehmann (1756) in John C. Greene, *The death of Adam: evolution and its impact on western thought* (Ames, Iowa, 1959), 60, and in Frank Dawson Adams, *The birth and development of the geological sciences* (London, 1938; reprinted New York, 1954), pl. 11. See also the exceptionally fine coloured engravings by F. H. Spörer in F. W. H. Trebra, *Erfahrungen vom Innern der Gebirge* (Dessau and Leipzig, 1785), in which evidence from mine-workings is explicitly used to infer structure in geological sections.

49. On industrial secrecy in mining, see Roy Porter, "The industrial revolution and the rise of the science of geology", in M. Teich and R. Young (eds), *Changing perspectives in the history of science* (London, 1973), 320-43. There is a small collection of unpublished mine sections in the Geological Society of London (mostly dating from the earliest years of the nineteenth century), which were probably presented in the first flush of enthusiasm for transcending considerations of secrecy by making the Society a 'Baconian' repository of mining records in the national interest. I imagine that far larger collections of such materials are preserved at the main Continental mining schools, but I have not had an opportunity to study them.

50. Westgarth Forster, *A treatise on a section of the strata, commencing near Newcastle-upon-Tyne, and concluding on the west side of the mountain of Cross Fell* (Newcastle, 1809), My Fig. 11 is reproduced from De la Beche's *Sections and views, op. cit.* (ref. 23), pl. 1.

51. John Whitehurst, *An inquiry into the original state and formation of the earth; deduced from the facts and the laws of Nature. To which is added an appendix, containing some observations on the strata of Derbyshire* (London, 1778). His sections are on pls 3, 4: my Fig. 6 is reproduced from pl. 3, Fig. 4. The earlier 'cosmogonical' sections of Steno (1669) are reproduced in Adams, *op. cit.* (ref. 48), 362, and Greene, *op. cit.* (ref. 48), 51, and in their original form in Rudwick, *op. cit.* (ref. 8), 67: compare with Descartes's section, reproduced in Helmut Hölder, *Geologie und Paläontologie in Texten und ihrer Geschichte* (Freiburg and München, 1960), 136. Hölder also reproduces two of Kircher's (1664) sections (Taf. 2, 3): compare with those of Scheuchzer (1731), also in Hölder (Taf. 4); and those of Moro (1740), in Adams, *op. cit.* (ref. 48), 370-1. See also those of Strachey (1725) and Catcott (1761), reproduced in Gordon L. Davies, *The earth in decay: a history of British geomorphology, 1578-1878* (London, 1969), pl. 3, 4; and those of Michell (1761), in Adams *op. cit.* (ref. 48), 417.

52. His drawings of a normal fault show the same masonry-like style. The analogy between strata and masonry is a natural one to make, in view of the regular bedding and jointing to be seen in many quarries. Dr Roy Porter has pointed out to me that the simile was already a *verbal* commonplace at this period: for example, "if a person was to see the broken walls of a palace or castle that had been in part demolished . . . . And in the same manner if a person was to view the naked ends and broken edges of the strata in a mountain . . . (etc.)", Alexander Catcott, *A treatise on the deluge* (London, 1761), 163.

53. John Farey, *General view of the agriculture and minerals of Derbyshire, with observations on the means of their improvement*, i (London, 1811): see his section through Derbyshire (pl. 5). Trevor D. Ford, "The first detailed geological sections across England, by John Farey, 1806-8", *Mercian geologist*, ii (1967), 41-49, gives re-drawn versions of the earlier sections, which Farey said he circulated widely in the form of hand-drawn copies. Some of Farey's later sections did show curved strata.

54. Farey, *op. cit.* (ref. 53), pls 3,4: my Fig. 22 reproduces a small corner of pl. 4.

55. William Smith, *A geological section from London to Snowdon showing the varieties of the strata, and the correct altitude of the hills, coloured*

to correspond with his geological map of England and Wales (London, 1817). Five further sections were published as separate sheets in 1819: for details, see Eyles, *op. cit.* (ref. 36). A category of visual products that would deserve more study is the *real* three-dimensional geological models that were constructed and marketed in the early nineteenth century.

56. For example in his *Mineralogische Reisen durch das Herzogthum Weimar und Eisenach und einige angränzende Gegenden, in Briefen* (Weimar, 1781-85) and in his *Practische Gebirgskunde* (ref. 14). A re-drawn version of one of his sections is in Hölder, *op. cit.* (ref. 51), 40. For a suggestive study of the engineering drawings of this period as a means of structural visual communication, see the original edition of Francis D. Klingender, *Art and the industrial revolution* (London, 1947), ch. 4, "Documentary illustration".

57. Cuvier and Brongniart, *op. cit.* (ref. 35): my Fig. 8 is reproduced from part of fig. 4. Their "coupe générale" is fig. 1. I do not mean to imply that these sections were totally original in style, but their relation to earlier German examples and their possible antecedents back towards those in the *Atlas minéralogique* need further research. Although I do not accept the 'catastrophist' historiography of Michel Foucault's *Les mots et les choses* (Paris, 1966; trns. as *The order of things* (London, 1970)), his interpretation of Cuvier's biological work, as penetrating the "surface" of the earlier epistemological "grid" in order to reveal "deeper" structures could be supported by an extension to Cuvier's *geological* work, of which these sections are in my opinion symptomatic.

58. Webster's sections in his 1814 paper (ref. 40) are on pl. 11. There was an uncoloured engraved section in the same style even in the first volume (1811) of the Geological Society's *Transactions,* illustrating a paper by Arthur Aikin (i, unnumbered pl.), and some simple coloured ones in the same volume as Webster's, illustrating a paper by J. F. Berger (ii, pl. 1). In the third and fourth volumes (1816 and 1817) there were many more, illustrating papers by MacCulloch, Aikin, Buckland, etc. A more detailed study of these illustrations would probably show that I have over-simplified the historical relation between the French and English sections; but I am not concerned here to establish 'firsts' so much as to suggest how the pattern of *standardized practice* developed.

59. My Fig. 12 is reproduced from Englefield, *op. cit.* (ref. 40), pl. 47. A slightly earlier but much less spectacular example, which may have given Webster his model, is Sir James Hall, "On the vertical position and convolutions of certain strata, and their relation to granite", *Transactions of the Royal Society of Edinburgh,* vii (1814), 79-108: see pl. 4.

60. My Fig. 23 is reproduced from De la Beche, *op. cit.* (ref. 23), pl. 5. Conybeare published a fine "Section from the north of Scotland to the Adriatic" with his "Report on the progress, actual state and ulterior prospects of geological science", *Reports of the British Association for the Advancement of Science, 1st & 2nd meetings 1831-2* (London, 1833), 365-414.

61. H. T. De la Beche, *Report on the geology of Cornwall, Devon and west Somerset* (London, 1839). His emphasis on these dangers of vertical exaggeration was motivated by theoretical concerns: for example, he

published a section of the whole earth, with mountains and oceans drawn to true scale, and commented, "how insignificant do our *tremendous* dislocations, *stupendous* mountains, and the like become, when we contemplate such a figure as that before us", and he suggested that "mere thermometrical differences beneath the earth's crust" could account for even the greatest features of its surface (*Sections and views* (ref. 23), 71 and pl. 40).

62. "Tableau théorique de la succession et de la disposition la plus générale en Europe, des terrains et roches qui composent l'écorce de la terre . . ." (Paris, undated), referring to Alexandre Brongniart, *Tableau des terrains qui composent l'écorce du globe* (Paris, 1829); William Buckland, *Geology and mineralogy considered with reference to natural theology* (London, 1836), ii, pl. 1, from part of which my Fig. 24 is re-drawn; Charles Lyell, *Elements of geology* (London, 1838), frontispiece, reproduced in Wilson, *op. cit.* (ref. 10), Fig. 52. In Brongniart's and Buckland's sections the vertical scale of the strata doubtless reflects primarily what was believed to be the relative average thicknesses of the successive formations; but it is difficult to avoid the conclusion that it also indicates the authors' conception of the relative time-scale that the generally tranquil deposition of the strata had occupied. If this inference is justified, the relatively small thickness assigned to the Tertiary strata, and the even more insignificant position of the so-called "diluvium", would confirm the impression gained from verbal documents that a time-scale of humanly unimaginable magnitude was accepted as a matter of course by geologists in the 1830s, and that the supposed "diluvial" event occupied no physically privileged position in earth-history.

63. L. Elie de Beaumont, "Recherches sur quelques-unes des révolutions de la surface du globe . . .", *Annales des sciences naturelles*, xviii (1829), 5-25, 284-416; xix (1830), 5-99, 177-240; also published in book form (Paris, 1830). My Fig. 9 is reproduced from part of pl. 3. The convention of dashed lines was used on this diagram to suggest his more speculative hypothesis that the most recent "révolution", which he believed had been responsible for the puzzling "diluvial" deposits, had been caused by the elevation of the Andes.

64. Saussure, *op. cit.* (ref. 13), ii (1786), pl. 5; compare i (1779), pl. 1. The work contains one remarkable iconographical experiment which was not, as far as I know, ever exploited further until modern times: a "vue circulaire" around the Glacier du Buet (i, pl. 8), which can only be compared with a modern panoramic photograph taken with a 180° fish-eye lens; the engraving is reproduced, but not well, in Richard J. Chorley, Anthony J. Dunn and Robert P. Beckinsale, *The history of the study of landforms, or the development of geomorphology*, i (London, 1964), 199.

65. In addition to my Fig. 1, examples of fine landscapes from *Campi Phlegraei* are reproduced in Hölder, *op. cit.* (ref. 51), Taf. 8, and in Fothergill, *op. cit.* (ref. 12), 65, 208, 209. The less successful pictures (*e.g.*, pls 9, 33), have not been reproduced.

66. Faujas de Saint Fond, *Recherches sur les volcans éteints du Vivarais et du Velay* (Grenoble and Paris, 1778): see pls 9, 10, 12. See also for example the set of engravings after drawings by Antonio de Bittio, published (without text) as *Basaltic mountains* (London, 1807), in which many of the

V

hexagonal columns have been 'rationalized' into square cross-sections; also the even later engravings in Scipion Breislak, *Atlas géologique ou vues d'amas de colonnes basaltiques* (Milan, 1818). My Fig. 14 is reproduced from the plate at p. 263 of Thos Pennant, *A tour of Scotland, MDCCLXXII, Part I* (London, 1774): Banks's "Account of Staffa" is at pp. 261-9. My Fig. 15 is reproduced from B. Faujas de Saint Fond, *Voyages en Angleterre, en Ecosse et aux Iles Hébrides; ayant pour objet les sciences, les arts, l'histoire naturelle et les moeurs . . .* (Paris, 1797). My charge of plagiarism is justified, I think, by a comparison of the incidental features of the two engravings.

67. For example, Fuseli referred to it as "that last branch of uninteresting subjects, that kind of landscape which is entirely occupied with the tame delineation of a given spot", and Gainsborough declined to paint "*real Views* from Nature in this Country" (*i.e.*, in England); quoted respectively in Kenneth Clark, "English romantic poets and landscape painting", *Memoirs and proceedings of the Manchester Philosophical Society*, lxxxv (1941-43), 103-20, and Luke Hermann, *British landscape painting in the eighteenth century* (London, 1973), 39-40. Twyman, *op. cit.* (ref. 18) emphasizes the low esteem in which topography was held by academic artists, and even argues that "the whole concept of the picturesque is at variance with the needs of topography" (12).

68. These were not mutually exclusive categories, nor were they completely separated from the social world of academic landscape painting. For example Gainsborough was happy to recommend Paul Sandby to his friends who wanted accurate topographical paintings of their country seats (see ref. 67). Sandby first encountered wild mountain scenery while attached as a draughtsman to the Highland survey after the '45 rebellion, and from 1768-97 was employed as principal drawing master at the Royal Military Academy at Woolwich; and he published successful topographical books, such as *The virtuosi's museum, containing select views, in England, Scotland, and Ireland* (London, 1778), which include landscapes. For the tradition of artist-naturalists on voyages see Bernard Smith, *op. cit.* (ref. 12), ch. 2.

69. My Fig. 16 is reproduced from Ayton and Daniell, *op. cit.* (ref. 17), ii, plate at p. 47. William Daniell was one of a family of topographical artists, who published many colour-plate books in the early nineteenth century on a variety of remote or exotic places. Some typical coastal views appended to naval charts are reproduced in Derek Howse and Michael Sanderson, *The sea chart. A historical survey based on the collections in the National Maritime Museum* (Newton Abbot, 1973).

70. My Figs 13, 17 are reproduced from Englefield, *op. cit.* (ref. 40), pls 46 and 25; similar cliff views are reproduced in Wilson, *op. cit.* (ref. 10), Figs 12, 15. For an example of a purely antiquarian drawing by Webster, see Englefield, pl. 43: see also the 'picturesque' view reproduced in Wilson, Fig. 17. Englefield's and Webster's drawings in this book to some extent suggest, *pace* Twyman (see ref. 67), that documentary accuracy could be combined successfully with the 'picturesque'.

71. My Fig. 18 is reproduced from pl. 16 of John MacCulloch, "On the parallel roads of Glen Roy", *Transactions of the Geological Society of London*, iv (1817), 314-92, pls 15-22. MacCulloch's drawings should be compared with the much cruder sketches of the same subject by Thomas

Dick Lauder, "On the parallel roads of Lochaber", *Transactions of the Royal Society of Edinburgh*, ix (1821), 1-64, pls 1-7; my Fig. 19 is reproduced from pl. 6. For an analysis of their (and later geologists') interpretations of these unusual horizontal terraces, see my "Darwin and Glen Roy" (ref. 43). The camera lucida of William Wollaston (who was another early member of the Geological Society of London) was a great improvement on the earlier camera obscura, because it was much less bulky and could be used in poor light. It would be interesting to know how generally it was used by early nineteenth century geologists.

72. See Hamilton, *op. cit.* (ref. 12), pl. 12, which shows a lava-flow in the 1760 eruption of Vesuvius flowing from a point low on the flank of the volcano and building a minor cone similar to a pre-historic cone nearby: this suggested that the accumulation of the whole volcano, including the minor cones, must have taken an extremely long period of time. My quotation is from *Campi Phlegraei*, 5.

73. A good early example is William D. Conybeare, "Descriptive notes referring to the outline of sections presented by a part of the coasts of Antrim and Derry", *Transactions of the Geological Society of London*, iii (1816), 196-216, pls 10-11*. A later example from Buckland's *Reliquiae diluvianae* (ref. 20) is reproduced in Chorley *et al.*, *op. cit.* (ref. 64), Fig. 24. The artistic antecedents of these geological cliff views (and other geological landscapes) are reflected in an evident reluctance to intrude any form of verbal labelling on to the drawing: the strata are identified by colours, and the landmarks often by 'flocks' of conventional 'birds', in both cases identified outside or above the landscape itself (see Fig. 20). The translucent interpretative overlays to the glacial landscapes in Louis Agassiz, *Recherches sur les glaciers* (Neuchatel, 1840; trans. as *Studies on glaciers*, ed. A. V. Carozzi, 1967) are a later example of the same tendency.

74. See for example Conybeare's 1816 cliff views (ref. 73). The last three (1819) of Smith's six engraved sections (ref. 55) were actually entitled "Section and view" as if to acknowledge their dual character.

75. For example, in the geological view which accompanied Richard Thomas's *Survey of the mining district of Cornwall* (London, 1819), the foreground is nominally a section in the same style as Smith's sections (which had the same publisher), but all the geological detail was in the landscape. Early examples of 'pure' geological landscapes of panoramic form are those illustrating articles by Nicholas Nugent and Thomas Weaver, in *Transactions of the Geological Society of London*, v (1821), pls 6, 7, 34. As usual the *Atlas minéralogique* (ref. 32) includes a much earlier precedent, although the landscape there bears only *verbal* labels indicating the surface *distribution* of the rocks: see feuille 76 (1769).

76. George Poulett Scrope, *Memoir on the geology of central France* (London, 1827). My Figs 20 and 21 are reproduced from his pl. 2 and 18 respectively. Part of another panoramic landscape (pl. 14) is reproduced in Wilson, *Lyell* (ref. 10), fig. 27; the other half in my article on "Poulett Scrope on the volcanoes of Auvergne: Lyellian time and political economy", *British journal for the history of science*, vii (1974), 205-42, which analyzes the theoretical meaning of Scrope's text and illustrations. Wilson also reproduces two of Scrope's non-panoramic landscapes (figs 28, 29). The importance of the colour on Scrope's landscapes can be

seen by comparing the originals with any of these reproductions or with the uncoloured versions in the second edition of his work, *The geology and extinct volcanos of central France* (London, 1858).

77. Lyell's illustrations were of the active volcano Etna in Sicily and the extinct volcanos near Olot in Spain (see ref. 21). A later example of his use of a 'theoretical' geological landscape is the frontispiece to his *Travels in North America* (London, 1845), which uses an imaginary aerial viewpoint to demonstrate the long-continued erosion of the Niagara Falls.

78. The major anomaly in this neat chronological scheme is the category of early 'cosmogonical' sections. However 'speculative' they may have been, these were unquestionably *causal* in intention, and I have argued that they were an important 'input' into the construction of more 'empirical' sections in the late eighteenth century. But these causal goals of 'cosmogony' were precisely what the dominant 'structural' tradition of the early nineteenth century rejected, and such theoretical ambitions were only gradually re-accepted in the 1820s as being respectably 'scientific'.

79. When I summarized this article for the Lyell Centenary Symposium, my interpretation was vigorously criticized by my friend Professor Dr R. Hooykaas for being too exclusively oriented towards British science and British sources. Being at present resident in his country, and therefore well able to detect provincial tendencies in some English-speaking historians of science, I am sensitive to this criticism, but I hope that the full version of my argument will make it plain that I have at least attempted to give equal weight to French evidence on this subject; I am well aware, however, that my treatment of German primary sources (and, for example, Italian and Scandinavian) is sketchy. My excuse must be that this article is intended primarily to stimulate further research, and not to say the last word on the subject; and I shall be glad if my interpretation can be corrected, for example, by a fuller study of sources relating to the central and northern European mining tradition. On the other hand, I am unrepentant about my emphasis on the crucial social role of the Geological Society of London in the development of the visual language of geology, because I do not think this is merely the result of linguistic or nationalistic short-sightedness. On the contrary, I think it is an unsurprising conclusion that the institution of the first specialist society and first specialist journal in geology should have played a major part in the visual aspect of the cognitive development of the science.

VI

Jean-André de Luc and Nature's Chronology

The theme of this volume is the 'age of the Earth'. The familiar phrase often implies a single numerical figure, which has certainly been of great and longstanding interest to cosmologists. But to Earth scientists what matters much more is to know what happened when, within that literally global timespan: in other words, to have a geochronology. However, early geologists got a very long way without quantifying deep time at all: they could and did work with an unquantified or so-called relative timescale, reconstructing events at least in the right temporal sequence. They clearly regarded it as desirable to put numbers on such a sequence, in order to give temporal precision to causal processes and to Earth history. But they were often reluctant to do so, for one very good reason: they knew they had little evidence for any such quantification, and they did not want to discredit their infant science by indulging in mere speculation.

Traditionally, however, this reluctance to quantify geohistory has been given a quite different explanation. It has been seen as a manifestation of a perennial conflict between science and religion, focussed in this case on a conflict between geology and Genesis. Early geologists, it was said, had avoided specifying the earth's timescale for fear of the Church, and those who had dared to do so had been persecuted for their temerity. But this is a historical myth, in the true sense of that word: not so much a false story, though it is that too, but a story that persists because it carries heavy ideological loading and is useful to those with various secularist agendas. The myth of intrinsic conflict between science and religion must be demythologized, as it has been by modern historians of the sciences (for example, Brooke 1991; Brooke and Cantor 1998). It was largely constructed in late nineteenth-century Europe (including Britain) and North America. It was a weapon in the hands of those who, often in the cause of professionalizing the sciences, were trying to wrest cultural and political power away from older social and intellectual elites, among which ecclesiastical elites had been prominent (Gieryn 1999, pp. 37–64; Moore 1979, pp. 19–122; Turner 1978). The myth survives here and there in Europe (including Britain) even today, especially in the rhetoric of some atheistic popular science writers. It was also revived in the United States during the twentieth century, owing to specific cultural conditions that gave as much political power to religious fundamentalists as to their secularist opponents (Numbers 1985, 1992).

Once the historically contingent character of the myth of intrinsic conflict is recognized, the way is open to be liberated from it. Only then can we get a clear view of what was going on in the past history of the sciences, including the Earth sciences. The traditional brief timescale for the Earth, epitomised by James Ussher's famous date of 4004 BC for Creation, has generally been regarded as utterly opposed to the immensely long timescale that Earth scientists now take for granted (Albritton 1980). In this paper I argue, on the contrary, that the short timescale was the direct progenitor of the long one, and facilitated its adoption. Historians no longer play the futile game of assigning praise or blame to people of the past for helping or hindering 'the progress of science'. But as a matter of history, the attempt to date human events even back to the Creation was one of the conceptual resources that were used in constructing an analogous scale for deeply prehuman time and geohistory. It was not the only such resource, but it was certainly an important one, and its role deserves to be recognised.

The Science of Chronology

A first clue to the affinity between the two timescales is given by the modern word 'geochronology' itself. It applies to the Earth the older concept of chronology, just as the word 'geohistory' applies the concept of history. Chronology was, and still is, a science within the practice of human historiography (the word 'science' is used here in the original pluralistic sense that only the anglophone world has abandoned, to denote *any* body of disciplined knowledge, about either nature or humanity). Chronologers analysed historical records, both texts and material artefacts (inscriptions, coins etc.), in order to reach accurate and reliable dates for historical events: first within some specific culture, but then correlating those dates on to a standard cross-cultural timeline. Chronology has ancient roots, but it first developed rigorous internal standards in the seventeenth century (Grafton 1975, 1991; North 1977); Ussher was just one of many scholarly chronologers, and not even the most distinguished. The goal of most early chronologers was indeed religious; it was to locate the great events of divine action, from Creation to Incarnation, in their context of world history. But in the course of doing so chronology itself was gradually secularized, as the histories of other cultures came to be treated in the same way as that of the ancient Jews and the Christian church. Chronology was eventually eclipsed in prestige by other kinds of historical work, but it never went extinct: its modern results are visible whenever, for example, dates BC are assigned to ancient Egyptian or Chinese events and artefacts.

Chronologers strove above all for precision based on good evidence critically evaluated (this admirable goal could readily be transposed into the later science of geochronology). However, working with fragmentary and often problematic texts, they found in practice plenty of scope for argument and controversy. For as they probed the records back in time from the familiar Romans and ancient Greeks, their sources became ever more sparse and enigmatic. Those chronologers, such as Ussher, who claimed to extend the timeline even back to a primal Creation recognized that they were pushing the science to its limits. It was here that their findings were most uncertain and controversial. Although they were all using much the same sources, there was no consensus, and Ussher's 4004 BC was just one of dozens of rival figures for the date of Creation. Given the central place of the Bible in the culture of Christendom, it is not surprising that the records extracted from Genesis were often given a privileged status, over and above what little was known about the early history of other cultures. But to justify that preference chronologers were obliged to compare and evaluate all the ancient texts, and hence to develop techniques of textual criticism that were later applied to the biblical documents themselves.

However, most work on chronology was focussed not on the date of Creation but on later history, where there was much more documentary evidence for constructing a reliable timescale. The great bulk of Ussher's work, like that of any other chronologer, dealt with the ordinary history of the last few centuries BC, with that of Greeks and Romans as well as Jews, not with the origin of the cosmos or even that of the human race.

This remained true of chronology a century and a half after Ussher's death. Chronology continued to flourish in the time of, say, Hutton, but as a scholarly discipline based on an array of texts that by then included those of ancient India and China. It was a science that tried to correlate the records of *all* human cultures and to condense them on to a single timeline of universal history. By around 1800, however, most chronologers had abandoned the attempt to extend the timeline back to the origin of the cosmos, restricting their science instead to the few millennia of recorded *human* history. For the growth of biblical criticism, particularly in the German universities, had led them to recognize that the Creation story in Genesis was the product of a culture profoundly different from their own, and that to treat it as a scientific text was misleading and inappropriate. So when naturalists became convinced on quite different grounds that there must have been a vast *prehuman* geohistory, there was no intrinsic conflict between them and the chronologers, or between geology and Genesis. Occasionally there was indeed forceful and even vehement argument; but only at specific times, in specific places, and above all in specific social settings. There was for example more argument in Britain than elsewhere in Europe, and more between savants and popularizers than among the scholars and naturalists themselves. Sweeping generalizations about perennial conflict, or alleging endemic ecclesiastical opposition, are not supported by the historical record.

Theories of the Earth

By around 1800, most naturalists with first-hand field experience of the relevant natural features shared a general sense of the vast though unquantifiable magnitude of deep time. They had no inhibitions about expressing that sense as concretely as their evidence allowed. They alluded almost casually to a million years or to thousands of '*siècles*' (the key French word could denote either centuries or indefinitely longer 'ages'). Such phrases expressed the time that seemed to be needed to account for the deposition of thick limestones and other stratified sediments, the erosion of deep valleys in mountain regions, the accumulation of piles of lava flows on the flanks of volcanoes, and so on (Ellenberger 1994, pp. 35–9; Taylor 2001). These 'guesstimates' usually remained vague and barely quantified, for the good and sufficient reason that naturalists knew they had no reliable evidence on which to base any firmer figures.

Significantly, the most striking exception proved the point by being highly controversial. Buffon's famously precise figures for the dating of his successive '*époques de la nature*' depended wholly on his hypothesis that the Earth had originated as an incandescent globe in space, and that it had cooled thereafter in the manner indicated by his experiments with small model balls (Buffon 1778; Roger 1962, 1989). In effect, the geophysics and the timescale came as a package: if Buffon's theory of global cooling was rejected, as it usually was on other grounds, his timescale necessarily collapsed with it. In fact his work often received the ultimate scientific dismissal of the time: it was a mere '*roman*', a novel, a piece of fanciful science fiction. But in any case Buffon's timescales, published and private, fell within the same range (from tens of thousands to a few millions of years) as the estimates of other naturalists. Even at the lower end, such figures went far beyond the traditional timescale of Ussher's century; Buffon kept the higher end to himself not out of fear of ecclesiastical authorities but because he knew he could not justify it scientifically with any concrete evidence.

Buffon's model of global cooling was rightly treated as just one example of what was usually called 'Theory of the Earth' (Taylor 2001). The phrase referred not to any specific theory about the Earth, but to a scientific *genre*, just as the novel, landscape and opera were artistic genres. 'Geotheory', as I suggest it should be called, was a genre that offered a Theory Of Everything about the earth, a terrestrial TOE; geotheorists proposed 'systems' or models of how the Earth works, which aimed at accounting for all its major features and all its causal processes. Geotheory always had a temporal dimension, of course; but that did not make it intrinsically *historical*, any more than, say, the theory of gravitation gave a historical explanation of planetary orbits. In the genre of geotheory, some specific 'systems' postulated a directional sequence, yet they were still scarcely geohistorical: in the case of Buffon's *Époques*, for example, the development was in effect determined or programmed from the start, by the physics of global cooling. Alternatively, some models proposed a cyclic or steady-state 'system', as in the case of Buffon's earlier one (1749), or more famously Hutton's. These were even less geohistorical, because the whole point of such models was to show that under unchanging laws of nature no period in past, present or future could be unique or even distinctive in character. Hutton made this explicit in his famous parallel between the endless changes or 'revolutions' of the Earth's surface and deep interior, and the equally endless revolutions of the planets around the sun (Hutton 1788, p. 304; Gould 1987, pp. 60–97). In fact, on the magnitude of time, Hutton was far from being an innovator; he was a typical 'natural philosopher' of the Enlightenment, who like Buffon took it for granted that in nature time is available without limit for explanatory purposes.

De Luc as a European Intellectual

It has been necessary to recall savants such as Buffon and Hutton, and far earlier ones such as Ussher, at some length, in order to provide an adequate context for understanding the origin of geochronology in the decades around 1800. For that origin is not to be found in Hutton's grand verbal gestures about infinite time and a 'system' without perceptible beginning or end. Nor is it to be found in the sense, which was quite widespread among Hutton's contemporaries, of an unquantifiably vast but *finite* geohistory. Instead, the future science of quantitative geochronology had its roots just where the word itself should lead us to expect: in transposing the pre-existing science of chronology from human history on to the Earth (Rappaport 1982, 1997). In initiating that process, one naturalist stands out: without lapsing into the discredited heroic style of history, Jean-André de Luc (or Deluc) (1727–1817) can be seen to have played a pivotal role. (The French naturalist Jean-Louis Giraud-Soulavie [1752–1813] also used the idea of 'nature's chronology' extensively in his natural history of southern France [Soulavie 1780–84], which was well known throughout Europe, but on this point his work had less impact than de Luc's.)

De Luc was well qualified to become Hutton's most formidable and persistent critic (Fig. 1). They were of the same generation, indeed they were born less than a year apart. De Luc's interests, like Hutton's, ranged far beyond what was later to become the science of geology; they were both Enlightenment 'philosophers', and to call either of them a scientist is deeply misleading and anachronistic. De Luc was a native and citizen of the Protestant city-state of Geneva (not yet incorporated into Switzerland). He gained a fine scientific reputation throughout Europe, particularly for his meteorology and not least for devising an accurate portable barometer that was widely used in plotting physical geography (De Luc 1772; Feldman 1990). In 1773, in middle age, he migrated to England, where he was elected to the Royal Society and appointed 'Reader' or intellectual mentor to the German-born Queen Charlotte, the wife of King George III (Tunbridge 1971). Like the royal family he lived at Windsor, just across the Thames from another immigrant, royal protégé and Fellow of the Royal Society, the great astronomer William Herschel. Throughout his long life de Luc wrote in his native French, which was then also the international language of the sciences, and indeed of the arts and diplomacy, just as much as English is today.

De Luc was certainly not a marginal figure in intellectual life, either in Europe as a whole or even just in England. But if he was so important at the time, it is legitimate to ask why he is not better known among historians and modern geologists. One reason is that his published work is voluminous, verbose and repetitious, and takes much effort to read and digest; but the same is often said of Hutton, and with as much truth. A more powerful reason is that de Luc was quite explicit about the religious motivation that underlay his scientific work; but this too is also true of Hutton (a fact that often dismays geologists who take the trouble to read the work of their father-figure, if they do so with secularist presuppositions). However, the crucial difference between them is that Hutton's theology was deistic, whereas de Luc's was quite orthodoxly Christian. Charlotte called him approvingly a 'proper philosopher' (*philosophe comme il faut*), because unlike many others he was not a sceptic in religious matters; he called himself unambiguously a 'Christian philosopher' (*philosophe Chrétien*) (Tunbridge 1971). And so, under the baneful influence of the myth of intrinsic conflict, de Luc has often been dismissed, even by historians who ought to have known better, as if he was in effect a modern American fundamentalist. He was not, and he deserves to be treated as seriously by modern geologists and historians as he was by his contemporaries, even if they do not share his religious beliefs.

Fig. 1. Jean-André de Luc: a portrait drawn around 1798 by Wilhelmine de Stetten and engraved by Friedrich Schröder (courtesy of the Bibliothèque Publique et Universitaire, Geneva).

De Luc's Geotheoretical Model

Like Buffon, Hutton and many others, de Luc devised his own Theory of the Earth or geotheoretical model. It first began to appear in 1778, when he published some letters that he had sent to his patron while he was travelling in the Alps. In a famous footnote, he tentatively proposed the term *'géologie'*, although not in its modern sense, for it was to denote the genre of geotheory as the terrestrial analogue of cosmology (De Luc 1778, pp. vii–viii). He then published no fewer than 150 more of his letters to the Queen, in six volumes, supporting his model with a mass of evidence based on his extensive fieldwork in central Europe and the Low Countries (De Luc 1779). All this work was published in Holland, which was a major centre of the international scientific book trade, and of course in French.

Over a decade later, as the political Revolution began to erupt in France, de Luc revised and amplified his earlier work. He sent a series of 31 papers to Paris, almost one a month, for publication in *Observations sur la Physique* (soon afterwards renamed *Journal de Physique*), which was one of the leading periodicals in Europe for the natural sciences (De Luc 1790–93). At the same time a set of four long letters to Hutton, criticizing the Scotsman's recent geotheoretical 'system', appeared in the English intellectual *Monthly Review*, making de Luc's own model well-known in his adopted country (De Luc 1790–91). Then, after the leading German naturalist Johann Friedrich Blumenbach of Göttingen had visited him in Windsor, de Luc sent him a set of seven letters, in which he promised to expound his 'system' more succinctly (in the event he was almost as prolix as before). Blumenbach had these essays translated for an important periodical edited in Gotha, the *Magazin für das Neueste aus der Physik* (De Luc 1793–96; Dougherty 1986). All but one were then re-translated, from German into English, for the new *British Critic* (De Luc 1793–95), before the whole set finally appeared in Paris, despite the state of war between France and Britain, in book form and in their original French (De Luc 1798). Much later, but still in wartime, his *Elementary Treatise on Geology* (1809) was published in Paris and London, in French and in English translation, making his ideas available at last in a final and somewhat briefer form.

This lengthy publication history underlines de Luc's prominence in the European scientific elite of his time, which remained highly international in outlook in spite of the wars sparked by the Revolution in France. His geotheoretical model changed and developed over the years, in part as a result of his continuing fieldwork, but for the purpose of this paper some enduring themes are of greatest importance. Unlike Buffon, Hutton and most other geotheorists, de Luc showed relatively little interest in offering causal explanations for events in the deep past. This was not because he doubted that they had had natural causes of some kind, but because he was much more concerned to establish the character of the Earth's *history*. He wanted to integrate the natural world, as it was being explored by naturalists such as himself, into the strongly historical perspective of his Christian beliefs, making the recent 'physical history' of the Earth the natural backdrop to the great events of human history.

Specifically, de Luc claimed that the Earth had undergone a radical 'Revolution' in the quite recent past, during which the previous sea floors had become dry land, and the previous land areas had sunk beneath the waves: in effect, an almost total interchange of continents and oceans. He thought this had probably been caused by sudden crustal collapse, rather than by the kind of deep-seated upheaval proposed by Hutton: the present continents would have emerged as a result of a global (eustatic) fall in sea level, after the deep collapse of the former continents. But exactly how it had happened was much less important to de Luc than the fact that it had happened. It was, he believed, a matter of contingent geohistory, an event for which there was compelling field evidence of many kinds. If such an event failed to fit the grand causal theories put forward by others such as Hutton, so much the worse for those theories. He criticized Hutton for spending too much time indoors and not enough in the field (De Luc 1790–1791, 2, p. 601). Certainly his own field experience was far wider than Hutton's; the latter had, for example, never seen the Alps or any other high mountains, or travelled elsewhere on the Continent (except to get his medical degree at Leiden).

De Luc's geotheoretical model can, I suggest, be usefully termed *binary* in character. His radical Revolution divided geohistory into two sharply contrasted periods, the Earth's '*histoire moderne*' and its '*histoire ancienne*', the familiar world of the present and an ancient or former world (De Luc 1779, 5, pp. 489, 505–6). Furthermore, since the earlier continents had collapsed out of sight, any evidence for whatever human life they might have sustained had become inaccessible. So in effect de Luc's binary model also divided geohistory into a prehuman and a human period. However, his two periods of geohistory were contrasted not only in relation to human life, but also in their timescales.

Timescales of Modern and Ancient Geohistory

De Luc argued that the 'ancient' or 'former' world (the key French word *ancien* bears both meanings) was of immense but unquantifiable antiquity. The literalism of the earlier tradition in chronology, in its application to the Creation story, was repudiated unambiguously. De Luc insisted on 'the enormous antiquity of the earth', adding that 'naturalists who have thought otherwise were not attentive observers' (De Luc 1790–93, **39**, p. 334). He reminded his readers how he had long rejected any attempt to compress the whole early history of the Earth into six literal days, as being 'in effect as much contrary to natural history as [it is] to the text that is to be explained'. Like other naturalists of his generation, de Luc had a good sense of how rock formations bore witness to successive periods of deposition over vast spans of time. He seems to have appreciated this increasingly over the years, probably as a result of seeing for himself the huge piles of fossiliferous strata exposed in various parts of Europe. About those along the south coast of England, for example, he exclaimed, 'What time must there have been for the formation of this pile of beds!' The timescale was 'indeterminate' or unquantifiable, but the successive periods of the former world could at least be put confidently into the correct order; the dating of this deep geohistory could be relative, but not absolute (De Luc 1790–93, **40**, pp. 282, 455).

Like other naturalists across Europe, de Luc also had a good sense of how the different formations were characterized by distinctive sets of fossils, suggesting a complex history of life. The limestones around Paris, he noted, 'contain shells that often vary from bed to bed … Their [the beds'] identity is recognized as much by these [fossil] bodies as by the nature of the rock' (De Luc 1790–93, **40**, pp. 281–2). On a larger scale he described how, on the south coast of England, shales were overlain by limestones, which were capped in turn by the Chalk; he noted that ammonites were confined to the lower and therefore older formations; and he interpreted that fact in terms of their extinction in the course of geohistory (De Luc 1790–93, **39**, pp. 458–60; Ellenberger and Gohau 1981). It was all rather crude by comparison with William Smith's more detailed stratigraphical work a decade or two later, and of course even more so in modern terms. Nonetheless it shows that de Luc had a clear conception of how an 'ancient history' of the world might be constructed from field evidence. Anyway, his strong Christian convictions certainly did not inhibit him from thinking in terms of humanly unimaginable spans of time. He did reconstruct geohistory as a sequence of seven vast periods which, like Buffon's 'epochs', corresponded explicitly to the seven 'days' of the Creation story in Genesis. He certainly ascribed important truth value to that story, but not in a literal sense, because he claimed that its primary purpose was religious and its significance theological. His seven periods of geohistory did not mark literal days, any more than Buffon's seven 'epochs'.

Compared to these vast tracts of deep time, de Luc's present world was quite young, and spanned only a brief period (it is worth noting that his final metaphorical 'day', like Buffon's, was not a divine sabbath rest as in the Genesis story, but the period of *human* dominance on earth). The major event that had brought the present world into being was only a few millennia in the past. De Luc identified it as none other than the event recorded in Genesis as Noah's Flood. That equation has been enough to provoke a knee-jerk reaction from some historians, who have promptly ridiculed or dismissed him (for an early and influential example see Gillispie 1951, pp. 56–66). But in fact his interpretation of the Flood story (De Luc 1798, pp. 287–337) was quite subtle, and far from the crass kind of literalism favoured by modern American fundamentalists. What mattered to him most was just one point, its date. He was concerned above all to show that the chronologers' date for the Flood, based on ancient textual records, was not contradicted by evidence from the world of nature. Only if the biblical documents were reliable *as human history*, he argued, could their authors be treated as trustworthy guides in moral, social and religious matters (De Luc 1779, **5**, pp. 630–46). Like many other intellectuals throughout Europe, de Luc believed that it was the repudiation of traditional moral norms, by fashionably deistic or even atheistic philosophers, that had led to the social catastrophe of the Revolution in France, which was lurching into its most radical, violent and regicidal phase even as he wrote his letters to Paris.

For de Luc's purposes it was not necessary for naturalists to achieve the kind of precision to which chronologers aspired, dating past events to a precise year. It was sufficient to show that there was natural evidence for a major physical change at around the right time, no more than a few millennia in the past. To establish this he made a close study of what he called '*causes actuelles*', physical causes that were 'actual' in the sense that is now obsolete in English but still current in French: they were *present* processes of change, processes that could be observed at work in the present world, such as erosion, deposition, volcanic activity and so on (Ellenberger 1987). On the basis of a mass of fieldwork he argued that many such actual causes cannot have been operating for an indefinitely long time, but must have started at a finite and relatively recent time in the past. He identified this as the time of the revolution that was marked by the emergence of the present continents *as land areas*, for only at that point could the varied terrestrial processes have begun to act on them.

One of de Luc's favourite examples was the delta of the Rhine, which he studied repeatedly during his travels in The Netherlands. The delta was manifestly growing in size by new deposition, at a rate that could be judged roughly from historical records of its former limits in Roman and medieval times. But its size was finite, so its historical rate of growth could be extrapolated back to an origin at some finite time in the past. Other examples were provided by lakes such as Lac Léman (the Lake of Geneva), where, at the far end from his native city, the

Rhône was clearly extending its delta year by year; yet if it had been doing so indefinitely it would long since have converted the whole lake into alluvial pastures. The screes below the nearby Alpine peaks were also growing yearly, by rock falls from above, but they too were of finite size. On the heathlands of the north German plain (which included both the homeland of his patron Charlotte and the domain of her Hanoverian husband George III), the thin cover of peat overlying sediments with marine fossils suggested likewise that the accumulation of peat from plant decay could not have been going on, and the region could not have been above sea level, for an indefinitely long time. (It will be clear from a modern perspective that many of de Luc's processes, in the regions he knew at first hand, had indeed started, or re-started, at a geologically recent time, namely after the last Pleistocene glaciation; but neither he nor any of his contemporaries had any reason to consider glacial ice as a possible causal agent outside the high mountain regions.)

Nature's Chronometers

Features such as these seemed to de Luc to provide several independent lines of evidence that 'the present world', the physical regime now observable on the present continents, dated from a relatively recent time in the past. They provided him with the possibility of constructing a 'physical chronology' based on *natural* evidence, to parallel and supplement the chronologers' narrative based on human textual evidence. More than that, they suggested how nature's chronology might even be quantified into nature's *chronometry*, at least for the period since the decisive revolution in which the present continents had emerged, if not for the far longer periods of the earth's 'ancient history' before that event.

In his earlier writings, de Luc invoked simple and even homely analogies to express this quite novel concept of geochronometry. For example, he argued that the heathlands of Hannover preserved the nearest there was to the pristine state of the continents:

> There are still many uncultivated lands there, which, like the teeth of a young horse, can give us some idea of the age of the world; I mean, of the date when the present surface took the form in which we know it today (De Luc 1779, **3**, pp. 11–12).

The analogy, more familiar in de Luc's day than in ours, was apt and illuminating. A horse's teeth, steadily worn down by its grassy diet, were an infallible and even quantitative guide to its age, making it proverbially inappropriate to 'look a gift horse in the mouth' to assess its age and its value before accepting it. So likewise the thickness of the peat might be used to estimate the date of the continent's emergence as a land mass; that the former was slight implied that the latter was recent.

The analogy with timekeeping became more explicit in a later letter to the Queen, in which de Luc referred to the growth of a delta such as that of the Rhine:

> This is the true *clepsydra* of the centuries, for dating the Revolution: time's zero is fixed by the unchanging sea-level and its degrees are marked by the accumulation of the deposits of the rivers, just as they are by the piling up of sand in our ancient instruments of chronometry. (De Luc 1779, **5**, p. 497).

Here was another apt analogy, although de Luc confused the dripping of a primitive water-clock or clepsydra with the trickling sand in an hour-glass; the latter was as familiar in his day as horses' teeth, being used on board ship and in the parson's pulpit as well as in the kitchen (where in the form of the humble egg-timer it survived well into the twentieth century). In de Luc's view the slowly accumulating deposits in a delta marked the passage of time since the continent emerged, just as reliably as the amount of sand in the lower half of an hour-glass marked the time that had elapsed since it was last up-ended.

De Luc later enhanced his argument by calling his physical features '*nature's chronometers*' (De Luc 1790–91, **2**, p. 580). Here the analogy was with the celebrated clockwork 'chronometers' (for which the word had been coined) that had recently enabled John Harrison to solve the outstandingly important practical problem of determining longitude at sea. De Luc conceded that by comparison with this supreme high-tech achievement of the eighteenth century his dating of natural features was far from precise. Nonetheless, the vivid metaphor made his goal of temporal accuracy unmistakeable. As he himself put it, referring again to the growth of the Rhine delta:

> There then is a true chronometer: one finds there the total operation [of the process of deposition] since the birth of our continents; one can see its causes and their progress, and one can distinguish the parts of all the products of known times. Doubtless there are too many causes of irregularity in this process for one to be able to count the centuries; but it is evident that their number cannot be considerable (De Luc 1790–93, **41**, p. 344).

So nature's own chronometers could provide at least the rudiments of a quantified chronology for the Earth itself, based on the observable rates of actual causes or ordinary natural processes.

De Luc himself applied his 'chronology of nature' only to the one problem that was his greatest concern: the dating of the boundary event that separated the present world, the world of human history, from the incalculably more lengthy former world of 'ancient history' represented by the formations of fossiliferous strata. Nonetheless it was a decisive move. For it showed by example how nature itself could provide the basis, however roughly, for its own quantitative chronometry, which in turn could give precision to nature's chronology, that is, for geochronology. It provided a template and a precedent for those who would later try to extend nature's chronology back in time beyond human history into prehuman geohistory.

De Luc's Legacy to Geochronology

De Luc's binary model for geotheory was immensely influential in the early nineteenth century, above all because it was adopted by the great French naturalist Georges Cuvier (Ellenberger & Gohau 1981). Cuvier used his skills as a comparative anatomist to show that most fossil bones belonged to terrestrial animals distinct from living species. He interpreted the contrast as the result of a mass extinction or 'catastrophe' in the quite recent past, a 'revolution' that corresponded closely to de Luc's (Rudwick 1997). Like de Luc, Cuvier remained uncertain about the cause of this event, being far more concerned to establish its sheer historicity. He agreed with de Luc that it had happened only a few millennia ago, but he widened the evidence for such a date by drawing on a vast multicultural range of human records. With a cultural relativism typical of the Enlightenment, he reduced the biblical narrative of the Flood to just one story among many, all of them more or less garbled but still perhaps with a kernel of historical truth (Cuvier 1812, pp. 94–106; Rudwick 1997, pp. 239–46). By treating them all as potentially valid evidence, Cuvier consolidated de Luc's tying of geohistory into human history, geochronology into chronology.

Ironically, however, de Luc's notion of nature's own chronometers, or of nature's chronology, was developed most effectively by a geologist who used it to refute the idea that there had been any catastrophic event whatever in the geologically recent past. A decade after de Luc's death (at the age of ninety), and more than two decades after Cuvier's innovative research on fossil bones, the London geologist Charles Lyell reacted against the Delucian ideas of his former mentor William Buckland at Oxford, and used de Luc's 'actual causes' to argue that no catastrophic event had disturbed the steady pace of physical change on earth, as far back in deep time as the record of the rocks could go. He could hardly adopt de Luc's phrase as it stood, without drawing unwelcome and embarrassing attention to his close affinity with the earlier naturalist, at least on this point. But under the guise of 'modern causes', as Lyell called them, de Luc's method for analysing and calibrating geohistory got a second wind, and became the basis for Lyell's own geotheoretical model, later dubbed uniformitarianism.

Lyell argued that a statistically uniform rate of organic change at the level of species could be used to construct a new natural chronometer. As in the case of 'actual causes', Lyell avoided adopting de Luc's term, but his idea was the same. His chronometer was to be based on the steadily increasing percentage of extant species, and the decreasing proportion of extinct ones, in the fossil molluscan assemblages of the successive Tertiary formations, which on this basis he termed Eocene, Miocene and Pliocene (Lyell 1830–33, **3**, pp. 45–61; Rudwick 1978). But his grand theory of constant faunal change soon foundered in the face of awkward empirical detail, and with it his bid to find a natural chronometer that could be applied to even earlier formations.

Conclusion

The aspiration to construct a quantitative geochronology remained, however, at least among a few geologists; and it can be traced through the rest of the nineteenth century, as in the work of John Phillips, and into the radiometric dating of the twentieth (Lewis 2001, Morrell 2001, Wyse Jackson 2001).

In taking that longer view, however, we should not overlook the decisive role of Jean-André de Luc, despised and rejected though he often has been, in transposing the ideas of chronology and chronometry from the human realm into the world of nature, and in working out how to interpret physical features as indices of geochronology. Furthermore, we should not ignore the historical evidence that the intellectual roots of this crucial transposition from culture into nature lay in de Luc's commitment to a religious tradition that had historicity at its heart. It was the science of chronology, powered originally by the desire to locate revelatory divine action within universal human history, that provided the model for de Luc's parallel 'chronology of nature'; and it was this in turn that others could later extend, as geochronology, into the depths of geohistory. Thus in the long run the traditional brief timescale exemplified by Ussher's 4004 BC was the historical template for the vastly longer timescale of modern geology. The perverse revival of a short timescale (or 'young Earth') by religious fundamentalists, in the radically changed epistemic circumstances of the modern world, should not blind us to its decisive importance in the earlier development of the earth sciences.

This paper is based on a talk given at the Geological Society in London in June 2000, during the millennial William Smith meeting, 'Celebrating the Age of the Earth'. Some of the research was done under grant no. SBR–9319955 from the (U.S.) National Science Foundation.

References

ALBRITTON, C. C. 1980. *The Abyss of Time: Changing Conceptions of the Earth's Antiquity after the Sixteenth Century.* Freeman & Cooper, San Francisco.

BROOKE, J. H. 1991. *Science and Religion: Some Historical Perspectives.* Cambridge University Press, Cambridge.

BROOKE, J. H. & CANTOR, G. N. 1998. *Reconstructing Nature: The Engagement of Science and Religion.* T. & T. Clark, Edinburgh.

BUFFON, G. LECLERC, COMTE DE. 1749. Histoire et théorie de la terre. *In:* BUFFON, *Histoire Naturelle, Générale et Particulière, avec la Description du Cabinet du Roi*, **1**, 63–203.

BUFFON, G. LECLERC, COMTE DE. 1778. Des Époques de la nature. *In:* BUFFON, *Histoire Naturelle, supplément*, **5**, 1–254.

CUVIER, G. 1812. Discours préliminaire. *In* CUVIER, *Recherches sur les Ossemens Fossiles de Quadrupèdes, où l'on Rétablit les Caractères de plusieurs Espèces d'Animaux que les Révolutions du Globe paroissent avoir Détruites*, 4 vols. Déterville, Paris.

DE LUC, J.-A. 1772. *Recherches sur les Modifications de l'Atmosphère*, 2 vols. Geneva.

DE LUC, J.-A. 1778. *Lettres Physiques et Morales sur les Montagnes et sur l'Histoire de la Terre et de l'Homme: Addressées à la Reine de la Grande-Bretagne.* De Tune, The Hague.

DE LUC, J.-A. 1779. *Lettres Physiques et Morales sur l'Histoire de la Terre et de l'Homme, Addressées à la Reine de la Grande-Bretagne*, 5 vols. in 6. De Tune, The Hague, and Duchesne, Paris.

DE LUC, J.-A. 1790–91. Letters to Dr James Hutton, F.R.S. Edinburgh, on his Theory of the Earth. *Monthly Review or Literary Journal Enlarged*, **2**, 206–27, 582–601; **3**, 573–86; **5**, 564–85.

DE LUC, J.-A. 1790–93. Lettres à M. de La Métherie. *Observations sur la Physique, sur la Chimie, sur l'Histoire Naturelle et sur les Arts*, **36**, 144–54, 193–207, 276–90, 450–69; **37**, 54–71, 120–38, 202–19, 290–308, 332–51, 441–59; **38**, 90–109, 174–91, 271–88, 378–94; **39**, 215–30, 332–48, 453–64; **40**, 101–16, 180–97, 275–92, 352–69, 450–67; **41**, 32–50, 123–40, 221–39, 328–45, 414–31; **42**, 88–103, 218–37; **43**, 20–38.

DE LUC, J.-A. 1793–95. Geological letters, addressed to Professor Blumenbach. *British Critic: A New Review*, **2**, 231–8, 351–8; **3**, 110–18, 226–37, 467–78, 589–98; **4**, 212–18, 328–36, 447–57, 569–78; **5**, 197–207, 316–26.

DE LUC, J.-A. 1793–96. Geologische Briefe an Hrn. Prof. Blumenbach. *Magazin für den Neueste aus der Physik und der Naturgeschichte*, **8** (4), 1–41; **9** (1), 1–123; **9** (4), 1–49; **10** (3), 1–20; **10** (4), 1–104; **11** (1), 1–71.

DE LUC, J.-A. 1798. *Lettres sur l'Histoire Physique de la Terre, addressées à M. le Professeur Blumenbach, Renfermant de Nouvelles Preuves Géologiques et Historiques de la Mission Divine de Moyse*. Nyon, Paris.

DE LUC, J.-A. 1809. *Traité Élémentaire de Géologie*. Courcier, Paris.

DOUGHTERTY, F. W. P. 1986. Der Begriff de Naturgeschichte nach J. F. Blumenbach anhand seiner Korrspondenz mit Jean-André DeLuc. Ein Beitrag zur Wissenschaftsgeschichte bei der Entdeckung der Geschichtlichkeit ihres Gegenstandes. *Berichte zur Wissenschaftsgeschichte*, **9**, 95–107.

ELLENBERGER, F. 1987. Les causes actuelles en géologie: origine de cette expression, la légende et la réalité. *Bulletin de la Société Géologique de France*, **3** (8), 199–206.

ELLENBERGER, F. 1994. *Histoire de la Géologie, **2**: La Grande Éclosion et ses Prémices, 1660–1810*. Technique et Documentation, Paris.

ELLENBERGER, F. & GOHAU, G. 1981. A l'aurore de la stratigraphie paléontologique: Jean-André de Luc, son influence sur Cuvier. *Revue d'Histoire des Sciences*, **34**, 217–57.

FELDMAN, T. S. 1990. Late Enlightenment meteorology. *In*: FRÄNGSMYR, T., HEILBRON, J. L. & RIDER, R. E. (eds) *The Quantifying Spirit in the Eighteenth Century*, University of California Press, Berkeley and Los Angeles, 143–77.

GIERYN, T. F. 1999. *Cultural Boundaries of Science: Credibility on the Line*. University of Chicago Press, Chicago.

GILLISPIE, C. C. 1951. *Genesis and Geology: A Study in the Relations of Scientific Thought, Natural Theology and Social Opinion in Great Britain, 1790–1850*. Harvard University Press, Cambridge, Mass.

GOULD, S. J. 1987. *Time's Arrow, Time's Cycle: Myth and Metaphor in the Discovery of Geological Time*. Harvard University Press, Cambridge, Mass.

GRAFTON, A. T. 1975. Joseph Scaliger and historical chronology: the rise and fall of a discipline. *History and Theory*, **14**, 156–85.

GRAFTON, A. T. 1991. *Defenders of the Text: The Traditions of Scholarship in an Age of Science, 1450–1800*. Harvard University Press, Cambridge, Mass.

HUTTON, J. 1788. Theory of the Earth; or an investigation of the laws observable in the composition, dissolution and restoration of the land upon the globe. *Transactions of the Royal Society of Edinburgh*, **1**, 209–304.

LEWIS, C. 2001. Arthur Holmes' vision of a geological timescale. *In*: LEWIS, C. L. E. & KNELL, S. J. (eds) *The Age of the Earth: from 4004 BC to AD 2002*. Geological Society, London, Special Publications, **190**, 121–38.

VI

LYELL, C. 1830–33. *Principles of Geology: Being an Attempt to Explain the Former Changes of the Earth's Surface by Reference to Causes now in Operation*, 3 vols. Murray, London.

MOORE, J. R. 1979. *The Post-Darwinian Controversies: A Study of the Protestant Struggle to Come to Terms with Darwin in Great Britain and America, 1870–1900.* Cambridge University Press, Cambridge.

MORRELL, J. B. 2001. Genesis and geochronology: the case of John Phillips (1800–1874). *In*: LEWIS, C. L. E. & KNELL, S. J. (eds) *The Age of the Earth: from 4004 BC to AD 2002*. Geological Society, London, Special Publications, **190**, 85–150.

NORTH, J. D. 1977. Chronology and the Age of the World. *In*: YOURGRAU, W. & BRECK, A. D. (eds) *Cosmology, History and Theology*. Plenum, New York, 307–33.

NUMBERS, R. L. 1985. Science and religion [in American history]. *Osiris* **1** (2), 59–80.

NUMBERS, R. L. 1992. *The Creationists*. University of California Press, Berkeley and Los Angeles.

RAPPAPORT, R. 1982. Borrowed words: problems of vocabulary in eighteenth-century geology. *British Journal of the History of Science*, **15**, 27–44.

RAPPAPORT, R. 1997. *When Geologists were Historians, 1665–1750*. Cornell University Press, Ithaca, New York.

ROGER, J. 1962. Buffon, Les Époques de la Nature: édition critique. *Mémoires du Muséum National d'Histoire Naturelle*, sér. C, **10**.

ROGER, J. 1989. *Buffon: Un Philosophe au Jardin du Roi*. Fayard, Paris.

RUDWICK, M. J. S. 1978. Charles Lyell's dream of a statistical palaeontology. *Palaeontology*, **21**, 225–44.

RUDWICK, M. J. S. 1997. *Georges Cuvier, Fossil Bones and Geological Catastrophes*. University of Chicago Press, Chicago.

SOULAVIE, J.-L. GIRAUD-. 1780–84. *Histoire Naturelle de la France Méridionale ou Recherches sur la Minéralogie du Vivarais [etc.]*, 7 vols. Belle, Nîmes, and Quillau, Paris.

TAYLOR, K. L. 2001. Buffon, Desmarest and the ordering of geological events in *époques*. *In*: LEWIS, C. L. E. & KNELL, S. J. (eds) *The Age of the Earth: from 4004 BC to AD 2002*. Geological Society, London, Special Publications, **190**, 39–49.

TUNBRIDGE, P. A. 1971. Jean André de Luc, F.R.S. (1727–1817). *Notes and Records of the Royal Society of London*, **26**, 15–33.

TURNER, F. M. 1978. The Victorian conflict between science and religion: a professional dimension. *Isis*, **69**, 356–76.

USSHER, J. 1650–54. *Annales Veteris Testamenti, a Prima Mundi Origine Deducti [etc.]*, 2 vols. Flesher, London.
WYSE JACKSON, P. N. 2001. John Joly (1857–1933) and his determination of the age of the Earth. *In*: LEWIS, C. L. E. & KNELL, S. J. (eds) *The Age of the Earth: from 4004 BC to AD 2002*. Geological Society, London, Special Publications, **190**, 107–19.

VII

Cuvier and Brongniart, William Smith, and the Reconstruction of Geohistory[1]

Introduction

François Ellenberger has given the second volume (1994) of his magnificent *Histoire de la Géologie* an appropriate subtitle: 'La grande éclosion et ses prémices, 1660–1810'.[2] Its implication is that the 'great flowering' of geology took place in the late eighteenth century and at the start of the nineteenth; and that its 'beginnings' reach back to the mid-seventeenth century. Most historians of geology would, I believe, broadly agree. This periodization itself has a long history: a century ago, for example, Karl von Zittel named the time around 1800 'das heroische Zeitalter der Geologie'.[3] What seemed to him to have made it a 'heroic period' were the feats of some naturalists in exploring dangerous and remote places: Horace Bénédict de Saussure reaching the peak of Mont Blanc, for example, and Peter Simon Pallas penetrating the remotest parts of Siberia. This did at least capture the sense in which *fieldwork* had gained a quite new prominence and prestige around that

[1] Translated, with some revision, from my 'Smith, Cuvier et Brongniart, et la reconstitution de la géohistoire', in J. Gaudant and G. Gohau (eds.), *De la géologie à l'histoire de la géologie*, 1997). This volume is dedicated to François Ellenberger, president of COFRHIGEO (the French Commission on the History of Geology), in celebration of his 80th birthday. I am indebted to Professor Ellenberger and other members of COFRHIGEO for valuable comments on the paper when I presented it in Paris in May 1995; the paper has also benefitted from discussions of an earlier version of this translation at the Cabinet of Natural History colloquium, Cambridge University, and at the History of Science and Technology colloquium at the Royal Institution in London. The research for this paper was supported by a Fellowship from the John Simon Guggenheim Memorial Foundation; by the National Science Foundation, grant no. SBR–9319955; by a President's Fellowship in the Humanities, University of California; and by a grant from the Academic Senate of the University of California San Diego.

[2] François Ellenberger, *Histoire de la géologie*, Tome 2, *La grande éclosion et ses prémices 1660–1810* (Paris: Technique et Documentation [Lavoisier], 1994).

[3] Karl Alfred von Zittel, *Geschichte der Geologie und Paläontologie bis Ende des 19. Jahrhunderts.* (München and Leipzig: R. Oldenbourg, 1899), p. 76.

time. Zittel's other criterion for 'the heroic period' was its rejection of large-scale theorizing and its emphasis on the value of collecting facts; as he put it, 'new ideas were no longer valued, but facts were' [man verlangte nicht mehr neue Ideen, sondern Thatsachen].[4] Here however Zittel accepted too readily the rhetoric of some of the savants of that period: he overlooked not only the heuristic value of 'theories of the earth' but also the way in which the search for so-called 'facts' was still guided by theoretical goals, albeit more limited ones.

Since Zittel's time, the establishment of a geohistorical dimension has often been regarded as an important factor – perhaps even as the key element – in the 'great flowering' of geology. That geohistorical dimension has generally been linked, in one way or another, with the adoption of *fossils* as a major criterion in reconstructing the history of the earth. This paper seeks to clarify what was involved in that new use of fossils.[5] (Since the word 'historical' is highly ambiguous, it is better to use the modern term *geohistorical* – although it is an anachronism – to denote ideas about the temporal history of the earth, as opposed either to human history or to non-temporal 'natural history'.)

In the older heroic tradition in the historiography of science, the main credit for introducing the effective use of fossils in stratigraphy was given either to William Smith or jointly to Georges Cuvier and Alexandre Brongniart. In fact the evaluation of this decisive new element in geological practice has often been confused by chauvinistic arguments about issues of priority. The rival claims of the Englishman and the two Frenchmen continued until recently to be championed with a vehemence that was more appropriate to the Napoleonic wars than to the century of the Entente Cordiale and the European Union. In this essay I hope to contribute to a more cordial evaluation, by putting some familiar material into a new perspective.

---

[4] Zittel, *Geschichte*, p. 76.

[5] Fossils had of course been treated as historical relics ever since the time of Robert Hooke a century earlier: see for example the classic paper by Cecil Schneer, 'The rise of historical geology in the seventeenth century', *Isis*, 1954, 45:256–68; also the important survey by Paolo Rossi, *I segni del tempo: storia della terra e storia delle nazioni da Hooke a Vico* (Milano: Giangiacomo Feltrinelli, 1979) [translated as *The Dark Abyss of Time* (Chicago: University of Chicago Press, 1984)]. I would argue however that this early 'historical' sense about natural objects was highly restricted in application compared to the later period with which this paper is concerned.

VII

The Reconstruction of Geohistory

Four Scientific Practices

First we should recognize the extent to which, around 1800, the study of fossils, and indeed of minerals, was still primarily a science of *specimens*.[6] In this respect 'mineralogy' – in the old broad sense that included the study of fossils – resembled the other two branches of natural history, botany and zoology. In all three sciences, of course, specimens had to be collected in the field. In practice, however, most major naturalists relied substantially on the specimens that peasants and provincial collectors (and, in the case of 'mineralogy', quarrymen and miners) offered for sale; or they studied those that travellers had collected and brought back from distant countries; or they delegated the task of collecting in the field to their students or their assistants. By contrast, the truly scientific work of a naturalist was considered to be the task of description, identification and classification; and this needed the resources of a collection, assembled *indoors* in a museum or 'cabinet' (and sometimes in its outdoor annexes, a botanic garden and a menagerie).

The resultant knowledge about fossils (or any other objects in natural history) was generally published in the form of descriptions and illustrations of the specimens that had been collected from a particular locality or region. Illustrated books on fossils were in effect mobile collections of *proxy* specimens.[7] They made the resources of particular museums or 'cabinets' available elsewhere. Thus for example the great illustrated volumes of *Merkwürdigkeiten der Natur* [Nature's remarkable objects] (1755–75), begun by Georg Wolfgang Knorr and completed after his death by Immanuel Walch, started with the spectacular fossils from the quarries in the lithographic stone of Solnhofen in Bavaria. Forty years later, the magnificent *Ittiolitologia Veronese* (1796) recorded some of the best collections of the equally fine fossil fish from Monte Bolca, in the hills behind Verona; soon afterward Barthélémy Faujas de Saint-Fond, Cuvier's senior colleague at the great Muséum d'Histoire Naturelle in Paris, described in his *Montagne de Saint-Pierre* (1799) all the fossils from the chalk mines near Maastricht in The Netherlands, including the monster that Cuvier later termed the mosasaur. And just after the turn of the century Jean-Baptiste de Lamarck, also at the Muséum, began describing the *Fossiles des environs de Paris* (1802–09) – in fact, mainly the mollusks from a

---

[6] Martin Rudwick, 'Minerals, strata and fossils', in N. Jardine, J.A. Secord and E.C. Spary (eds.), *Cultures of natural history* (Cambridge: Cambridge University Press, 1996), 266–86.
[7] Mark Hineline, *The visual culture of the earth sciences, 1863–1970*. Ph.D thesis (San Diego: University of California San Diego, 1993).

prolific quarry at Grignon near Versailles – in the early volumes of the Muséum's new *Annales*.[8]

This way of studying fossils, centred on collections from specific localities or regions, detached them in effect from the rocks in which they were found, or at least from what would now be called their stratigraphical context. Fossils were treated in practice as discrete specimens, or as collections of specimens, which were studied and identified in a museum, just like dried plants and animal skeletons or shells. And like those other objects of natural history, fossils were arranged and stored according either to their place of origin or to their place in a taxonomy. Apart from its assigned name, only the locality where a fossil had been found was generally recorded; at most, a note such as 'ten feet down' might be added. Like animal and plant specimens, almost the only information that fossils brought with them, as it were, from the field was locational, and hence merely two-dimensional or geographical.

Another branch of 'mineralogy' was, by contrast, thoroughly grounded in fieldwork, and profoundly three-dimensional in its practice. This was the science which, partly through the teaching of Abraham Gottlob Werner at the Bergakademie [mining school] at Freiberg in Saxony, had become universally known as *Geognosie*. It is important however to recognize that geognosy was primarily a *structural* science, not a geohistorical one. As one might expect from its context in mining practice and mining education, it was primarily concerned with the structural relations of rock-masses. Then as now – as Ellenberger has aptly put it – anyone concerned with mining 'lives in a three-dimensional lithological world' [vit dans un monde lithologique à trois dimensions].[9] In clarifying the three-dimensional relations of the 'Gebirge', 'terrains' or 'formations' in any region, their structural *order* was of primary importance: it was essential to know whether one rock-mass lay above or below others. That structural order was recognized as being – often, at least – the result of the temporal order in which they had been formed (*how* they had been

---

[8] J.C. Emmanuel [*sic*] Walch, *Sammlung der Merkwürdigkeiten der Natur und Naturgeschichte der Versteinerungen zur Erläuterung der Knorr'scher Sammlung*, 4 vols., (Nürnberg, 1755–75). [Giovanni Serafino Volta], *Ittiolitologia Veronese del Musei Bozziano ora annesso a quello di Conte Giovambattista Gazola e di altri gabinetti di fossili Veronesi con la versione latina*, (Verona, 1796). Barthélémy Faujas-Saint-Fond, *Histoire naturelle de la Montagne de Saint-Pierre de Maestricht* (Paris, An 7 [1799]). Jean-Baptiste de Lamarck, 'Mémoires sur les fossiles des environs de Paris, comprenant la détermination des espèces qui appartiennent aux animaux marins sans vertèbres, et dont la plupart sont figurés dans la collection des vélins du Muséum', *Annales du Muséum d'Histoire Naturelle* 1802–09, *1–14* [39 short papers, the later ones with plates].

[9] Ellenberger, *Histoire de la géologie*, vol. 2, p. 247.

5   *The Reconstruction of Geohistory*

formed, by sedimentation, precipitation or some other process, was more problematic). But the geognostic order of formations – in modern terms, the stratigraphical sequence – was primarily a matter of structure, not geohistory. Significantly, Werner entitled his seminal booklet on geognosy a *Kurze Klassifikation* [Short classification] (1787): like any other kind of natural history, it dealt primarily with the 'species' of formation [Gebirgsarten], and with their classification into larger categories.[10]

For those larger categories, Werner simply adopted the kind of classification that by his time had become consensual throughout Europe. The lowest kinds of formation, which he called 'uranfänglich Gebirgsarten' [apparently primitive kinds of formation], were those that, for example, Giovanni Arduino had called the 'monti primari'. The overlying formations that he called 'Flötzgebirge' [bedded formations] were Arduino's 'monti secondari'.[11] Of course all naturalists understood the principle of superposition, and recognized that the Primary formations were older than the Secondary, and, within the Secondary, that lower formations were older than those that overlay them. But as Werner himself explained, when he visited Paris in 1802 and was interviewed by the editor of the *Journal de Physique* (perhaps the most widely distributed periodical for the earth sciences at the time), geognosy was above all a science of structural features observable in the field.[12]

Within this well established practice of geognosy, the relative neglect of fossils was due, at least in part, to a matter so practical that it has often been overlooked. Werner and other geognosts were concerned mainly with the exploration and exploitation of mineral resources. Most mineral ores of economic value were found in the Primary formations and in the veins that traversed them; those rocks were totally lacking in fossils, by empirical observation and hence also by definition. That in itself was bound to make the science of fossils a relatively marginal matter for most geognosts. Even in the Secondary formations, where fossils were often abundant, they seemed almost equally marginal to the main business of geognosy;

---

[10] Abraham Gottlob Werner, *Kurze Klassifikation und Beschreibung der verschiedenen Gebirgsarten* (Dresden, 1787), reprinted almost unchanged from a journal article published the previous year; the latter is reproduced in facsimile, with an English translation, in Alexander M. Ospovat (trans. and ed.), *Abraham Gottlob Werner. Short classification and description of the various rocks* (New York, Harper, 1971).

[11] Ezio Vaccari, *Giovanni Arduino (1714–1795): il contributo di uno scienziato veneto al dibattito settecentesco sulle scienze della terra* (Firenze: Olschki, 1993). This fine monograph supersedes most earlier research on Arduino, but an excellent summary of his work is in Ellenberger, *Histoire de la géologie*, vol. 2, pp. 258–65.

[12] Jean Claude de Lamétherie, 'Idées de Werner sur quelques points de la géognosie: extraits de ses conversations', *Journal de physique, de chimie, de l'histoire naturelle et des arts* An 11 [1802], 55:443–50.

the acceptance of their organic origin – which was universal by the mid-eighteenth century – tended in practice to detach them from the study of the rocks in which they were embedded. As 'accidental' or 'extraneous fossils', it seemed that they could contribute little to the definition and classification of the formations in which they were found.[13]

In effect, therefore, there were two distinct traditions of practice around 1800, in what would now be called the earth sciences; but neither was truly geohistorical in orientation. On the one hand, the study of fossils was a science of specimens, pursued primarily indoors in museum collections, with the aid of illustrated monographs on the fossils from particular localities or regions. It was an ordinary part of natural history, dealing with description, identification and classification; its link with field study was virtually restricted to the spatial dimension of geographical distribution. On the other hand, the study of rock formations in geognosy was based on work in the field and its underground extension in mines; but it was primarily concerned with the structural relations of three-dimensional rock-masses, and it had a strong practical bias toward those in the cores of mountain regions, which contained most of the useful minerals but few if any fossils.

A third kind of investigation, the long established genre of 'the theory of the earth', had recently been given the new name *géologie*.[14] It proposed models or 'systems' for the causal development of the whole earth, but they were deeply ahistorical. True to their Cartesian origins, they were genetic, or – better still – *epigenetic* in character. Each 'system' posited a set of initial conditions, combined it with a set of physical principles, and then derived a sequence of stages through which the earth must have passed, and through which it would have to pass in the future. That sequence of stages took place within time, of course, but it represented a programmed unfolding of physical states; it was directly analogous to the programmed development of an embryo into an adult organism (or, in contemporary terms, to its 'epigenesis' or its *evolutio*).[15]

---

[13] This point is well argued by Rachel Laudan, *From mineralogy to geology: the foundations of a science, 1650–1830* (Chicago: University of Chicago Press, 1987), on pp. 146–7.

[14] Jean-André de Luc [later, Deluc], *Lettres physiques et morales sur les montagnes et sur l'histoire de la terre et de l'homme: addressées à la Reine de la Grande-Bretagne* (La Haye, 1778), pp. vii–viii. Dennis R. Dean, 'The word "geology",' *Annals of science*, 1979, 36:35–43, cites still earlier occurrences of the word and its cognates; but semantic continuity – though not identity – with modern usage seems to date from Deluc's famous footnote.

[15] Jacques Roger, 'La théorie de la terre au XVIIe siècle', *Revue d'histoire des sciences*, 1973, 26:23–48; Roger, 'The Cartesian model and its role in eighteenth-century "theory of the earth"', in Thomas M. Lennon, John M. Nicholas and John W. Davis (eds.), *Problems*

7   *The Reconstruction of Geohistory*

The 'theories of the earth' that were proposed during the 'long' eighteenth century (to which Ellenberger's subtitle refers) were bewildering in their diversity, but all of them shared this general character. It made no difference whether the timescale proposed was quite short by the standards of modern geology, as in Buffon's celebrated *Époques de la nature* (1778), or unimaginably vast. It made only a minor difference whether the sequence envisaged was a directional development or, as in James Hutton's *Theory of the Earth*, a cyclic system with 'no vestige of a beginning, no prospect of an end'.[16] What all these 'systems' lacked was any significant element of the *contingency* that would have marked a truly geohistorical narrative, any sense of the unpredictable complexity and particularity of history. In each system specific observations were indeed added to the programmatic description of the Earth, but they were annexed to the theory merely as 'notes justificatives' (as in Buffon) or as 'proofs and illustrations' (as in Hutton), rather than being treated as the primary evidence on which a geohistory should be reconstructed.[17]

In any case, the younger generation of naturalists was beginning to reject the entire genre of the 'theory of the earth' as misconceived, or at best premature. The very diversity of the speculative 'systems' on offer suggested that in 'geology' too many theories were chasing too few facts. As Cuvier commented in 1807, 'it has become almost impossible to mention its [geology's] name without provoking laughter' [il est devenu presqu'impossible de prononcer son nom sans exciter le rire].[18]

Around 1800, therefore, neither the indoor study of fossil and mineral specimens nor the outdoor practice of geognosy was geohistorical in orientation; and 'geology',

---

of Cartesianism (Montreal and Quebec: McGill-Queen's University Press, 1982), pp. 95–125; David R. Oldroyd, 'Historicism and the rise of historical geology', *History of Science*, 1979, *17*:191–213, 227–57.

[16] George Louis Leclerc, count de Buffon, 'Des époques de la nature', in Buffon, *Histoire naturelle* [20], *Supplément* (1778) 5:1–254; a modern edition is in Jacques Roger, 'Buffon: Les Époques de la Nature; édition critique', *Mémoires du Muséum National d'Histoire Naturelle*, 1962, sér. C, *10* [clii + 343 pp.]. James Hutton, 'Theory of the earth; or an investigation of the laws observable in the composition, dissolution and restoration of the land upon the globe', *Transactions of the Royal Society of Edinburgh*, 1788, *1*(2):209–304, on p. 304.

[17] See note 16 and, for Hutton, the book-length version of his work, *Theory of the earth, with proofs and illustrations in four parts*, 2 vols. (Edinburgh, 1795) [facsimile reprint, Codicote: Wheldon and Wesley, 1959].

[18] Georges Cuvier [et al.], 'Rapport de l'Institut National (classe des sciences physiques et mathématiques), sur l'ouvrage de M. André, ayant pour titre: Théorie de la surface actuelle de la terre', *Journal des mines*, 1807, *21*:413–30, on p. 417.

or the search for an all-embracing causal 'theory of the earth', was concerned with essentially 'programmed' and therefore ahistorical changes, and in any case its highly speculative character had made it deeply suspect.

I believe that, by contrast, the origin of a truly geohistorical dimension is to be found in the transfer into natural history of concepts from the *human* sciences, and specifically from human historiography. In the later eighteenth century, the language of historians and 'antiquarians' (in modern terms, archaeologists) was beginning to be deployed consistently by at least a few naturalists: they wrote of the 'epochs', 'documents', 'archives' and 'monuments' *of nature*.[19] Buffon used such metaphors for cosmogonical purposes of his 'theory of the earth'; but others applied them to the much more empirical analysis of the geohistory of specific regions. Nicolas Desmarest, for example, explained the different 'epochs' of the extinct volcanos of Auvergne (1779), and Jean Louis Giraud-Soulavie used a wide variety of historical metaphors to describe the complex history of Vivarais (1780–84).[20] Geohistorical analyses of specific regions such as these were closely and intentionally analogous to the careful investigations of contemporary antiquarians and local historians, based on human archives and monuments. They were far removed from the global ambitions of the authors of the 'theory of the earth', just as the archival work of local historians was far removed from the speculative 'conjectural history' of Enlightenment philosophes.

William Smith as a Geognost

In the context of these four distinct traditions of practice, it is now possible to define more precisely what was distinctive or novel about Smith's work and that of Cuvier and Brongniart.

---

[19] Rhoda Rappaport, 'Borrowed words: problems of vocabulary in eighteenth-century geology', *British Journal of the History of Science*, 1982, *15*:27–44; Gabriel Gohau, *Les sciences de la terre aux XVIIe et XVIIIe siècles: naissance de la géologie* (Paris: Albin Michel, 1990). Some of these metaphors date back to the time of Hooke, but their use became far more powerful and significant in the late eighteenth century.

[20] Nicholas Desmarest, 'Extrait d'un mémoire sur la détermination de quelques époques de la nature par les produits des volcans, et sur l'usage qu'on peut faire de ces époques dans l'étude des volcans', *Observations sur la physique, sur l'histoire naturelle, et sur les arts*, 1779, *13*:115–26. Jean Louis Giraud-Soulavie, *Histoire naturelle de la France méridionale, ou recherches sur la minéralogie du Vivarais, du Viennois, du Valentinois, du Forez, de l'Auvergne, du Velai, de l'Uségeois, du Comtat Venaissin, de la Provence, des Dioceses de Nismes, Montpellier, Agde, &c...*, 6 vols (Nîmes and Paris, 1780–84).

Smith was a practical man, as much as any Continental geognost; but unlike most of them his early employment was in the one important part of the mining industry – namely coal – that was concerned not with Primary but with Secondary formations. Later he worked as a land surveyor in parts of England underlain by Secondary formations some of which contained fossils in exceptional abundance. Such activities brought him into contact with local fossil collectors, and hence with the practice of natural history based on museums and cabinets. These two circumstances made his experience distinct from that of most other geognosts. Yet geognosy was what he was practising, even if he was unaware of it.

As a canal engineer, Smith was concerned with the structural or geognostic relations of a complex sequence of varied formations, many of which were deceptively similar (limestones, clays) yet distinct in position. By 1796 he had found that – to use his own terms – he could distinguish the 'strata' by their 'characteristic' fossils; the fossils could be used empirically to trace their outcrops across the countryside from one exposure to another. By 1801 he had traversed much of England in the course of his work as a land surveyor, and he published a prospectus for a map that would plot 'the natural order of the various Strata' and their outcrops across the whole country. In 1802 his patron Sir Joseph Banks, the president of the Royal Society of London, exhibited a manuscript version of the map, already 'in a very considerable state of forwardness', at the great agricultural show on the estate of the Duke of Bedford at Woburn, and thereafter at his own house in London.[21] But a succession of practical and financial problems delayed the publication of the map, and its brief textual explanation, until 1815; and Smith's album of illustrations of the fossils that characterized the formations appeared even later (1816–19).[22]

The story of Smith's map, and of its basis in his empirical use of 'characteristic fossils', is well known. But chauvinistic arguments about his priority and originality have obscured the fact that his work fell clearly into the tradition of geognostic description: like any geognost, Smith had found by careful fieldwork the structural

---

[21] The quotations are from the review of Smith's work in Joan M. Eyles, 'William Smith, Sir Joseph Banks and the French geologists', in Alwyne Wheeler and James H. Price (eds.), *From Linnaeus to Darwin: commentaries on the history of biology and geology* (London: Society for the History of Natural History, 1985), 37–50.

[22] William Smith, *A delineation of the strata of England and Wales with part of Scotland...* (London, 1815); *A memoir to the map and delineation of the strata of England and Wales, with part of Scotland* (London, 1815); *Strata identified by organized fossils, containing prints on coloured paper of the most characteristic specimens in each stratum* (London, 1816–19).

order of the formations in a specific region. However, the sheer magnitude of the region covered by Smith's map – the whole of England and Wales – was remarkable; the map plotted the outcrops of the formations with great precision, at least in the areas he knew best and had surveyed most thoroughly; and it distinguished a large number of separate formations, some of them only a few metres in thickness. His unusual cartographical convention – each outcrop coloured with a *graded* intensity – highlighted the three-dimensional structure of the formations, and thereby accentuated the geognostic character of the map; but the format of the map itself was not radically novel (Figure 1).[23]

What was most original about Smith's work was of course his use of 'characteristic fossils' as a new – or rather, a greatly refined – criterion in the identification and classification of formations.[24] Fossils were not Smith's sole criterion: some of his formations contained few fossils or none at all, and like any geognost he also used superposition and lithology. But in a lowland region in which the land was thickly covered with soil and vegetation, and exposures of the rocks were generally small and scattered, those traditional criteria were inadequate, and his use of fossils was decisive.[25] Paradoxically, however, this most original feature of Smith's work was also the most invisible: for until he published his illustrations of the fossils, their value could only be appreciated, *or checked*, by those who inspected his collections in person at his house. But those who did so found his fossils ranged on shelves that corresponded to the formations. Thus in effect the structural order of geognosy, determined by outdoor fieldwork, had been brought indoors and replicated in a fossil collection in a cabinet: those two traditions were thereby united, effectively for the first time (Figure 2).[26]

Smith's use of fossils extended the practice of geognosy decisively, and enabled it to be applied much more effectively to regions of Secondary rocks. Yet his work

---

[23] The latter claim is clearly justified in the light of François Ellenberger, 'Recherches et réflexions sur la naissance de la cartographie géologique, en Europe et plus particulièrement en France', *Histoire et Nature*, 1985, no. 22/23:3–54. Fig. 1 is reproduced from Smith, *Map and delineation* (1815), by permission of the British Library, London.

[24] A *crude* fossil-based distinction between, for example, the older and newer Secondary formations – those with and without fossils such as ammonites and belemnites – was already widely recognized and understood; but Smith's distinctions were far more detailed and precise.

[25] Compare Rachel Laudan, 'William Smith: stratigraphy without palaeontology', *Centaurus*, 1976, *20*:210–26; and Joan M. Eyles, 'William Smith: great discoverer or mere fossil collector?', *Open earth*, 1979, *2*:11–13.

[26] Fig. 2 is reproduced from Smith, *Strata identified* (1816–19), by permission of the University Library, Cambridge.

Fig. 1.   A part of William Smith's *Map and delineation of the strata* (1815), showing the area around Bath in south-west England. A black-and-white reproduction cannot do justice to the striking colours of the original, but it does show how the edges of the successive formations or 'strata', indicated by stronger tints, accentuated the three-dimensional or geognostic *structure* of the region. The darkest areas are the coalfields, with collieries marked as crosses. By permission of the Syndics of Cambridge University Library.

remained essentially geognostic: it was concerned only with the *structural* order of the formations, and its goals were hardly geohistorical at all. Smith used fossils as diagnostic characters for the identification of three-dimensional rock-masses, not as natural archives for the reconstruction of past events. It is true that he later

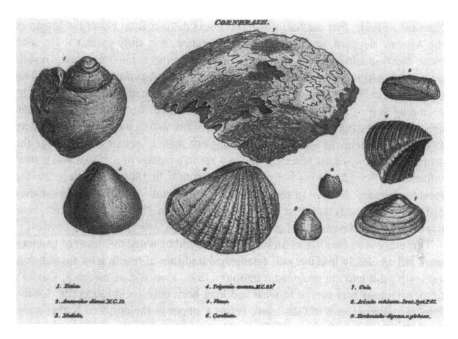

Fig. 2. William Smith's illustrations of the 'characteristic' fossils of the Cornbrash, published in his *Strata identified by organized fossils* (1816–19). The set of fossils from each formation was printed on coloured paper, matching the colour chosen to denote the outcrop of that formation on his map. Thus his specimens were integrated with the geognostic order, as in the arrangement of his own collection in London, but rendered mobile in this published proxy form. The Cornbrash limestone was one of the thinnest of the formations distinguished by Smith, yet its fossils helped him to trace its outcrop across England for more than 200 miles. By permission of the Syndics of Cambridge University Library.

applied to fossils the epithets – by that time conventional – of 'medals' and 'antiquities'; and in private he related his own ideas to the speculative 'theory of the earth' proposed by a much earlier compatriot, John Strachey. But such analogical and theoretical considerations played no perceptible role in his practical work as a land surveyor or in the production of his map.

Cuvier as a Natural Antiquarian

In the same year that Smith first wrote a private note about the relation between 'strata' and their fossils, Cuvier (1796) claimed, at a meeting of the new Institut

National in Paris, that the fossil mammoth was distinct from either the Indian or the African elephant, and therefore, by inference, was truly extinct.[27] He soon made fossil quadrupeds his own major research project; his preliminary results were published in journals such as the *Magasin encyclopédique* and the *Bulletin* of the Société Philomathique; and they were soon translated into foreign journals, becoming known throughout Europe and even in America. In 1800 he issued an appeal under the prestigious auspices of the Institut, asking naturalists everywhere to send him specimens – or at least drawings of them as proxies – to enlarge the scope and validity of his research; and as a result a stream of pictorial and textual information flowed into his house at the Muséum.[28] In 1804 he began to use the recently founded *Annales* of the Muséum as a medium for publishing his studies of fossil quadrupeds and their living relatives; and in 1812 he reissued all these papers as his *Recherches sur les ossemens fossiles* [Researches on fossil bones].[29]

That story is as familiar as Smith's; but it should be noted that most of Cuvier's work fell as clearly into the well established tradition of research on fossil bones as Smith's did into the geognostic tradition. In its aims and its methods, Cuvier's fossil osteology was similar to what had long been done by, for example, Johann Friedrich Blumenbach in Göttingen, Petrus Camper in the northern Netherlands, Louis Daubenton in Paris, and William Hunter in London. It was indoor work on

---

[27] Georges Cuvier, 'Mémoire sur les éspèces d'elephans tant vivantes que fossiles, lu à la séance publique de l'Institut National le 15 germinal, an IV', *Magasin encyclopédique*, 1796, 2e année, *3*:440–45.

[28] Cuvier, 'Extrait d'un ouvrage sur les éspèces de quadrupèdes dont on a trouvé les ossemens dans l'intérieur de la terre, addressé aux savants et aux amateurs des sciences. Imprimé par ordre de la classe des sciences mathématiques et physiques de l'Institut National, du 26 brumaire an 9', *Journal de physique*, An 9 [1801], *52*:253–67. Martin Rudwick, '*Recherches sur les ossemens fossiles*: Georges Cuvier et la collecte d'alliés internationaux', in Claude Blanckaert et al. (eds.), *Le Muséum au premier siècle de son histoire* (Paris: Muséum National d'Histoire Naturelle, 1997), 591–606. Contemporary translations are listed in Jean Chandler Smith, *Georges Cuvier: an annotated bibliography of his published works* (Washington, DC: Smithsonian Institution Press, 1993).

[29] Cuvier, *Recherches sur les ossemens fossiles de quadrupèdes, où l'on rétablit les caractères de plusieurs espèces d'animaux que les révolutions du globe paroissent avoir détruites*, 4 vols (Paris, 1812). Rudwick, *Georges Cuvier, fossil bones and geological catastrophes* (Chicago: University of Chicago Press, 1997) includes translations of Cuvier's most important early papers on fossils and of the 'Discours préliminaire' to the *Ossemens fossiles*; the accompanying commentary assesses Cuvier's geological work in more detail than is possible here. See also Dorinda Outram, *Georges Cuvier: vocation, science and authority in post-revolutionary France* (Manchester: Manchester University Press, 1984), pp. 141–60.

specimens that had been collected by others; it was aided by comparisons with the skeletons of extant species and with published or unpublished illustrations – proxy specimens – elsewhere; it, in turn, was published in the form of densely illustrated monographs. Cuvier's impressive grasp of comparative anatomy, backed by the unparalleled resources of the Muséum, enabled him to convince most other naturalists that extinction was a genuine natural phenomenon; but in doing so he was merely closing a debate that others had started. Cuvier brought new rigour to the comparative anatomy of fossil bones, just as Smith did to the use of fossil shells in the geognosy of Secondary formations; but most of his work was not radically original in its methods, any more than Smith's.

There are however two crucial exceptions to that claim. First, Cuvier adopted at an early stage the language of nature's 'documents', 'archives' and 'monuments', and he exploited it with unprecedented thoroughness. In a public lecture at the Institut as early as 1798, he explicitly likened his own work to that of an antiquarian; and he suggested how certain fossil bones could not only be identified and classified – the normal task of the natural historian – but also be *reconstructed* into a living animal that had since vanished from the face of the earth.[30] This was in effect to claim the possibility of compiling a true *history* of life on earth in the distant past; and Cuvier argued that to try to do so was no *more* speculative than the 'systems' of self-styled 'geologists' (such as his senior colleague Faujas). He himself later became seriously involved in antiquarian research, when he acted as a consultant on natural history for the French edition of the celebrated *Asiatick researches* of British orientalists.[31] And he continued to use the historical metaphor throughout his career: in the 'preliminary discourse' to his *Fossil bones* he famously introduced his work by styling himself 'antiquaire d'une espèce nouvelle' [a new kind of antiquarian].[32] This was no empty claim, because he had for the first time extended the truly geohistorical method of naturalists such as Desmarest and Soulavie to the organic realm, and had begun to reconstruct a detailed history of life on earth (Figure 3).[33]

---

[30] This lecture is translated, and the original manuscript is transcribed, in Rudwick, *Georges Cuvier* (1997), pp. 35–41, 285–90.

[31] A. Labaume, [ed.]. 1805. *Recherches asiatiques, ou mémoires de la Société établie au Bengale pour faire des recherches sur l'histoire et les antiquités, les arts, les sciences et la littérature de l'Asie...*, 2 vols (Paris, 1805). This contained a selection of the papers, for example on the Sanskrit literature of India, that had been published by the Asiatick Society in Calcutta.

[32] Cuvier, 'Discours préliminaire', p. 1, in *Ossemens fossiles* (1812), vol. 1.

[33] Fig. 3 is reproduced from Cuvier, *Ossemens fossiles* (1812), vol. 3, 7e mémoire, unnumbered pl., by permission of the University Library, Cambridge. Cuvier's reconstructed

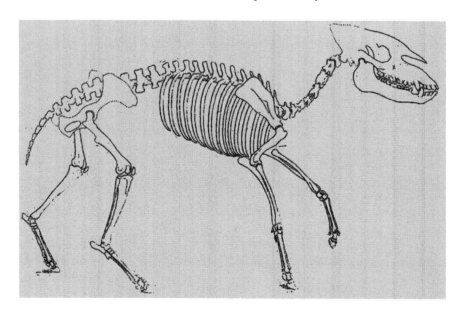

Fig. 3. Georges Cuvier's reconstruction of the skeleton of the palaeotherium, one of the extinct mammals whose bones were found in the gypsum quarries around Paris. This lively drawing, published in his *Researches of fossil bones* (1812), exemplifies his ambition to be an 'antiquarian' or *historian* of nature, by reconstituting the contingent vanished past from the surviving 'archives' of nature. Note his careful depiction (with dotted lines) of those bones that had *not* yet been found, and which were therefore still conjectural. By permission of the Syndics of Cambridge University Library.

The second way in which Cuvier's work embodied a novel dimension was in its incorporation of geognosy. Although he was familiar with Werner's practice even before he arrived in Paris in 1795, Cuvier's early progress reports of his own work on fossil bones show no sign that he appreciated the relative geognostic position – let alone the relative geohistorical age – of his various specimens: like his contemporaries and predecessors, he treated them all as relics of a single undifferentiated 'former world' [ancien monde], which had been destroyed by 'some kind of catastrophe' [un catastrophe quelconque].[34]

---

skeletons later became the basis for the new genre of scenic reconstructions of the life of past periods: Martin Rudwick, *Scenes from deep time: early pictorial representations of the prehistoric world* (Chicago: University of Chicago Press, 1992).

[34] Cuvier, 'Espèces d'éléphans', (1796), p. 444.

However, one particular set of specimens soon led him to recognize that not all fossil bones were alike in this respect. These were the specimens from the gypsum quarries in what were then the suburbs of Paris, for example at Montmartre; they came from the fossil localities closest to his home at the Muséum, and they were abundant in public and private collections in Paris. At first he identified them as belonging to some species of dog; but as he studied a larger number of specimens – and as *better* specimens were discovered year by year – he concluded that the fossils were *more* unlike living mammals than any of his other fossil species, including even the 'animal de Paraguay' that he had named the megatherium and the 'animal de Ohio' that he later named the mastodon. At the same time, he saw for himself that the Parisian fossils came from beds of solid rock, not from the loose silts and gravels [*terrains meubles*] that yielded the bones of large fossil mammals such as mammoths. To reconstruct the history of life thus required distinguishing its separate and successive periods; but that in turn required identifying the different geognostic positions of his various specimens. The indoor naturalist had to go outdoors.

It is well known that Cuvier tackled that problem in joint fieldwork with his friend Alexandre Brongniart.[35] Although Brongniart had been trained as a pharmacist, and his scientific preference was probably for zoology, his employment shifted him inexorably in the direction of mineralogy. During the Revolution he was one of the first mining engineers appointed by the new Mines Council [Conseil des Mines], and by the time Cuvier was installed at the Muséum, Brongniart was teaching at the Mining School [École des Mines] in Paris. In 1800 he reluctantly accepted the directorship of the state porcelain works at Sèvres just outside Paris, because he was newly married and needed the salary. So he had his own reasons for surveying the Paris region: in fact he started fieldwork there as early as 1801, looking for new sources of ceramic materials, particularly kaolin.

In the wartime economy, a better knowledge of German 'mineralogy' had long been recognized as an urgent necessity for the French mining industry, and Werner's concept of geognosy became still better known in France when he himself visited Paris in 1802.[36] It is almost certain, however, that Brongniart's understanding of geognosy was modified significantly by his visit to England later in 1802, during the brief Peace of Amiens. In London he met Sir Joseph

---

[35] Louis de Launay, *Une grande famille de savants: les Brongniart* (Paris: G. Rapilly, 1940), remains the best biographical source.

[36] See note 12.

Banks and attended the meetings of the Royal Society and the scientific soirées at Banks's house.[37] It is probable that he saw the preliminary version of Smith's map, for which Banks was trying to collect subscriptions; he may even have met Smith himself, although Smith had not yet moved to London with his fossil collection. In any case Smith's map of the Secondary formations of England could well have suggested to Brongniart that the formations around Paris could be surveyed in the same way, using their equally abundant fossils to trace formations across the countryside. The idea of a geognostic map cannot have been unfamiliar to him, nor the fact that – in broad terms – different Secondary formations contained different fossils; but the idea of paying such *close* attention to fossils was probably novel to Brongniart, and Smith's claim that they could be used to identify such a large number of distinct formations may even have been surprising.

In any case, Brongniart seems to have started systematic fieldwork around Paris soon after his return, and at the latest by 1804. He was joined by Cuvier at least from time to time; it was the only substantial fieldwork that Cuvier ever did. A preliminary version of their joint paper on the Paris 'basin' was read at the Institut in 1808, with the explanation that 'certain circumstances oblige us to present this summary today' [quelques circonstances nous obligent de présenter aujourd'hui cet abrégé]. With some exaggeration they also described the map they exhibited (it was not published until 1811) as 'a first [*sic*] attempt at a mineralogical map in which each kind of formation is depicted in a particular colour' [un premier essai de cartes minéralogiques dans lesquelles chaque sorte de terrain est enluminée d'une couleur particulière].[38] When John Farey, an English mineral surveyor like Smith, read their article in translation, he sprang to Smith's defence. He inferred that the unexplained 'circumstances' were that the French naturalists had heard that Smith's map was now nearing publication, and that they wanted to

---

[37] Eyles, 'Smith, Banks and the French geologists' (1985). Eyles's inferences about Brongniart's visit to London, for which she conceded that her evidence was largely circumstantial, are confirmed by Brongniart's manuscripts in Paris; I hope to publish this material elsewhere.

[38] Cuvier and Brongniart, 'Essai sur la géographie minéralogique des environs de Paris', *Annales du Muséum d'Histoire Naturelle*, 1808, *11*:293–326; the article was also printed in *Journal des mines*, 1808, *23*:421–58. The full version, amplified by massive local details and illustrated by the map and detailed sections, was published under the same title in *Mémoires de la Classe des Sciences Mathématiques et Physiques de l'Institut Impérial de France*, 1811, *1810*:1–278, 2 pls, 1 map; it was reissued unchanged in the first volume of Cuvier's *Ossemens fossiles* (1812).

steal the credit that was due to him.[39] Even if Farey's suspicions were justified, however, it is important to note the limited extent to which in fact the French memoir followed Smith's example.

Cuvier and Brongniart entitled their monograph a 'géographie minéralogique', and their map a 'carte géognostique'. Both phrases indicate unambiguously that they regarded their work as a contribution to geognosy. What was novel about it was that it applied standard geognostic methods to a region of flat-lying formations with abundant fossils, lying *above* all the formations that Werner and his students had described elsewhere in Europe, and even above most of those on Smith's map. However, several of the formations that Cuvier and Brongniart described contained no fossils at all; like Smith they could have done some of their mapping just by using the conventional criteria of superposition and rock-type. They did indeed note the total contrast between, for example, the fossils of their lowest formation (the Chalk) and the next fossil-bearing one (the Coarse Limestone or *calcaire grossier*); but that kind of *general* use of fossils was already common, and they did not need to borrow it from Smith. In fact, only *within* one particular formation (the Coarse Limestone again) did they mention 'fossiles caractéristiques', and claim to have used them to trace several distinct sets of subordinate beds over large distances; only here did they claim that fossils were 'a sign of recognition that hitherto has not deceived us' [un signe de reconnaissance qui jusqu'à présent ne nous a pas trompés].[40]

A far more important feature of the French memoir owed nothing to Smith's work. As might be expected from Cuvier's participation, the conventional geognosy of the successive formations, described in order from bottom to top, was systematically interpreted in *geohistorical* terms, from earliest to most recent: the fossils were treated as nature's documents or archives. For example, Cuvier and Brongniart were not the first to notice that some Parisian formations contained fossil mollusk shells of what are now exclusively freshwater genera; but they

---

[39] John Farey, 'Geological remarks and queries on Messrs Cuvier and Brogniart's [sic] memoir on the mineral geography of the environs of Paris', *Philosophical magazine*, 1810, *35*:113–39. An alternative possibility, suggested to me by Rod Home, is that one of the authors was currently a candidate for a Parisian scientific position, and needed to demonstrate his qualifications by having his research in the public domain without delay: similar phrasing was usual in such circumstances. Of the two, it was probably Brongniart who was looking for preferment, and he was in fact appointed Professor of mineralogy at the Collège de France that year; Cuvier by contrast was already loaded with all the positions and honours anyone could reasonably ask for.

[40] Cuvier and Brongniart, 'Environs de Paris' (1808), pp. 307–8.

elevated that observation into great significance, by claiming it showed that long periods of marine conditions had alternated with equally tranquil periods of freshwater (lacustrine or fluviatile) conditions. They then used that geohistorical reconstruction to suggest that the most recent 'catastrophe' (which Cuvier believed had caused the extinction of the mammoths) had been just the last in a succession of similar natural events, which in earlier times too had caused occasional abrupt changes in the environment, with consequent sudden changes in the fauna.

Above all, however, both the tranquil periods and the putative sudden events were integrated in effect into a long and complex geohistorical narrative. There had been marine periods: for example, the Chalk and, much later, the Coarse Limestone that provided Lamarck with most of his fossil shells. There had been freshwater periods: for example, the gypsum beds that contained Cuvier's strangest mammals and, much later, the Freshwater Limestone [*calcaire d'eau douce*]. And there had been periods unrepresented by any sediments: for example after the Chalk and, much later, before the superficial 'detrital silt' [*limon d'aterrissement*], with its bones of mammoths and other extinct species. Finally, that last deposit helped bridge the gap between geohistory and human history: Cuvier and Brongniart concluded that 'although very modern in comparison with the other [formations], it is still anterior to historical times' [quoique très-moderne en comparaison des autres, elle est encore antérieure aux tems historiques] (Figure 4).[41]

Nothing could have been more truly *historical* in character than this detailed narrative reconstruction, for it shared all the contingency of human history. Cuvier and Brongniart presented a geohistory of the Paris region that was as complex and – even in retrospect – as unpredictable as the bewildering twists and turns, the war and peace, the sudden *coups d'état* and quieter interludes, of the Revolutionary and Napoleonic politics they had both lived through in the past two decades. Furthermore, just as the politics of those years had been different, though related, in other parts of Europe, so the geohistory that Cuvier and Brongniart offered for the Paris region could be expected to be related, but not identical, to that of other regions. It followed that the equivalent formations elsewhere would be need to be analyzed in equal detail, before a full history of even this relatively recent (post-Chalk) period could be written. Cuvier and Brongniart therefore offered their monograph explicitly as an *exemplar* of what should be done by other naturalists in other regions.

---

[41] Cuvier and Brongniart, 'Environs de Paris' (1808), p. 457. Fig. 4 is reproduced from Cuvier and Brongniart, 'Géographie minéralogique' (1811), pl. 1, fig. 1, by permission of the University Library, Cambridge.

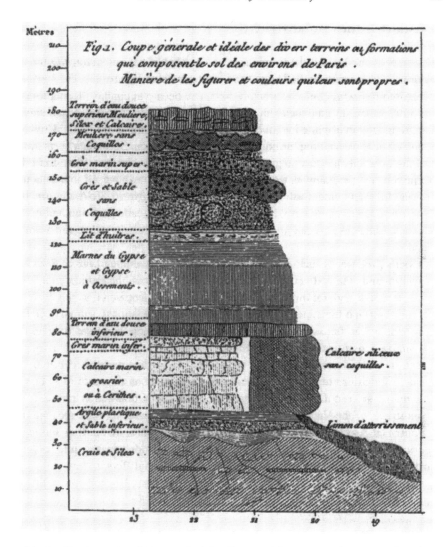

Fig. 4. The 'general and ideal section' that illustrated the monograph on the Paris region by Georges Cuvier and Alexandre Brongniart (1811); it was generalized from several carefully measured traverse sections. Resting irregularly on a foundation of Chalk with flints (*Craie et Silex*) is a pile of varied formations. Some are noted as 'marine' (*marin*) in origin, others as 'freshwater' (*d'eau douce*), in both cases on the evidence of their fossils. Even younger than the uppermost formation is the 'detrital silt' (*limon d'aterrissement*) on the floors of the valleys excavated in the other formations. This diagram, although in format a conventional geognostic section, was thus made the basis for a distinctively *geohistorical* interpretation in terms of a contingent sequence of marine and freshwater periods, and periods of erosion. (Two of the formations are depicted side-by-side as lateral equivalents; although well attested by the authors' fieldwork, this was so unexpected that they left it without satisfactory explanation.) By permission of the Syndics of Cambridge University Press.

A Research Agenda for the Tertiary

In a report to the Institut the previous year, Cuvier had set out an *agenda* for the sciences of the earth: he had urged the importance of integrating the study of fossils into a new geognostic practice focussed on the Secondary formations.[42] Five years later, he concluded the 'preliminary discourse' to his *Fossil bones* with an agenda that made its *geohistorical* goals even more explicit.[43] Cuvier's agenda was a research program focussed on the more recent Secondary formations (such as those of the Paris region), because those were the ones that could be interpreted most reliably in geohistorical terms, and – above all – that could connect the past history of the earth to the *present* world. To focus on the obscure and complex Primary formations, as the geognostic tradition had done, was, he argued, like neglecting the whole of French history since the Roman period.

Cuvier's prescriptive agenda was printed in the same volume of *Fossil Bones* as the revised and enlarged version of the joint monograph on the Paris region, so it was soon noted and followed all over Europe. For example, Giovanni Battista Brocchi described the superbly preserved fossil mollusks of the similarly young formations on the flanks of the Apennines, down the whole length of Italy, in his *Conchiologia fossile subapennina* (1814). Thomas Webster gave an account of the 'Freshwater formations of the Isle of Wight' (1814), finding distinct echoes of the Parisian alternations of marine and freshwater conditions in the south of England. Brongniart's student Constant Prévost described the 'Constitution physique et géognostique' of the Vienna basin (1820), again with formations that recalled the Parisian ones. And Brongniart himself gave an account of the *Terrains de sédiment supérieurs du Vicentin* (1823), finding similar formations in the hills behind Vicenza and elsewhere in northern Italy.[44]

---

[42] Cuvier, 'L'ouvrage de M. André' (1807), pp. 422–5.

[43] Cuvier, 'Discours préliminaire' (1812), pp. 111–16.

[44] Giovanni Battista Brocchi, *Conchiologia fossile subapennina con osservazione geologiche sugli Apennini e sul suolo adiacente*, 2 vols (Milano, 1814). Thomas Webster, 'On the freshwater formations in the Isle of Wight, with some observations on the strata over the chalk in the south-east part of England', *Transactions of the Geological Society of London*, 1814 [ser. 1] 2:161–254, pls. 9–11. Constant Prévost, 'Sur la constitution physique et géognostique du bassin à l'ouverture duquel est située la ville de Vienne en Autriche', *Journal de physique*, 1820, 91:347–67, 460–73. Brongniart, *Mémoire sur les terrains de sédiment supérieurs calcareo-trappéens du Vicentin, et sur quelques terrains d'Italie, de France, d'Allemagne, etc., qui peuvent se rapporter à la même époque* (Paris, 1823). The earlier of these publications describe research begun *before* Cuvier's 'Discours' appeared,

With impressive monographs such as these, the relatively recent period represented by the younger Secondary formations – or the 'Tertiary', as they were soon being called – rapidly became the most reliably reconstructed part of geohistory. It was also clearly perceived as the most important part, in that it linked the more distant and obscure periods with the directly observable present world. So it was no surprise that, two decades after Cuvier issued his clarion call for research on the Tertiary, Charles Lyell used those same formations as the testbed for his neo-Huttonian 'theory of the earth', in the culminating volume (1833) of his *Principles of Geology*.[45]

Conclusion

The monograph by Cuvier and Brongniart on the Paris basin, published in full in 1811, exemplified decisively the fusion of three of the four research traditions that were outlined earlier in this essay, and foreshadowed the incorporation of the fourth. It utilized the collections of fossil shells in the public and private museums of Paris, which Lamarck was already describing and classifying; and Cuvier worked within the same museum practice when he studied fossil bones. But he also turned all those specimens – shells and bones alike – from traditional natural history into materials for geohistory, by adopting the methods and outlook of antiquarians and local historians, taking seriously the analogy between fossils on the one hand and human documents, archives and monuments on the other. Independently, Brongniart's geognostic orientation, derived in part from his practical concern with ceramics, was greatly enriched when he learnt from Smith's work how fossils could be used to help in a geognostic survey; for Smith had already combined a close study of fossil collections with geognostic fieldwork, to help identify the structural order of the various formations. Then, in their collaborative fieldwork, Cuvier and Brongniart integrated that enriched geognosy with the geohistorical perspective: the structural sequence of formations and their fossils was turned into a temporal sequence or narrative of ecologically distinctive periods punctuated by occasional sudden changes. And so, finally, even 'the theory of the

---

but show clear signs of being influenced by his agenda and by his and Brongniart's Parisian fieldwork.

[45] Charles Lyell, *Principles of geology, being an attempt to explain the former changes of the earth's surface, by reference to causes now in operation*, 3 vols (London, 1830–33). See Martin Rudwick, 'Introduction' to the facsimile reprint of *Principles* (Chicago: University of Chicago Press, 1990–91), vol. 1, pp. [vii]–[lviii].

earth' began to seem capable of redemption from an almost useless past, if – as Cuvier suggested programmatically – it could in the future be grounded in an assemblage of similarly detailed reconstructions of regional geohistories. '*Géologie*' might yet become a science about which one would laugh no more.[46]

---

[46] The semantic shift in the word 'geology' in just these years, by which it became both respectable and comprehensive as a label for what would now be termed the earth sciences, is a linguistic trace of the gradual incorporation of the 'theory of the earth' into empirical science. Its adoption in 1807 by the *Geological* Society of London is particularly significant, since the founders of that body – the first of its kind – were explicitly averse to the earlier style of theorizing.

VIII

Researches on Fossil Bones: Georges Cuvier and the Collecting of International Allies[1]

In the first decades of its new life under a new name, the Muséum d'Histoire Naturelle became the world centre for all the natural history sciences. From the start, its professors aimed to make it a storehouse of the natural riches of the entire globe, not just of France itself or even of France's conquered territories and overseas possessions. In that respect they were just continuing what Buffon had intended for the Jardin du Roi, in the cosmopolitan ethos of the Enlightenment.[2] It is too easy, however, to celebrate the internationalism of the Muséum in its early years, without analyzing how international contacts really worked, in an age of increasingly nationalistic sentiment and, of course, at a time of major military conflict throughout Europe and beyond it.

More generally, we know very little about the day-to-day practice of the museum sciences, either in Revolutionary Paris or indeed at any past time or place. By contrast, we can now get a fairly clear impression of the practice of the laboratory sciences in the past, thanks to the research of certain modern historians. We can imagine what it would have been like to watch Faraday in early 19th-century London, or Lavoisier in late 18th-century Paris, or even Boyle in late 17th-century London, actually at work in their respective laboratories.[3] But not all scientific knowledge is created in laboratories, although physicists (and historians of physics)

---

[1] First published as '*Recherches sur les ossements fossiles*: Georges Cuvier et la collecte des alliés internationaux', in Claude Blanckaert, Claudine Cohen, Pietro Corsi and Jean-Louis Fischer (eds.), *Le Muséum au premier siècle de son histoire*, Paris (Muséum National d'Histoire Naturelle), 1997, 591–606. The research for this paper was supported by the U.S. National Science Foundation (grant no. DIR–9021695) and by a grant from the Academic Senate of the University of California San Diego.

[2] See Marie-Noëlle Bourget, 'La collecte du monde: voyage et histoire naturelle (fin XVIIème siècle – début XIXème siècle)', in Claude Blanckaert *et al.*, *Le Muséum*, 1997, 163–96.

[3] See David Gooding, *Experiment and the making of meaning*, Dordrecht (Kluwer), 1990; Frederic L. Holmes, *Lavoisier and the chemistry of life*, Madison, Wisconsin (University of Wisconsin Press), 1985; Steven Shapin and Simon Schaffer, *Leviathan and the air-pump*, Princeton, New Jersey (Princeton University Press), 1985.

often speak as though it is. Much scientific knowledge, in the past and today, is of course made in the field and in museums.

In this paper I explore the practice of one museum science, in an international context, by taking a specific example from the early years of the Muséum d'Histoire Naturelle. My example is Georges Cuvier, and specifically the research that culminated in 1812 in the four volumes of his great *Researches on fossil bones*. It was this research that gave authority to the broader claims that Cuvier made in the famous *Preliminary discourse* that he prefixed to that work.[4] Without his detailed analyses of specific fossil quadrupeds, neither Cuvier's opinions on the 'revolutions of the globe', nor his claims to be a 'new species of antiquarian', would have had much impact on the world of science, however persuasive his verbal rhetoric.

When Cuvier arrived in Paris in 1795, and got himself appointed *suppléant* to Mertrud at the Muséum, he brought some internationalism with him. As a Montbéliardois, he was no Frenchman by birth, but a subject of the Duke of Württemburg. He had received his higher education at the Karlsschule in Stuttgart, where he had had to become fluent in German.[5] As a result he gained access to what was then the second most important body of scientific literature, as well as the first. This was an advantage enjoyed by few if any of his colleagues at the Muséum, most of whom were as complacent in their command of the premier scientific language as anglophone scientists are today. Cuvier later made himself fluent in Italian, and seems to have been able to read English, if not to speak it, without difficulty. And of course his classical education also enabled him to read the still substantial scientific literature in Latin. As Giovanni Fabbroni of Florence commented later, with flattering exaggeration, 'All languages are familiar to you.'[6]

Soon after Cuvier joined the Muséum, his unplanned excursion into fossil anatomy, which in the event delayed his *Animal kingdom* until after *Fossil Bones* was completed, was precipitated by two events that brought him crucial specimens from *outside* France. First, a Republican official on a visit to Spain sent the

---

[4] Cuvier, *Recherches sur les Ossemens Fossiles de Quadrupèdes*, Paris (Déterville), 4 vols, 1812. On the *Discours préliminaire*, see Claudine Cohen, 'Stratégies et rhétorique de la preuve dans les *Recherches sur les ossements fossiles de quadrupèdes* de Cuvier', in Blanckaert *et al.*, *Le Muséum*, 1997, 523–39; the text is available in a convenient modern edition, ed. Pierre Pellegrin, Paris (Flammarion), 1992.

[5] On Cuvier's career, see particularly Dorinda Outram, *Georges Cuvier: vocation, science and authority in post-Revolutionary France*, Manchester (Manchester University Press), 1984; also Howard Negrin, *Georges Cuvier: administrator and educator*, New York (dissertation, New York University), 1977.

[6] Fabbroni to Cuvier, 26 December 1801 (Institut de France, MS 3223/36).

Institut National pre-publication copies of some engravings from Madrid. They were of the large fossil bones that had been found in 1789 near Buenos Aires in Spanish America, and of the skeleton that Juan-Bautista Bru had reconstructed and mounted in the Gabinete Real in Madrid. Cuvier, the most appropriate (and the youngest) member of the scientific First Class at the Institut, was asked to report on these illustrations. He concluded that this 'huge beast' or *Megatherium* (as he named it) was quite unlike any living animal; but he claimed that it was clearly an edentate, and related most nearly to the lowly sloths and anteaters.[7] However, to reach that sensational conclusion Cuvier required more than just a wide knowledge of comparative anatomy, impressive though his already was. It could only have been reached by a naturalist with access to one of the finest zoological collections in the world: few if any other museums, apart from the Hunterian in London, contained at that time the range of skeletons of rare and exotic animals that Cuvier's detailed comparisons required.[8]

Second, in the wake of the Revolutionary army's victories in the Netherlands, the great artistic and scientific collection of the Stadhouder in the Hague had been plundered for the benefit of the French Republic. As a result, Cuvier found at his disposal in the Muséum many rare new specimens, the incidental products of the Dutch commercial empire. Among these, the skulls of elephants from both Ceylon and Cape Colony were of decisive importance. They enabled Cuvier to transform a plausible conjecture, already suggested by several older naturalists, into a persuasive argument supported by detailed evidence. This was that living elephants belonged to two distinct species, the Indian and the African; and, much more importantly, that the fossil mammoth was different from either. That conclusion, reported in Cuvier's first paper to the Institut National, was again only possible because of the resources of the Muséum.[9]

---

[7] Cuvier, 'Notice sur le squelette d'une très-grande espèce de quadrupède', *Magasin encyclopèdique*, 2e année, 1: 303–10, 2 pls, 1796; this is item 22 in Jean Chandler Smith, *Georges Cuvier: an annotated bibliography of his published works*, Washington DC and London (Smithsonian Institution), 1993. On Cuvier's research on the megatherium, see R. Hoffstetter, 'Les rôles respectifs de Bru, Cuvier et Garriga dans les premières études concernant *Megatherium*', *Bulletin du Muséum d'Histoire National*, 31: 536–45, 1959; and José López Piñero, 'Juan Bautista Bru (1740–1799) and the description of the genus *Megatherium*', *Journal of the history of biology*, 21: 147–63, 1988.

[8] On the relation between the Hunterian collection and those of the Muséum, see Philip R. Sloan, 'Le Muséum de Paris vient à Londres', in Blanckaert *et al.*, *Le Muséum* (1997), 607–34.

[9] Cuvier, 'Mémoire sur les éspèces d'elephans tant vivantes que fossiles', *Magasin encyclopédique*, 2e année, 3: 440–45, 1796; 'Mémoire sur les éspèces d'elephans vivantes

4   *Researches on Fossil Bones*

Cuvier's interpretations of the megatherium and the mammoth were so significant that he decided to make fossil anatomy his top priority for research. His work on the anatomy of the molluscs and other invertebrates took second place; his plans for surveying and restructuring the entire *Animal Kingdom* were shelved; and he set his course towards massive *Researches on Fossil Bones*.

Cuvier's appreciation of the material resources of the Muséum was probably crucial in his decision to decline Berthollet's invitation to him to join the scientific and cultural team that accompanied Bonaparte's military expedition to Egypt: as he recalled many years later, 'I was [already] at the centre of the sciences and in the midst of the finest collection.'[10] That decision implied an ambition to make a career by a novel and almost uncharted route, namely as a purely *museum* naturalist. In any case, within a few years his duties as *Sécrétaire Pérpetuel* to the First Class at the Institut made him, as he punned to Fabbroni, 'pérpetuellement fixé' in Paris.[11] So the only fieldwork he could do in pursuit of his research on fossil bones was to collaborate with Alexandre Brongniart on their great survey of the 'géographie minéralogique' of the Paris basin. Luckily for Cuvier, Paris happened to be situated in an area of decisive importance for his research: the gypsum quarries at Montmartre were yielding some of the best and most puzzling of all the fossil bones at his disposal from anywhere in the world. But even here, it is not clear how much of the fieldwork was done by Cuvier: it seems likely that it was mainly Brongniart who got mud on his boots, although it was Cuvier who wrote the draft of their joint paper.[12]

Anyway, that Parisian material was the exception: for the rest of his research on fossils Cuvier was almost completely dependent for his specimens on what the Muséum already contained, or what he could acquire for it. Later, when his governmental appointments in educational administration entailed travelling to the south of France, Italy and the Netherlands, Cuvier took the opportunity to examine fossils in the museums of all the towns he visited. But by then he was so fixed in his ways as a museum naturalist that he seems to have done little if any true fieldwork.

---

et fossiles', *Mém. Inst. Nat. Sci Arts, sci. math. phys.* 2: 1–22, pls. 2–6, 1799 (Smith, *Bibliography*, items 23, 35).

[10] Cuvier, autobiographical fragment (Institut de France, MS 2598/3), p. 39; printed in Flourens, *Receuil des éloges historiques*, 1856–7, vol. 1, p. 185. See Outram, *Georges Cuvier*, pp. 61–3.

[11] Fabbroni to Cuvier, n.d. [1803], quoted in Outram, *Georges Cuvier*, p. 67.

[12] The MS is in Bibliothèque Centrale du Muséum (MS 631). The paper was published as 'Essai sur la géographie minéralogique des environs de Paris' in *Journal des mines*, 23: 421–58, 1808; and, with minor alterations, in *Annales du Muséum*, 11: 293–326, 1808 (Smith, *Bibliography*, items 234, 241).

So even his travel was used, when he was off-duty, to extend the collections of the 'virtual museum', as I will call it, in his head and in his files of notes.

Here it is important to recognize an intrinsic feature of palaeontological material, which constrains research today as much as it did in Cuvier's time. In the case of botanical and zoological material, duplicate specimens are generally available, unless a species is unusually rare or valuable. Much of the ordinary international traffic among natural history museums is in the loan or exchange of such duplicates. This means that the resources of any one museum can be replicated in others, to the extent that that is required for the needs of research or public display. But for fossils the case is often quite different. Vertebrate fossils in particular are generally rare and fragmentary, and often poorly preserved. Therefore those specimens that are unusually complete or well preserved have exceptional scientific value. In palaeontological collections there is quite routinely a premium, as it were, on particular specimens; this is a situation that is only rarely paralleled in botanical or zoological collections (except of course in the case of type specimens). As a result, those in charge of collections of fossils are naturally reluctant to lend, let alone to give away in exchange, their best and most informative specimens. This was as true when the Muséum was founded as it is today.

In modern museums the problem is alleviated by modern technologies of casting. These, in effect, create highly satisfactory 'proxy specimens',[13] which can be exchanged between museums without risk to the precious originals. Therefore palaeontologists can often do much of their research without having to visit distant museums in person. In Cuvier's time, however, the techniques of casting were crude and somewhat unsatisfactory, failing to reproduce many of the finer features of form and texture of the original. They were used *faute de mieux*, but in any case the sending of casts, or indeed of duplicate specimens when those were available, was fraught with risks. Particularly in the wartime conditions in which Cuvier did most of his fossil research, sending parcels of heavy specimens by wagon or diligence was expensive and hazardous: Cuvier's correspondence with Adriaan Camper in The Netherlands, for example, is full of anxious enquiries about the arrival or non-arrival of such consignments.[14]

---

[13] I borrow the term 'proxy' in this context from Mark Hineline, *The visual culture of the earth sciences, 1863–1970*, San Diego (dissertation, University of California San Diego), 1993.

[14] Cuvier's side of this correspondence is in the Universiteitsbibliotheek in Amsterdam; Camper's is in the Muséum and the Institut de France: see the summaries in Bert Theunissen, 'De briefwisseling tussen A.G. Camper en G. Cuvier', *Tijdschrift der Geschiedenis der Geneeskunde, Natuurwetenschappen, Wiskunde en Techniek*, 3: 155–77, 1980.

Another kind of proxy specimen, however, could be sent more safely, more cheaply and more swiftly. These were drawings of particular specimens. Cuvier's manuscript papers, now preserved in the Bibliothèque Centrale of the Muséum, are full of manuscript drawings of specific bones, which he received from his network of international contacts; and it is clear from his correspondence, most of it now preserved in the library of the Institut de France, that many of these drawings were sent in response to Cuvier's detailed requests for pictures of particular specimens. Likewise, when he did travel, he made his own drawings of the most significant specimens in other museums, as proxies that he could take back to Paris.

Drawings, in fact, were the main currency of international exchange in Cuvier's research on fossil vertebrates. Often they were simply sketches in pencil or ink, but they were usually drawn with impressive accuracy and shaded with considerable skill. Frequently, however, they were water-colour paintings, often made with astonishing virtuosity, either by professional artists but in many cases by the naturalists themselves. The practical function of all these drawings, as proxy specimens, put a premium on an almost *trompe-l'oeil* style of extreme naturalism.[15] Of course it was a fortunate accident of art history that this was a period when such a naturalistic style, for example in still-life paintings, was highly valued for its own sake in the world of art. It was equally fortunate for science, though no accident, that it was also a period when competence in the graphic arts was a taken-for-granted accomplishment among educated people of both sexes, and when aspiring naturalists in particular realised the importance of skills in drawing. These were skills that Cuvier himself possessed – or rather, had developed early in life – to an outstanding degree.

Once he had decided to make fossil quadrupeds his main research project, Cuvier lost no time in ensuring that his intention was widely known. His papers on the megatherium and on the species of living and fossil elephants were quickly published in summary in the *Magasin encyclopédique*, and were duly noticed and translated in journals elsewhere in Europe. But as always in the natural history sciences, images were more informative and persuasive than any text. In the case of the megatherium, a copy of Bru's engraving of the reconstructed skeleton was published with Cuvier's preliminary paper; although crude in quality and much reduced in size, this was better than nothing. In the case of the elephants, Cuvier

---

[15] The illustration (Fig. 1) at this point in the original French version of this article is omitted here, because it is reproduced, with more detailed explanation, in the next article in the present volume (chap. X, fig. 4).

used pre-publication copies of the plates for his forthcoming memoir, engraved from his own outline drawings of the skulls, to send to several correspondents outside France. In this way, his crucial evidence was seen by savants elsewhere in Europe, without waiting three years until the memoir was published by the Institut. Meanwhile his ambitious research project was summarized in 1798 in the bulletin of the *Société philomathique* and translated into German and English.[16] It seems to have been read by many naturalists spread across Europe; certainly several of them wrote to comment on it.

Cuvier's decisive bid for international recognition came, however, with his programmatic paper to the Institut in 1800.[17] This was presented as a mere extract from 'a work on the species of quadrupeds of which the bones are found in the earth's interior'. It was explicitly addressed 'to savants and to amateurs of the sciences': both to the rather few scientific specialists on fossil bones, and to the many serious collectors of such specimens. Cuvier set out his claim that the fossil bones represented a whole fauna of species distinct from any living animals. He posed the problem of deciding whether the fossil species had (a) somehow become extinct, or (b) been transformed in the course of time, or (c) merely migrated elsewhere on earth. Of these three alternatives, his own strong conviction was, of course, that they had all become extinct by some kind of catastrophe in the geologically recent past. But that high-level theoretical issue depended for its resolution on the collection of far more empirical evidence: only with far more and better specimens could he confirm his initial impression, and consolidate his bold claim, that all the fossil bones were distinctly different from those of any living species.

Cuvier therefore made an explicit appeal for international collaboration. He did not ask for the loan of specimens, for that would have raised suspicions that Paris was planning further cultural plunder and would never return them. Instead he appealed explicitly for proxy specimens, and in particular for accurate drawings. In return he promised to send his authoritative identifications of the specimens portrayed, and to give his collaborators full acknowledgement in what he intended

---

[16] Cuvier, 'Extrait d'une mémoire sur les ossemens fossiles de quadrupèdes', *Bulletin des sciences, Société Philomathique* [2e sér.] 1: 137–9, 1798 (Smith, *Bibliography*, item 44; translations are listed as items 50, 59 and 84).

[17] Cuvier, 'Extrait d'un ouvrage sur les éspèces de quadrupèdes dont on a trouvé les ossemens dans l'intérieur de la terre, addréssé aux savants et aux amateurs des sciences', *Journal de physique*, 52: 253–67, 1801. This was reprinted from the booklet published by the Institut in 1800; it was translated into Italian, with extracts in German and English (see Smith, *Bibliography*, items 60, 98, 105, 106, 125).

to be a massive and magisterial publication. 'This reciprocal exchange of information [lumières]', he claimed, 'is perhaps the noblest and most interesting commerce that men can have'.[18] He carefully set out his own credentials for the task, not only by referring to the results he had already achieved, but also by listing the savants who had already assisted him. Finally, but not least important, he tacitly warned off any potential competitors, by emphasizing that his own project was already far advanced. In particular, in the essential and expensive matter of illustrations, he mentioned that no fewer than fifty plates of his drawings had already been engraved and were ready for publication.

The text of Cuvier's programmatic paper reveals that he was starting out with a quite small number of informants. He was careful to acknowledge no fewer than eight of his Parisian colleagues: among them were Daubenton, his chief French predecessor in the study of fossil bones; Déodat de Dolomieu, whose geological theorizing was closest to his own; and even, tactfully, Faujas de Saint-Fond, his chief rival at the Muséum in the field of geology. Cuvier also mentioned three major Parisian collectors, on whom he depended for some of his best specimens; chief among them was 'citizen Drée', the temporarily democratized Marquis de Drée, Dolomieu's brother-in-law. Outside Paris, however, Cuvier had built up few sources of specimens or information on fossil bones, as a simple distribution map shows (Fig. 1).

Apart from two 'colleagues' in newly French territories – Johann Hermann in Strasbourg and Gotthelf Fischer in Mainz – only eight 'foreign naturalists' were listed, and in at least two cases they, not Cuvier, had taken the initiative. Christian Wiedemann in Braunschweig seems to have contacted him initially to get permission to reproduce Cuvier's illustrations in his own periodical; and Adriaan Camper was encouraged to contact Cuvier after Auguste de Candolle of Geneva visited Franeker in the northern Netherlands and saw the magnificent collection that the Dutchman had inherited from his father, the great anatomist Petrus Camper. Cuvier seems to have contacted Johann Blumenbach of Göttingen, his most experienced predecessor and greatest potential rival in the field, only just before he announced his ambitious research plans. Cuvier's only other contacts in German-speaking Europe were Autenrieth and Jaeger in Württemberg. In the Netherlands, Cuvier was already in contact with another comparative anatomist, Sebald Brugmans of Leiden, even before the younger Camper put himself on the scene; and in the Italian states Cuvier mentioned his contacts with Fabbroni in Florence and Giovanni-Battista Fortis in Bologna. He mentioned no informant in the anglophone world, but that is hardly surprising in view of the war.

---

[18] *Ibid.*, p. 266.

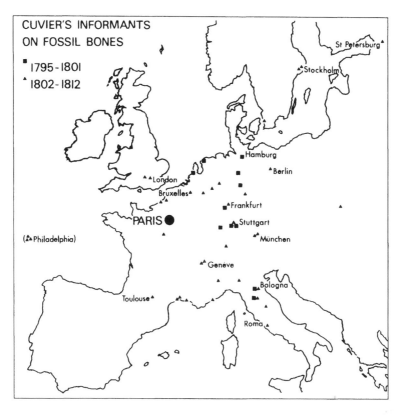

Fig. 1. Distribution map of Cuvier's informants on fossil bones, showing the location of those he mentioned in his programmatic paper of 1800 (black squares), and those who responded to that appeal for collaboration, in time for their assistance to be incorporated in the first edition (1812) of his *Researches on fossil bones* (small black triangles).

Cuvier's appeal for international collaboration was printed by the Institut National as a separate booklet; doubtless the fact that Cuvier had become one of the secretaries of the First Class helped in getting it this unusual treatment. The booklet was certainly distributed widely, both to individuals and to scientific bodies throughout Europe and even in North America. Fischer, whom Cuvier had asked to translate his full work in due course, sent him a list of German savants to whom further copies should be sent. A German translation was published in Jena, and was responded to by savants as far afield as Stockholm and eastern Galizia. Fabbroni promised to distribute a translation among his Italian friends, and Camper told Cuvier he would get a Dutch translation published in the Netherlands. A copy reached Sir Joseph Banks in London, probably during the brief Peace of Amiens, and must surely have been shown around at the Royal Society and at Banks's own scientific soirées. Another copy was sent to the American Philosophical Society in Philadelphia, and a summary in English was published in a medical journal in New York. In addition to all this, Cuvier's paper was also printed in Lamétherie's widely circulated *Journal de physique* as well as in the *Magasin encyclopédique*, both of which must have given the paper still further publicity.

Cuvier's voluminous correspondence, and his files of material for his work on fossil bones, show that his appeal for international collaboration was highly successful. Letters and drawings poured in from all over Europe and beyond it. A simple distribution map of those whose correspondence with him has survived can only give a minimal impression of his network; but it shows an impressive enlargement compared to the period before he published his appeal for collaboration (Fig. 1).[19] His informants now ranged from Stockholm in the north to Rome in the south, and from St Petersburg in the east to Philadelphia in the west.

Of course the status and value of the information that Cuvier received from these sources, and doubtless from many others for which the documentary evidence has not survived, was very uneven. Some correspondents simply sent him a drawing of a striking bone found in their vicinity, or informed him of specimens in some private collection that he might not have heard about. Others, more knowledgeably, sent drawings of specimens that they knew to be of particular significance, or they discussed his work in relation to their own 'theories of the earth'. But however modest the information, Cuvier was careful to mention it later in print; and that acknowledgement, coming from the world centre of scientific research, was doubtless ample reward for many of these scattered 'amateurs' of science. They did not even have to wait until 1812, when at last the *Researches on fossil bones* was published; for as early as 1804 Cuvier's papers on specific fossil animals began to appear in the *Annales du Muséum*, and in each memoir he carefully mentioned his informants by name.

After the bold cultural plunder of the Revolutionary wars, the Muséum received relatively few new solid specimens of fossil bones for its collections, apart from the steady stream of those found close by, at Montmartre and in what were then other suburbs of Paris. But in the years of the Consulate and Empire Cuvier was almost deluged with proxy specimens. Often the correspondent had taken

---

[19] This map, and the verbal summaries above, are based on the letters received by Cuvier, now preserved in the library of the Institut de France. I am indebted to the staff of the library for helpful assistance in consulting these manuscripts.

verisimilitude to the limit, by sending a naturalistic painting of, say, a large mammoth tooth or even a piece of tusk at natural size; in such cases Cuvier would carefully redraw the picture at a reduced size suitable for publication. Sometimes the drawing sent did not show the requisite diagnostic features, and he would write back to ask for another drawing from a different angle; or a drawing came without a clear indication of scale, and he would ask for precise measurements. But many of these drawings were so good, and so adequate for his purposes, that he simply cut them out and mounted them, along with his own, to compose a plate of figures ready for the engraver. The engraver's craft then had an inevitable consequence that was probably not unwelcome to Cuvier. Whatever the style and medium of the drawings he had been sent, they were all rendered uniform on the final plate: Cuvier had in effect *appropriated* them all into his own project.[20]

The results of Cuvier's appeal 'to savants and to amateurs of the sciences' gave him the decisive advantage of a network of correspondents throughout Europe and even beyond. These were his international allies, whose practical assistance bolstered the authority of his claims. Like any true alliance, however, the collaboration benefitted both sides. As Cuvier's scientific reputation expanded, so his public acknowledgement of collaboration became an increasingly valuable form of recognition for many of those who supplied him with information and proxy specimens.[21]

Cuvier's correspondents provided him with a vast enlargement of the resources of the Muséum, if not in fossil specimens then at least in drawings of them. By appropriating these proxies into a virtual museum that was wholly under his own control, they too became in a sense his international allies, as much as the naturalists who had sent them to him.[22] For without this enlargement of his resources, his work would have been far less complete in its description of the osteology of the fossils, and therefore far less persuasive evidence for his major theoretical claims. Without the proxy specimens elicited by his appeal for international assistance, Cuvier's claim to have identified a whole fauna of species

---

[20] The concept of 'appropriation' is borrowed from Robert Marc Friedman, *Appropriating the Weather*, Ithaca, New York (Cornell University Press), 1989. The illustrations (Figs. 3, 4) at this point in the original French version of this article are omitted here, because they are reproduced, with more detailed explanation, in the next article in the present volume (chap. X, figs. 5, 6).

[21] See Outram, *Georges Cuvier*.

[22] The notion of inanimate entities, in this case fossil specimens, as 'allies' in the agonistic field of scientific persuasion, is borrowed from Bruno Latour: see for example his *Science in Action*, Cambridge, Mass. (Harvard University Press), 1987.

VIII

distinct from any living animals would have been no better than the highly speculative 'systems' or 'theories of the earth' that he so deplored in geological science.[23] Still less would there have been such plausibility in his further, famous and sensational interpretation of his claim, namely that this fauna of previously unknown species demonstrated 'the existence of a world anterieur to ours, destroyed by some kind of catastrophe'.[24]

By the time the *Researches on Fossil Bones* was published, less than twenty years after the foundation of the Muséum, Cuvier's work had focussed world attention on fossil quadrupeds on to the concentric circles of France, Paris, the Muséum, its Gallery of Comparative Anatomy, and of course Cuvier himself. That focus was due above all to the vast collection of individual fossil bones he had assembled: not only the real specimens that the Muséum contained, but also and decisively the proxy specimens – drawings and casts – that his international contacts had sent him at his request. He had concentrated this great virtual museum into one spot on earth, in one corner of the Jardin des Plantes, next to the street that now bears his name. In writing the memoirs that were reprinted in the *Fossil bones*, but even more decisively in assembling the plates that illustrated them, Cuvier appropriated specimens from around the world – both real and proxy – into the service of his own project. Once his work was published, all these specimens became his powerful international allies: engraved images of them acted as further proxies for those that remained in Paris, and spread his conclusions persuasively outwards through the whole scientific world.

---

[23] See for example Cuvier, Haüy and Lelièvre, 'Rapport de l'Institut National (Classe des sciences physiques et mathèmatiques), sur l'ouvrage de M. André, ayant pour titre: Théorie de la surface actuelle de la terre', *Journal des mines*, 21: 413–30, 1807 (Smith, *Bibliography*, item 209). The report was written by Cuvier: the MS is in the library of Institut de France (MS 3160).

[24] Cuvier, 'Espèces d'éléphans', 1796, p. 444.

IX

Georges Cuvier's paper museum of fossil bones

INTRODUCTION

In recent years, historians of science have reconstructed the practice of some of the laboratory sciences of the past, giving a vivid impression of what it would have been like to follow certain individual scientists as they conducted their experiments. But not all scientific work is done in laboratories, although modern scientists often speak as though it is; nor was it in the past, although historians of the physical sciences often talk in the same way. Much scientific knowledge, particularly in the natural history sciences, has of course been created in the field and in museums. Compared to laboratory life, however, those "places of knowledge" are both almost *terra incognita* in historical terms. Leaving fieldwork aside, this paper examines a historical case of the practice of scientific work in a museum, to try to understand how new knowledge was generated there. Specifically, the paper is designed to show how both the practice and the knowledge were visual in character throughout. This will be no surprise to the scientists who now do this kind of research, but many historians, trained exclusively on textual sources, still seem to find it difficult to comprehend.[1]

The naturalist Georges Cuvier arrived in Paris in 1795, at the age of 25, soon after the end of the Terror had made the city a highly congenial environment for any ambitious young man of science; and he was appointed to a junior position at the recently reformed Muséum d'Histoire Naturelle. His meteoric rise to prominence in subsequent years, in Paris and therefore in the international world of the sciences, was the result of hard political work as well as exceptional scientific talents and opportunities. It involved the careful cultivation of allies and patrons, and discreet campaigns against rivals and critics, as well as the full use of the unrivalled resources of the Muséum and of his own outstanding skills as an anatomist.[2]

In 1798 Cuvier had his portrait painted, and chose a design that epitomised his ambitions for a career as a museum naturalist (Figure 1).[3] He is portrayed working indoors, in respectable city clothes. On his table are various zoological specimens, awaiting his treatment of them as raw material for his comparative anatomy. But in fact they are anything but raw

52   CUVIER'S PAPER MUSEUM OF FOSSIL BONES

Figure 1. A portrait of Georges Cuvier, painted in 1798, three years after
he had arrived in Paris and been appointed to a junior position at the
Muséum d'Histoire Naturelle.

bits of nature. They have already been processed in various ways: labelled and catalogued,
no doubt, but also visibly preserved in jars of alcohol, and perhaps dissected. A compound
microscope stands symbolically in the background: symbolic, because for Cuvier's purposes
it was less effective than a simple lens, as a means to explore the details of animal anatomy.
Further in the background are the bound volumes of books or periodicals that represent the
existing body of relevant knowledge; and Cuvier looks conventionally up towards heaven for
inspiration, as he sits poised to add to that knowledge by his own writing.

The portrait is rich in meaning, but it omits some of the crucial phases in the process of
making knowledge in a museum. It shows the specimens, and the manuscript and printed
accounts of them: the relatively raw materials, and the relatively stable end products.
But Cuvier is not shown examining his specimens, nor is he shown transferring what
he sees on to paper in the form of visual images and verbal notes. Furthermore, his
work appears almost solitary and individual: he is tacitly building on the work of other
naturalists, as represented by the volumes in the background, but he is not seen in any
kind of interaction with them.

This paper will try to fill in those crucial gaps, using as evidence Cuvier's own research files and correspondence, now conserved in Parisian archives, as well as the published record of his and others' work. The paper focuses on what became his main research project, just at the time this portrait was painted; ironically, this was research that diverted him for over a decade from what the portrait showed him doing. He was diverted on to the comparative anatomy of fossil animals, which in turn led him into the then dubiously scientific field for which the term "geology" had recently been coined. Cuvier's work on fossil bones will, however, highlight some important issues about the practice of museum science more generally.[4]

CUVIER AND THE MUSÉUM

Research on fossil bones could logically be said to begin in the field, where specimens are found. But Cuvier himself was not there, or only rarely. Most prominent naturalists around 1800 did not spend a lot of time in the field, once their careers and reputations were established. That was certainly the case with Cuvier. He turned down an invitation to join the team of savants attached to Napoleon's expedition to Egypt because, as he put it later, "j'étais au centre des sciences et en milieu de la plus belle collection": he was already in the world centre of the sciences, Paris, and in the midst of the finest relevant collection, at the Muséum d'Histoire Naturelle.[5] He did not need to go into the field, because the specimens came to him, if they were not already around him.

Cuvier's unanticipated research project on fossil bones entailed exploiting the riches of the Muséum in Paris. At the time, there was no other museum in which any naturalist could have found a fuller collection of material for the comparisons that were needed. It is not clear how extensive the collections of fossil bones were, before Cuvier started work on that kind of material; but for skeletons of living vertebrate species the Muséum was the best in the world.[6] This gave him an inestimable advantage in his work on fossils: a material factor that historians who work exclusively with textual sources too easily overlook.

What Cuvier did with fossil bones was explicitly and necessarily based on an actualistic method. The known present was the key to understanding the unknown past; living mammals were obviously the best basis for determining the affinities of fossil forms. For example, in his very first foray into the vexed problem of fossil bones, Cuvier claimed that a large fossil skeleton recently found near Buenos Aires and mounted in the royal museum in Madrid was that of a gigantic edentate, previously unknown and probably extinct, which he named the "megatherium". But he could never have perceived that surprising affinity, or made it convincing to other naturalists, if he had not had access to the Muséum's unparalleled array of the skeletons of the whole known range of living sloths and anteaters. Likewise, in Cuvier's other and equally striking early paper on fossil bones, he claimed to show decisively that the well known "mammoth" bones found in Siberia did not belong to the same species as either the Indian or the African elephant, as was commonly supposed. But that striking conclusion, which had major implications for the history of the earth, was only made possible by the recent arrival in Paris of cultural trophies from the conquered Netherlands, including elephant skulls originally from Dutch colonies at the Cape and in Ceylon. Without that acquisition of new specimens for the Paris museum, the identity of the Siberian bones might have remained, as it had long been, ambiguous and controversial.[7]

The museum as an indoor place of knowledge became a necessity for Cuvier for other reasons. His official duties as "secrétaire perpétuelle" of the natural-scientific "First Class" at the Institut National (the Revolutionary replacement of the old Académie Royale des Sciences) soon made him "perpétuellement fixé" in Paris, as he punned to an Italian colleague.[8] When later his duties as an educational bureaucrat entailed travelling around Europe, it was museums that he made a point of visiting in his spare time, not the field localities where the museum specimens had been found. In the case of his one and only piece of substantial field research – his joint work with the mineralogist Alexandre Brongniart (1770–1847) on the rock formations and fossils of the Paris region – the localities were close enough to home for the fieldwork to be done at weekends. But even here, he delegated the collection of new specimens to an assistant: "mon homme", my man, as he called him in his rare allusions to this now anonymous and almost "invisible" technician.[9] This was a rational strategy, however. Fossil bones, important and exciting though they were, were too rare for a naturalist to be likely to find any on a brief weekend trip. So it made good sense to pay someone to keep watch on the men working in the gypsum quarries around Paris, such as those at Montmartre, and to reward them for what they found and handed in.

THE PROCESSING OF SPECIMENS

Many new specimens arrived in this way at Cuvier's working space at the Muséum. Further specimens were lent to him, not only from other institutions such as the École des Mines, but also by amateur collectors in Paris. Other specimens were already there, of course, in the existing collections at the Muséum. But crucial data about them might be frustratingly incomplete, owing to the less than efficient habits of earlier curators. For example, Cuvier described one important new fossil species, a pygmy hippopotamus, solely on the basis of specimens he obtained from a large block of rock that he had found in a store room, with nothing to show where it had come from. In this particular case he also mentioned how he – or possibly a now invisible technician – had used chisels, files and gravers to extract the bones and teeth from their hard matrix.[10]

Such allusions to his working methods are very rare, however, in Cuvier's published papers. In one other instance it is clear that he wanted to emphasise the skilled action involved. This concerned a unique little specimen from the gypsum quarries, which was of exceptional importance because its dentition suggested to him that it was close to the living opossums, an affinity with major geological implications. He described how he had used a sharp steel needle delicately to chip away the concealing rock, revealing the hidden marsupial bones that confirmed his provisional identification and vindicated the anatomical principles on which the prediction had been based.[11] His procedure was the fossil equivalent of his equally delicate manual work in dissecting the soft anatomy of living molluscs; both are examples of the tactile skills that are easily forgotten, if historical figures are treated only as the writers of texts.

Having processed his specimens by excavating them from their matrix or piecing them together from fragments, the next stage was to transform them into "inscriptions", or images and notes on paper.[12] In fact, Cuvier's research practice tied visual and verbal elements into a tightly integrated whole, as epitomised by a typical item from his files (Figure 2).[13] These drawings were made with his usual high standard of subtle shading, to give a convincing representation of the three-dimensional specimen. His artistic talents are another important aspect of scientific practice that it is all too easy to gloss over historically. Unlike the collection of specimens in the field, such drawings were not delegated to an assistant, at

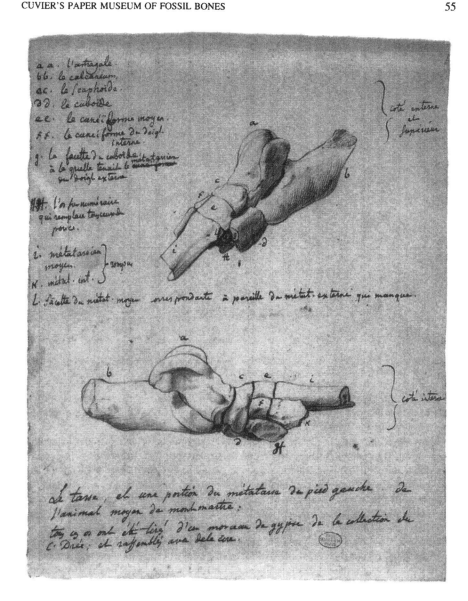

Figure 2. Cuvier's drawings of an assemblage of fossil bones from the Paris gypsum beds, after he had extracted it from its matrix, with notes identifying the bones as those of the ankle region of an unknown "animal de Montmartre", which he later named the "paléotherium magnum".

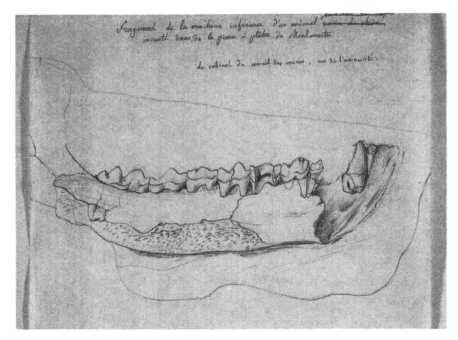

Figure 3. Cuvier's sketch of a fossil jaw from the Paris gypsum, recording an early stage in extracting it from its matrix, with two of his three successive interpretations of it (top right).

least in the earlier part of his career, and many of his published illustrations are recorded firmly as his own ("Cuvier del.").

Many other drawings in Cuvier's files are the museum equivalents of rough laboratory notes on an experiment. In the case of one particular specimen from the Paris gypsum, he made a series of drawings at successive stages as he extracted it from its matrix. His first drawing also records the development of his process of understanding it (Figure 3).[14] Initially he called it the lower jaw of an animal "voisin du chien" (related to the dog). Later, he crossed that out and reinterpreted it as "semblable au tapir" (similar to the South American tapir), an inference that made the find much more puzzling and important. Later still (but not visible in Figure 3) he reinterpreted it again, as belonging to the "paléotherium magnum", one of the species of the extinct genus which by that time he had defined on the basis of this and many other specimens. So in this case a single image served as the enduring basis for a developing interpretation: a complex role that is inadequately captured by the static metaphor of an inscription.

Cuvier was not limited, however, to the fossil bones brought to his working space by his man in the Parisian quarries, or lent to him by other institutions and amateur collectors in Paris. In principle, museums and collectors elsewhere in Europe could also have loaned him their specimens. But it is clear from Cuvier's correspondence that they were loath to do so, partly out of fear that the French state might expropriate them, as it had often done during the Revolution. Besides, there were real risks of damage or loss in transit, particularly in the wartime years in which Cuvier did much of his best research. A safer alternative was

IX

for other museums and collectors to send casts of their specimens, and sometimes they did so. But the casting technique, adapted from the *cire-perdu* (lost wax) process used in sculpture and jewellery, was crude by modern standards: it often failed to show the more subtle features of a bone, such as the scars of the muscle attachments which were crucial for any reconstruction of the animal's anatomy. Anyway, casts were as bulky and almost as heavy as real specimens, and therefore as expensive and troublesome to transport.

Cuvier's collection of specimens, on which he could found his publications and his theoretical conclusions, was therefore enlarged far more substantially by the inflow to Paris of "proxy specimens" in the form of accurate drawings on paper. They were proxies because they purported to stand in for the original specimens, as effective substitutes for them; they functioned as "mobiles", for they could be shifted far from the museums in which the original specimens remained.[15] For example, one fossil tooth was sent to Cuvier in proxy form by Blumenbach in Göttingen, but the real specimen depicted in the drawing had come originally from the famous locality of Big Bone Lick on the Ohio, in what is now Kentucky; it was in effect doubly mobilised (Figure 4).[16] Since few naturalists could draw as well as Cuvier, many proxies were made by professional artists, commissioned by the local collector or curator. The effectiveness of such images, as accurate proxies, was greatly enhanced by the fact that a *trompe l'oeil* style was much in vogue in the art world at this time, for example in still-life painting.

Proxy pictures poured into Cuvier's working space at the Muséum from informants all over Europe and even beyond; they were the product of the social network that he actively constructed to help in his research project. He rewarded his informants by giving their specimens authoritative identifications, and less tangibly by the prestige and credit that they gained from being publicly acknowledged as having assisted the great naturalist in Paris. Cuvier himself added to this collection of proxies when he visited other museums in the course of his bureaucratic travels: he drew all their best specimens to take back to Paris in proxy form. So his "paper museum", or "virtual" collection of proxy specimens, came to be far richer than even the rich collections of real specimens at the Muséum in Paris.[17]

FROM DRAWINGS TO ENGRAVINGS

In 1802 the Muséum launched its own periodical, *Annales du Muséum*, with a generous allowance of engraved plates. Two years later, in the third volume, Cuvier began to publish an astonishing torrent of specialised papers on the bones of particular fossil mammals. He made his actualistic method quite explicit: if the osteology of a living mammal was poorly known, as for example with the tapir, he would publish a paper on it before describing what he regarded as the remains of its fossil relatives. In this way the skeleton of the living South American tapir became persuasive evidence for his striking claim that other tapirs had formerly roamed France itself.[18]

Some of Cuvier's papers on fossil bones depended heavily or even exclusively on proxies rather than real specimens. One of his earliest, which has already been mentioned, was based entirely on pictures. A set of proof copies of some unpublished engravings had been sent to Paris from Madrid, showing some fossil bones from South America and their reconstruction as a complete skeleton. Cuvier was asked to report on the prints, and claimed – without seeing the specimens themselves – that they depicted the bones of a huge and probably extinct sloth, which he named the "megatherium". A later paper on what he named

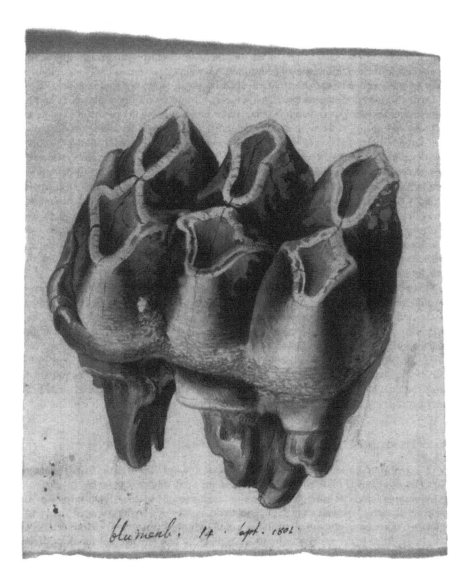

Figure 4. A drawing of a fossil molar tooth of the "Ohio animal" from North America, sent to Cuvier in 1801 by Johann Friedrich Blumenbach of Göttingen; Cuvier later defined and named the animal as the "grande mastodonte".

the "ptérodactyle", which he interpreted strikingly as a flying reptile, was based likewise on a proxy picture of the only known specimen, which had been drawn many years earlier, before the specimen itself was mislaid during the removal of its museum to a new site.[19]

Usually, however, Cuvier had a rich mixture of real and proxy specimens on which to work. In preparing each paper, he first assembled his own drawings of relevant bones together with the proxies that he had been sent from elsewhere. He then cut out all the pictures to be published, and pinned them to a background sheet of stout paper, to make the most economical use of the expensive copper plate on which they would later be engraved (Figure 5).[20] Such sets of mounted drawings were usually sent away to professional engravers. But at an early stage in his research Cuvier got the prominent engraver Simon Charles Miger (1736–1820) to teach him the craft, and a few of his published pictures – such as those based on the drawings reproduced here as Figure 2 – are his own engravings and are marked accordingly ("Cuvier del. et sc."). Like a composer who tries to learn about the instruments he is writing for, even if he does not play them expertly, Cuvier evidently wanted to understand at first hand what was entailed in the process of turning a drawing into an engraving.

There are several contrasts between Cuvier's sets of mounted drawings and the published plates that his engravers derived from them. Most noticeably, the drawings are all reversed by the engraving process, so that they no longer match the real specimens except as mirror images (Figure 6).[21] This was not unavoidable. In other contexts – for example, when making prints of topographical scenes – engravers were accustomed to using tracing paper to reverse the drawings they were given; they would then engrave from the reversed images, so that the original orientation would be restored in the final prints. But in the natural history sciences the extra stage was usually omitted, presumably to save labour costs and perhaps to avoid possible further degradation of the image. In the case of fossil bones the loss of veracity was not very important: a real specimen of a left femur, for example, would emerge from the engraving process looking like a right femur, which was its mirror image anyway. So the reversal was a visual convention that was apparently just taken for granted; it received no comment in Cuvier's published work.[22]

A second contrast between drawings and engravings was the relative degradation of the representation of shading, and hence of the *trompe l'oeil* illusion of three-dimensional solidity. The subtle pencilled shading of Cuvier's drawings was unavoidably replaced by the relatively crude "grid" of parallel or crosshatched lines that does duty for shading in an engraving.[23]

A third contrast is more significant for the historian. The varied styles of the original drawings – some by Cuvier himself, some by his informants or their artists – were all rendered into a uniform style on the engraved plate. This shows how proxy images, although indeed "mobiles", were anything but "immutable".[24] In effect, it demonstrates how Cuvier appropriated the work of other naturalists into his own project, in a strong sense of that verb. In this particular case, he even put his own name in the usual position at the foot of the plate, as its artist, thereby tacitly claiming that they were all his own drawings, which in fact they were not.

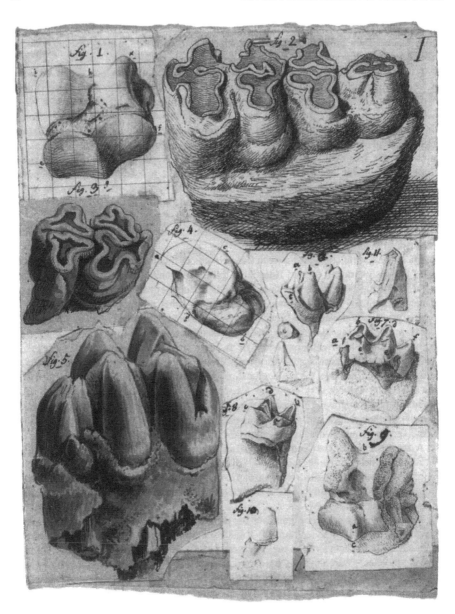

Figure 5. Drawings of fossil hippopotamus bones and teeth, assembled by Cuvier ready for engraving. Some drawings were by Cuvier himself, others by his informants or their professional artists; hence the diversity of styles. Those marked with a ruled grid had been reduced in size to fit on the plate.

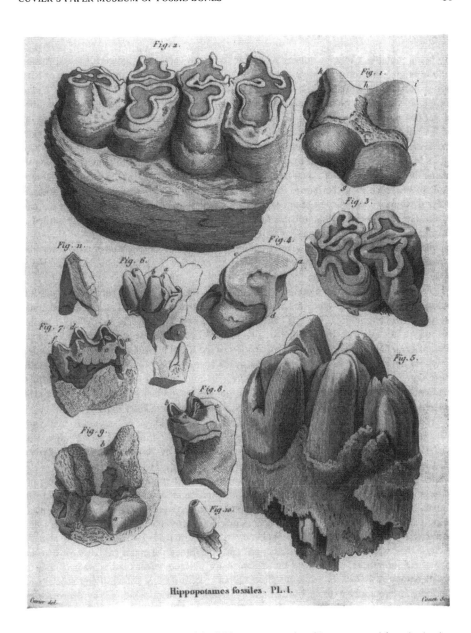

Figure 6. One of Cuvier's published plates of fossil hippopotamus teeth and bones, engraved from the drawings reproduced in Figure 5. He claimed tacitly that they were all his own drawings ("Cuvier del.", bottom left), although in fact they were not.

IX

PUBLISHING FOSSIL BONES

By being mutable, Cuvier's images were able to be the vehicles of powerful visual rhetoric (in the proper and non-pejorative sense of that word): they made his interpretations of fossil bones persuasive. Usually this was the rhetoric of plain factuality. For example, Cuvier took a set of images of the skulls of bears, living and fossil, drawn by himself and others at various scales, and had them all re-drawn at exactly the same scale. This enabled one of his published plates to make an interpretative point persuasively, because it highlighted the distinctive size and character of the putatively extinct cave bear, the largest of them all.[25] In another case, by contrast, the visual rhetoric depended on deliberately reducing the factuality. In one of Cuvier's first papers, which was mentioned earlier, he reduced the huge skull of the fossil megatherium to the same size as those of two species of living sloths, in order to support his startling claim about their zoological affinity (Figure 7).[26] In the case of the famous specimen that Johann Jakob Scheuchzer a century earlier had identified as "a man a witness of the deluge", Cuvier debunked that conceit in typical Enlightenment manner. He established it as a giant amphibian, making his case persuasive by publishing a plate showing the fossil alongside the skeletons of a fish, frog and salamander, engraved at deliberately different scales.[27]

Cuvier's most striking piece of visual rhetoric was deployed to support his startling interpretation of the little specimen, mentioned earlier, that he identified as a fossil marsupial. He published engravings showing the fossil before and after he had excavated it to find the marsupial bones he had predicted (Figure 8).[28] In fact this was a dramatically staged presentation: Cuvier did his delicate excavation of the fossil in front of competent witnesses, to whom he had made his risky prediction in advance. So his published account, including these drawings, was designed to convert his readers into "virtual witnesses" of that fulfilled prediction, which in turn helped explicitly to put his anatomical principles on to the same prestigious level as the predictive physics of his patron Pierre-Simon Laplace.[29]

That last example illustrates how, in all Cuvier's papers on fossil bones, his engravings were integrated tightly with his texts. The integration was not as close as he might have wished: like all naturalists at this period, he was constrained by the technicalities of engraving (just as modern palaeontologists are, by collotype printing). His illustrations could not be printed on the same page as the relevant explanatory text: engravings (like collotypes) had to be printed separately, on different paper, and bound in at the end of each article. But still, abundant cross-references made the integration as tight as it could be. Rather than the plates being the illustrations of the texts, as historians customarily and revealingly put it, it would be better to say that the texts were explanations of the images. The pictures were primary, because they were proxies for the specimens that constituted the empirical basis for Cuvier's theoretical inferences. They were the evidence for his zoological claims, namely that the fossils belonged to species that were almost all distinct from any known living species, and therefore inferentially extinct. Further than that, they were also indirect evidence for his even bolder geological claims, namely that the putative mass extinctions had been caused by sudden physical events that deserved to be called catastrophes.[30]

It was the images, tightly bound to their textual explanations, that constituted the publicly accessible evidence on which all Cuvier's claims could be judged by other naturalists. The published images mobilised his collection of real specimens supplemented by the proxies or virtual specimens that he had been sent. Out of that rich total collection, the specimens he chose to publish were of course those to which he attached particular scientific significance.

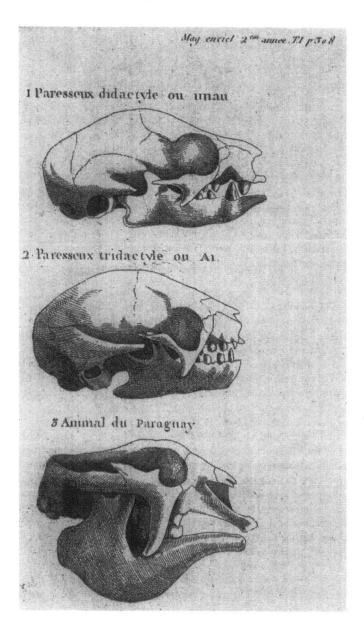

Figure 7. Cuvier's drawings of the skulls of two living species of sloths, with the far larger skull of the fossil megatherium ("animal de Paraguay") reduced to the same size on paper in order to highlight its osteological similarity and inferred affinity to the edentates.

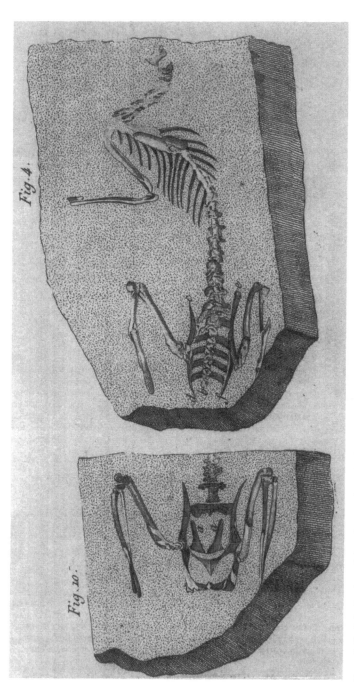

Figure 8. Cuvier's drawings of the unique specimen that he interpreted as a fossil opossum, shown before ("fig. 4", right) and after ("fig. 10", left) he had dramatically revealed its predicted marsupial bones ("a, a" on "fig.10").

So his published images constituted a new paper museum, which was distributed wherever the volumes of the *Annales* were sent. A few years later, Cuvier assembled offprints of all his research articles, rearranged them in a systematic order, supplemented them with new material, and reissued them as his great four-volume *Recherches sur les ossemens fossiles*, researches on fossil bones (1812). In that form they were spread even more widely, all over the scientific world. They constituted a paper museum, now assembled in a way that maximized the scientific meanings that Cuvier attached to them.[31]

Cuvier's museum practice was thus structured from start to finish around the production of pictorial images, tightly integrated with textual explanations of them. Most of his images purported to be authentic proxies for original specimens. His theoretical inferences would have had little impact on the international network of other naturalists, let alone the general literate public, if they had not been seen to be founded on his vast museum resources, much of it a paper museum of proxy specimens. Those images had converged on Paris, and on Cuvier's working space at the Muséum d'Histoire Naturelle, from all over the civilised world and in some cases from beyond it. But with the publication of his work that movement was reversed, and the images were disseminated from the Muséum to naturalists everywhere, carrying with them the persuasive power of Cuvier's interpretations. At least in Cuvier's case, and perhaps more generally, museum science entailed a complex and subtle visual practice. Anything further from the purely verbal and textual practices described by many historians of science would be hard indeed to imagine.

ACKNOWLEDGEMENTS

This paper is based on a talk given at the meeting on "Drawing from nature", organised by the Society for the History of Natural History and held at the Natural History Museum in London in April 1999. Earlier versions were given at the Universities of California (San Diego), Cambridge, Oxford and Sydney, and at a joint meeting of the Palaeontographical Society and the History of Geology Group of the Geological Society; I am grateful to those who gave me valuable comments on all these occasions. I am also grateful to the librarians and staff at the Bibliothèque Central du Muséum National d'Histoire Naturelle, and at the Bibliothèque de l'Institut de France, both in Paris, for giving me access to their Cuvier archives; and to the former, for permission to reproduce images from manuscripts in Cuvier's research files as Figures 2–5. Figures 6–8 are reproduced from printed sources by courtesy of the Syndics of Cambridge University Library.

NOTES

[1] For the historical analysis of laboratory practice, see for example Holmes (1985) on Antoine-Laurent Lavoisier, Gooding (1990) on Michael Faraday, and Shapin and Schaffer (1985) on Robert Boyle. The notion of "laboratory life" comes from the classic study by Latour and Woolgar (1979); Ophir and Shapin (1991) drew attention to the importance of studying "places of knowledge".

[2] Cuvier's active construction of his career is described and analysed by Outram (1984).

[3] Figure 1 is reproduced from Bultingaire (1932), frontispiece.

[4] Rudwick (1997a) traces the course of Cuvier's research on fossil bones, giving the more important texts in translation, and also charts his changing perception of the new field of "geology". Rudwick (1997b) covers some of the same ground as the present paper, but focusses on his institutional context and his network of informants.

[5] The quotation is from an autobiographical fragment in the Bibliothèque de l'Institut de France, MS 2598/3, 39, printed in Flourens (1856–57: vol. 1, p. 185); see also Outram (1984: 61–63).

[6] John Hunter's collection in London, its only rival, was virtually inaccessible during Cuvier's most creative period of research on fossils; see Sloan (1997) for a comparison between the Hunterian and the Muséum. On the

66   CUVIER'S PAPER MUSEUM OF FOSSIL BONES

early history of the Muséum, see also the other essays in Blanckaert *et al.* (1997).

[7] Cuvier (1796a, 1796b); see Rudwick (1997a: 13–32, with Cuvier's texts in translation).

[8] Cuvier to Giovanni Fabbroni, [1803], printed in Outram (1982: 210).

[9] Shapin (1989) introduced the term "invisible technicians" to denote those who operated the laboratories of savants such as Robert Boyle, but whose role was effaced from the published record; see also Shapin (1994: 355–407).

[10] Cuvier (1804e: 112); reprinted in Cuvier (1812: vol. **2**, 6e mémoire, p. 14).

[11] Cuvier (1804f: 286); reprinted in Cuvier (1812: vol. **3**, 10e mémoire, p. 10; translated in Rudwick (1997a: 68–73)).

[12] See Latour (1990) for the concept of "inscriptions" in scientific work. Latour has been second to none in emphasising the role of visual materials, yet regrettably his chosen metaphor implicitly privileges the textual over the pictorial.

[13] Figure 2 is reproduced from a sheet in the file "Paléotheriums et Anoplotheriums" in MS 628, Bibliothèque Central, Muséum National d'Histoire Naturelle, Paris. The drawings were later engraved – by Cuvier himself – and published in Cuvier (1804d: pl. 41, figs 1,2), and reissued in Cuvier (1812: vol. **3**, 3e mémoire, pl. 4).

[14] Figure 3 is reproduced from a drawing in the file "Paléotheriums et Anoplotheriums" in MS 628, Bibliothèque Central, Muséum National d'Histoire Naturelle, Paris. It was later engraved and published in Cuvier (1804c: pl. 23, fig. 1), and reissued in Cuvier (1812: vol. **3**, 1e mémoire, pl. 1).

[15] Hineline (1993) proposed the term "proxy" for a wide range of visual images used in this way in the practice of the earth sciences in the later nineteenth and twentieth centuries; for "mobiles", see Latour (1990).

[16] Figure 4 is reproduced from a drawing annotated by Cuvier "blumenb. 14 sept. 1801", now in the file "Grande mastodonte: matériaux nouveaux" in MS 630(2), Bibliothèque Central, Muséum National d'Histoire Naturelle, Paris. It was later engraved and published in Cuvier (1806b: pl. 49, fig. 5), and reissued in Cuvier (1812: vol. **2**, 10e mémoire, pl. 1).

[17] On Cuvier's international network of correspondents, many of whom supplied him with proxies, see Rudwick (1997b, and particularly the map, fig. 2). On the exchange of credit involved in his correspondence, see Outram (1982). The notion of a "paper museum" derives from the famous *Museo Cartaceo* assembled by Cassiano dal Pozzo (1588–1657): see Haskell (1993).

[18] Cuvier (1804a, 1804b), reprinted in Cuvier (1812: vol. **2**, 3e mémoire; vol. **3**, 8e mémoire), respectively.

[19] Cuvier (1796a); Cuvier (1809), reprinted in Cuvier (1812: vol. **4** (part 5), 5e mémoire). See Hoffstetter (1959) and López Piñero (1988) on the megatherium, and Padian (1987) on the pterodactyl (later, pterosaur).

[20] Figure 5 is reproduced from a set of drawings in the file "Hippopotame: matériaux anciens et nouveaux, et ostéologie" in MS 628, Bibliothèque Central, Muséum National d'Histoire Naturelle, Paris. It was later engraved and published: see Figure 6.

[21] Figure 6 is reproduced from Cuvier (1812: vol. **2**, 6e mémoire, pl. 1), published originally in Cuvier (1804e: pl. 9).

[22] Nor has the convention been much discussed historically, except by Gould (1996), who points out how the short cut could not be taken in the case of pictures of gastropods: the representation of coiled shells as sinistral where the real specimens were dextral would have entailed serious error in scientific content.

[23] Ivins (1953), in his classic study of the history of prints as a medium of visual communication, analysed the effects of this "grid of rationality" on engravings of sculptures and paintings.

[24] On "immutable mobiles", see Latour (1990); notwithstanding his chosen adjective, Latour's work has in fact demonstrated the indefinite mutability of scientific images in the course of their deployment.

[25] Cuvier (1806a: pl. 19), reprinted in Cuvier (1812: vol. **4** (part 4), 3e mémoire, pl. 2).

[26] Figure 7 is reproduced from Cuvier (1796a: unnumbered plate, p. 308).

[27] Cuvier (1809: pl. 30), reprinted in Cuvier (1812: vol. **4** (part 5), 5e mémoire, unnumbered plate).

[28] Figure 8 is reproduced from Cuvier (1804f: pl. 19), reprinted in Cuvier (1812: vol. **3**, 10e mémoire, unnumbered plate).

[29] Shapin (1984) used the concept of "virtual witnessing" to denote the intended effect of highly detailed and circumstantial accounts of experiments, for example those of Boyle; he described them as the "literary technologies" that persuaded Boyle's readers that it was as if they themselves had been present to witness

the veracity of his reporting.

[30] On Cuvier's use of his empirical evidence in the service of these broader conclusions, see the interpretative sections in Rudwick (1997a).

[31] The first volume of Cuvier (1812) opens with his long "Discours préliminaire", setting out his broader conclusions in a non–technical style, as a prelude to all his detailed papers on fossil bones; see Rudwick (1997a: 173–252), which includes a translation.

REFERENCES

BLANCKAERT, C., COHEN, C., CORSI, C. and FISCHER, J.–L. (editors), 1997 *Le Muséum au premier siècle de son histoire*. Paris: Éditions du Muséum National d'Histoire Naturelle. Pp 687.

BULTINGAIRE, L., 1932 Iconographie de Georges Cuvier. *Archives du Muséum d'Histoire Naturelle*, 6e série, **9**: 1–12.

CUVIER, G., 1796a Notice sur le squelette d'une très–grande espèce de quadrupède inconnue jusqu'à présent, trouvé au Paraguay, et déposé au cabinet d'histoire naturelle de Madrid. *Magasin encyclopédique*, 2e année, **1**: 303–310.

CUVIER, G., 1796b Mémoire sur les éspèces d'elephans tant vivantes que fossiles, lu à la séance publique de l'Institut National le 15 germinal, an IV. *Magasin encyclopédique*, 2e année, **3**: 440–445.

CUVIER, G., 1804a Description ostéologique du tapir. *Annales du Muséum d'Histoire Naturelle* **3**: 122–131.

CUVIER, G., 1804b Sur quelques dents et os trouvés en France, qui paroissent avoir appartenus à des animaux du genre de tapir. *Annales du Muséum d'Histoire Naturelle* **3**: 132–143.

CUVIER, G., 1804c Sur les espèces d'animaux dont proviennent les os fossiles répandus dans la pierre à plâtre des environs de Paris. Premier mémoire: restitution de la tête. *Annales du Muséum d'Histoire Naturelle* **3**: 275–303.

CUVIER, G., 1804d Sur les espèces d'animaux dont proviennent les os fossiles répandus dans la pierre à plâtre des environs de Paris. Troisième mémoire: restitution des pieds. *Annales du Muséum d'Histoire Naturelle* **3**: 442–472.

CUVIER, G., 1804e Sur les ossemens fossiles d'hippopotame. *Annales du Muséum d'Histoire Naturelle* **5**: 99–122.

CUVIER, G., 1804f Mémoire sur le squelette presque entier d'un petit quadrupède du genre de sarigues, trouvé dans le pierre à plâtre des environs de Paris. *Annales du Muséum d'Histoire Naturelle* **5**: 277–292.

CUVIER, G., 1806a Sur les ossemens du genre de l'ours, qui se trouvent en grande quantité dans certaines cavernes d'Allemagne et de Hongrie. *Annales du Muséum d'Histoire Naturelle* **7**: 301–372.

CUVIER, G., 1806b Sur le grande Mastodonte, animal très–voisin de l'éléphant, mais à mâchelières hérissées de gros tubercles, dont on trouve les os en divers endroits des deux continens, et surtout près des bords de l'Ohio, dans l'Amérique Septentrionale, improprement nommé Mammouth par les Anglais et par les habitans des États–Unis. *Annales du Muséum d'Histoire Naturelle* **8**: 270–312.

CUVIER, G., 1809 Sur quelques quadrupèdes ovipares fossiles conservés dans des schistes calcaires. *Annales du Muséum d'Histoire Naturelle* **13**: 401–437.

CUVIER, G., 1812 *Recherches sur les ossemens fossiles de quadrupèdes, où l'on rétablit les caractères de plusieurs espèces d'animaux que les révolutions du globe paroissent avoir détruites*. Paris: Déterville. 4 vols.

FLOURENS, M. J. P., 1856–1857 *Receuil des éloges historiques lus dans les séances publiques de l'Académie des Sciences*. Paris: Académie des Sciences. 3 vols.

GOODING, D., 1990 *Experiment and the making of meaning: human agency in scientific observation and experiment*. Dordrecht: Kluwer Academic. Pp 310.

GOULD, S. J., 1996 Left snails and right minds, in *Dinosaurs in a haystack: reflections in natural history*. New York: Harmony.

HASKELL, F. (editor), 1993 *The paper museum of Cassiano dal Pozzo*. London: Olivetti. Pp 288.

HINELINE, M. L., 1993 *The visual culture of the earth sciences 1863–1970*. San Diego: University of California San Diego. Pp 272. (Doctoral dissertation: University Microfilms no. 94 20724).

HOFFSTETTER, R., 1959 Les rôles respectifs de Bru, Cuvier et Garriga dans les premières études concernant *Megatherium*. *Bulletin du Muséum National d'Histoire Naturelle* **31**: 536–545.

HOLMES, F. L., 1985 *Lavoisier and the chemistry of life: an exploration of scientific creativity*. Madison: University of Wisconsin Press. Pp 565.

IVINS, W.M., Jr, 1953. *Prints and visual communication*. London: Routledge and Kegan Paul.

LATOUR, B., 1990 Drawing things together, in LYNCH, M. and WOOLGAR, S. (editors), *Representation in scientific practice*. Cambridge, Massachusetts: MIT Press.

LATOUR, B. and WOOLGAR, S., 1979 *Laboratory life: the social construction of scientific facts*. Beverly Hills: Sage. Pp 271.

LOPEZ PIÑERO, J.M., 1988 Juan Bautista Bru (1740–1799) and the description of the genus *Megatherium*. *Journal of the history of biology* **21**: 147–163.

OPHIR, A. and SHAPIN, S., 1991 The place of knowledge: a methodological survey. *Science in context* **4**: 3–21.

OUTRAM, D., 1982 Storia naturale e politica nella correspondenza tra Georges Cuvier e Giovanni Fabbroni. *Ricerche storiche* **13**: 412–440.

OUTRAM, D., 1983 Cosmopolitan correspondence: a calendar of the letters of Georges Cuvier (1769–1832). *Archives: Journal of the British Records Association* **16**: 47–53.

OUTRAM, D., 1984 *Georges Cuvier: vocation, science and authority in post–revolutionary France*. Manchester: Manchester University Press. Pp 299.

PADIAN, K., 1987 The case of the bat-winged pterosaur: typological taxonomy and the influence of pictorial representation on scientific perception, in CZERKAS, S. J. and OLSON, E. C. (editors) *Dinosaurs past and present*. Seattle: University of Washingon Press.

RUDWICK, M.J.S., 1997a *Georges Cuvier, fossil bones, and geological catastrophes: new translations and interpretations of the primary texts*. Chicago and London: University of Chicago Press. Pp 301.

RUDWICK, M.J.S., 1997b *Recherches sur les ossements fossiles*: Georges Cuvier et la collecte des alliés internationaux, in BLANCKAERT, C., COHEN, C., CORSI, C. and FISCHER, J.-L. (editors), *Le Muséum au premier siècle de son histoire*. Paris: Éditions du Muséum national d'Histoire naturelle.

SHAPIN, S., 1984 Pump and circumstance: Robert Boyle's literary technology. *Social studies of science* **14**: 481–520.

SHAPIN, S., 1989 The invisible technician. *American scientist* **77**: 554–563.

SHAPIN, S., 1994 *A social history of truth: civility and science in seventeenth–century England*. Chicago: University of Chicago Press. Pp 483.

SHAPIN, S. and SCHAFFER, S., 1985. *Leviathan and the air–pump: Hobbes, Boyle, and the experimental life*. Princeton: Princeton University Press. Pp 440.

SLOAN, P., 1997 Le Muséum de Paris vient à Londres, in BLANCKAERT, C., COHEN, C., CORSI, C. and FISCHER, J.-L. (editors), *Le Muséum au premier siècle de son histoire*. Paris: Éditions du Muséum national d'Histoire naturelle.

X

Encounters with Adam, or at least the hyaenas: Nineteenth-century visual representations of the deep past

Adam is indeed dead in the flesh – at least until the final resurrection – but in visual representations he is alive and well and on view in every major museum of natural history.[1] Of course he is not named Adam; but *ha'adam*, The Man, the origin and fount of us all, is presented to our view under the form of *Homo habilis* or *H. erectus*, standing upright (more or less), knapping a flint or carving a carcase, or gazing out from a cave mouth to survey an artist's impression of the East African landscape of the early Pleistocene.

Often this transmuted Garden of Eden is but the last in a sequence of dioramas that depict various phases in the history of life, far back beyond The Man. For example, *Tyrannosaurus rex*, that perennial children's favourite, lords it over the lesser dinosaurs of the Wild West of the Cretaceous. Or our view dips underwater, like visitors to a dolphinarium, into the Jurassic seas of Dorset or Württemburg, to watch a school of ichthyosaurs darting after their squid-like prey. Far further back in time, a steamy swamp scene shows us the Everglades of the Carboniferous, with gigantic tree ferns in place of mangroves. And still further back we peer, like scuba divers in the Caribbean, at the coral reefs of Silurian Shropshire or Gotland, and see the trilobites and nautiloids going about their unspectacular business among the coral clumps. Reversing our stroll through the museum gallery, and following these scenes in their real geohistorical order, we are transported on a Wellsian time machine, or perhaps more appropriately like Dr Who, through an updated Genesis One. It is a creation narrative as powerful – and as mythopoeic – as its original:

> And Evolution said, "Let the waters bring forth swarms of living creatures." So Evolution created the great sea monsters and every living creature that moves. And Evolution said, "Let the

earth bring forth creeping things and beasts of the earth." And it was so. Then Evolution said, "Let us make humankind in our image." So Evolution created humankind in her own image.[2] Evolution has merely replaced *'elohim*. In the implicit message of the dioramas the plural actions of natural selection melt into an agency as overarching, and imaginatively almost as personal, as the regally plural God of the Genesis narrative. Such sequences of reconstructed scenes do not logically entail a theory of evolution; but, to borrow a term from the physical sciences, they certainly help to give that theory *Anschaulichkeit*, conceivability or picturability. They make the prehuman and barely-human past, its initial otherness and its gradual approach towards familiarity, conceivable and imaginable. They thereby create the imaginative groundwork for an evolutionary interpretation of the scenes to become, if not logically compelling, at least as plausible and persuasive as any scientific theory can be.

These sequences of dioramas in our modern museums, with their *trompe d'oeil* backdrops merging forwards into fully three-dimensional reconstructions of extinct animals and plants, have themselves descended from ancestors that were equally ambitious in conception although more modest in their means. Those ancestors are to be found not in museums but in books, as sequences of two-dimensional pictures of the prehistoric past.[3] The changing style and iconography of these visual representations reveals more than their designers intended, precisely because their form was (as it still is) so greatly underdetermined by the available evidence. It is customary to refer disparagingly to what is 'an artist's impression' of the scene. But it is the scientist who must decide how to extrapolate beyond the fragmentary evidence, to create a scene that will carry conviction; the artist merely translates that vision on to paper or into plaster and plastic, in an operation that is no less creative for being under the constraints of the scientist's instructions. The genre of the reconstructed scene demands explicitness; it forces the scientist to reveal judgements of likelihood, or mere hunches and prejudices, that can discreetly be left implicit in more formal media of publication. Above all, therefore, reconstructed scenes give us a unique insight into how successive generations of scientists (and their collaborating artists) have visualized the long aeons of 'deep time' that lie beyond human history or even the origins of our humanity. In this brief essay I want to explore the historical origins of this visual genre, which has become such a powerful means of rendering the remote past imaginable, and its evolutionary explanation persuasive.[4] I start at a point at which the genre can be seen to have been already well established, and then trace it backwards in search of its origins.

The World before the Deluge

One of the earliest full sequences of scenes from the time-machine, and certainly one of the most influential, was due to the prolific scientific popularizer Guillaume Louis Figuier (1819–94). Figuier was the author of profusely illustrated books on an amazing variety of topics: among others, *Great Discoveries of Modern Science* (1851), *Marvels of Modern Times* (1860), *The Earth and the Seas* (1864), *The Vegetable World* (1865), *Fish, Reptiles and Birds* (1868), *The Human Race* (1872), and so on and on to *Metropolitan Railways* (1886), not to mention *The Day after Death, or Our Future Life According to Science* (1871). Most of these books went through several editions in French; they were also translated into English, German and other languages, and thereby reached a literate audience that was almost global.

The *World before the Deluge* (1863), which contains Figuier's time-machine series, was no exception. Already in its sixth edition by 1872, it had by then appeared in English (published in both London and New York), German, Spanish (published in Mexico), Danish and probably other languages too. With such a spread of editions, it was immensely influential in conveying to a wide reading public – not least the young people to whom it was explicitly directed – a sense of the spectacular implications of the burgeoning sciences of geology and palaeontology.[5] Figuier borrowed most of his illustrations – some 300 engravings of fossils – from a respectable academic source, namely the two-volume *Elementary Course on Palaeontology and Stratigraphical Geology* (1849–52) by Alcide d'Orbigny (1802–57), who until his death had been professor of palaeontology at the Natural History Museum in Paris.[6] But Figuier added new illustrations of his own, to bring d'Orbigny's fossils to life: a magnificent sequence of twenty-five (in later editions, thirty) full-page 'ideal views' of the past history of the earth and its living inhabitants. These were drawn under his direction by the young painter and illustrator Edouard Riou (1833–1900), who, at the very same period, was also taking his imaginative time-machine and his artistic style in the other direction, in his illustrations for the science fiction of Jules Verne.

Figuier's great temporal panorama began even before the beginnings of life, with a view of the original condensation of water on to the surface of the hot primitive earth, a scene of torrential rain, lightning and boiling seas. It continued with a view of the Sun breaking through clouds over a Silurian seascape, with trilobites and nautiloid shells thrown up on the shore. After a similar Devonian scene came a view of Carboniferous marine life that at first glance could have been of a crowded Victorian aquarium, together with a view of a steamy

Carboniferous coal swamp in a style that has remained canonical for such reconstructions to the present day. And so the panorama continued, still in a style that is unmistakably continuous with that of our modern museums, through Triassic and Cretaceous, Eocene and Pliocene. The last 'ideal view' in that style was a chilly scene of Europe during the Quaternary Ice Age, with mammoths and cave-bears, woolly rhinoceros and the so-called Irish elk.

After that scene, however, came one entitled 'The Deluge of the North of Europe'. This, and the title of the book itself, are a reminder that the 'Deluge', although long abandoned by geologists as a global event or major agent of change, remained vivid in the imagination of the reading public even in the later nineteenth century. But it was Figuier's business as a popularizer to keep abreast of the latest science. So what he depicted in visual form (and described in his text) was an interpretation of the Deluge that was quite widely current among scientists in the 1860s and even the 1870s. A mass of swirling waters was shown submerging a sub-Arctic landscape of conifers, with huge icebergs being swept along, dropping their erratic boulders – unseen – as the ice melted, scratching the bedrock with their embedded stones and churning up the surface of the ground. A localized 'Deluge of the North of Europe' could thus explain the problematical Diluvium or Drift deposits of that region; and it did so with impeccable naturalism, since such a Deluge seemed a plausible physical consequence of the Ice Age, if it had indeed ended abruptly.

That Deluge, however, was clearly *not* the one that was recalled in Genesis and other ancient literatures. On this point Figuier adopted a standard compromise of the period. He affirmed the historicity of the Deluge of Genesis, but restricted its effects to Mesopotamia; he attributed it naturalistically to a sudden elevation of the land to the north, accompanied by a catastrophic volcanic eruption that had produced Mount Ararat. Riou illustrated this 'Asiatic Deluge' with an apocalyptic scene in the style of John Martin's celebrated painting of *The Deluge*, which had recently been exhibited in Paris: he showed an imaginary prehistoric city (and a couple of elephants) being overwhelmed by swirling waters and torrential rain.

Between these two distinct and contrasted Deluges, the origins of humankind had somehow to be accommodated. In his later editions, Figuier claimed that the discovery of a human jaw in the gravels near Abbeville in 1863 had at last precipitated a consensus among the scientists, that the human beings who had crafted the well-known prehistoric stone axes must have lived among the equally well-known extinct mammals of the Pleistocene.[7] So he got Riou to draw a new 'ideal view' that would embody that latest discovery.

'The Appearance of Man': an engraving by Edouard Riou from a later edition of Louis Figuier's *World before the Deluge* (1866), in which the co-existence of early human beings and the mammals of the Pleistocene was first depicted.

'The Appearance of Man' depicted a tribal group dressed in the skins of animals. The women were grouped domestically in a cave mouth (and were as topless as on a modern French beach); the men were equipped with stout flint-headed axes, and stood ready to ward off a hostile nature and hunt it down for food and clothing. For with the incongruous exception of a fine thoroughbred horse and a solitary deer, nature was not only wild but hostile: bears, hyaenas, rhinoceros and mammoths threatened the very survival of humanity. A deep ditch separated the tribal group from this hostile environment; more symbolically, it was a chasm that divided the human world from the natural. For Figuier's human beings, although primitive in time, and simple in tools, clothing and shelter, were no primitives in any other sense: they were unmistakably white and European, and wholly modern in physical appearance.[8]

When Figuier's book was first published, however, this co-existence of early human beings with the mammals of the Pleistocene was still highly controversial. In the early editions, his interpretation of the origin of humankind was even more traditional, as its visual representation clearly showed. The original version of the engraving just described was strikingly different in character: it depicted not a tribal group outside a cave, but a primal human family in a verdant temperate landscape.[9] Man, woman and child were surrounded by the sheep, goats, cattle and horses that they would soon domesticate, with only a distant deer to suggest the wilder nature that might need hunting rather than herding. To match this pastoral economy, the man was equipped not with a flint axe, but with no more than a staff. The scene could almost have come from the hand of a Poussin or a Claude; in style and tone, this Garden of Eden was in striking contrast to the long sequence of strange scenes from the deeper past, which so vividly illuminated the body of the book. The Deluge of Figuier's title stood here as a symbolic barrier between the immense panorama of deep time and its brief human epilogue, between 'the ancient world' of his subtitle and the civilized human world described in so many of his other books. The Arcadian scene of his original 'Appearance of Man' portrayed the not-so-distant origins of that world of humanity; the whole of the preceding history of the earth, despite its manifest diversity as portrayed in all the other 'ideal views', could in this perspective be lumped together as *the* 'ancient world'.

This contrast between Figuier's earlier and later editions neatly captures the historical moment at which the beginning of humanity came to be depicted not only as the final episode of a long sequence that stretched back beyond the beginning of life, but also as an episode not wholly different in character from the deeper past. Figuier may not have been persuaded by the evolutionary theories that a certain Englishman had just brought back to the forefront of public debate; but whether he knew it or not, and whether he willed it or not, his great series of 'landscapes of the ancient world' must surely have helped to make such theories literally *anschaulich*, picturable, and therefore potentially plausible to his vast and almost worldwide audience.

Landscape representations

Needless to say, however, Figuier was not the author of the genre. Fortunately for our phylogenetic task, he himself named the direct ancestor of his visual representations: it was *The Primitive World in Its Different Periods of Formation* (1847), by the Austrian palaeobotanist Franz Unger (1800–70).[10] By the time Figuier borrowed from his work, Unger had become professor of botany in Vienna, but when he first published his series of 'landscape representations' he held a similar but provincial post in Graz. Just as Figuier's 'ideal views' were published in a popular book, so likewise Unger's work had its origin in a lecture course to a group of amateurs in Graz. As a professional scientist, Unger claimed he had had serious scruples about 'crossing from strictly scientific research into fantasy'; but the work of his illustrator, the topographical artist Josef Kuwasseg (1799–1859) of Graz, convinced him that the exercise need be no more speculative than any other portrayal of the unobservable past. The landscapes that formed the *raison d'être* of the book thus embodied – and, for the first time, explicitly – the partnership between artist and scientist that has remained characteristic of the genre to the present day. Unger ensured that their work would become widely known in the scientific world, by publishing it with the text in both German and French; and the list of subscribers, who had enough confidence in Unger's abilities to risk their money beforehand, included naturalists from most of the European countries and even one (Asa Gray) from North America.

Unger, more cautiously than Figuier, began his series of fourteen scenes not with the pre-biotic world – for which there was still no unquestioned evidence in the stratigraphical record – but with a scene from the Transition period, using the older name for what had only recently been differentiated into Silurian and Devonian.[11] The series continued in a sequence that Figuier was to borrow with little substantive change, through the Carboniferous and the New Red Sandstone periods, the Oolite and the Chalk, and so into the Tertiary with scenes of Eocene and Miocene life. But Kuwasseg's lithographs were far superior to Riou's engravings, not only artistically but also in how they portrayed the fauna and flora of each period. Working under Unger's close supervision, Kuwasseg produced scenes that Unger could proudly claim were more authentic than any earlier 'landscape representations' had been. Rather than crowding into one scene a set of reconstructions of all the characteristic fossils found in a given group of strata, Unger got his artist to portray an ecologically plausible view of the *community* of organisms that had lived at each period in the past. Being a botanist, it is not surprising that his scenes gave the plant life more prominence than the animals; but the animals were shown in their likely ecological relations, as predators and prey, carnivores and herbivores, and so on.

Unger rounded off his series with two scenes that brought the immensely long history of life into the human epoch. Unlike Figuier's series, there was no portrayal of any Deluge, even a prehuman and periglacial one. The penultimate scene was of the glacial period on the edge of the Alps – it might have been a scene near Graz itself. In the foreground were cave bears, one with its paw on a pile of huge mammoth bones; in the middle distance, a herd of bison stood peacefully at the edge of a lake not far from the snout of a glacier. That scene, like Riou's later adaptation of it, was of a world still without human beings. By contrast, Unger's final scene depicted the culmi-

The Ice Age in Europe: a lithograph by Joseph Kuwasseg, forming the penultimate 'landscape representation' in Franz Unger's *The Primitive World in Its Different Periods of Formation* (1847). (Reproduced by permission of the Syndics of Cambridge University Library.)

The origin of humankind, as depicted by Kuwasseg for the final scene in Unger's *Primitive World*. (Reproduced by permission of the Syndics of Cambridge University Library.)

nation of the history of life in a style similar to Riou's *original* drawing for Figuier. As the sun rose on the dawn of humankind, a family group in primal nakedness – the man with no more than a staff in his hand – surveyed a subtropical Eden in which only some playful horses were visible to represent the untamed animal world, and the most prominent natural object was a large palm tree – as it were, a secular Tree of Life.

Unger claimed that his 'landscape representations' were superior to their forerunners; in fact, such forerunners are hard to find. As we push our search for the origins of the genre back into the earlier nineteenth century, the evidence becomes fragmentary and problematic. There was certainly a tradition of such visual representations before Unger's were published, but his book may well represent the first time such scenes were published as a *sequence* stretching all the way from the beginnings of life to the beginnings of humanity. The significance of that step, for creating the imaginative infrastructure for a developmental – and later, an evolutionary – view of the living world, needs no emphasis.[12]

Still, even the reconstruction of a scene from a *single* period in the prehuman past represents an imaginative achievement, the magnitude of which is only masked from us by its familiarity. One of Unger's likely models decorated the title-page of a short monograph (1836) on the then recently discovered 'colossal skull' of the *Dinotherium*, an extinct elephant from the Tertiary of Hesse. The authors were two Hessian geologists: August Wilhelm von Klipstein (1801– 94), who had recently been appointed to the chair of mineralogy at Giessen (thereby becoming a colleague of the more famous professor of chemistry, Justus von Liebig), and Johann Jakob Kaup (1803–73), a curator at the grand-ducal museum at Darmstadt. Their monograph consisted of a short descriptive text, with seven plates of maps, geological sections and drawings of the fossils themselves. Outside this conventional series of illustrations, however, were two others: a sketch of the workmen, supervised by a geologist, excavating the gigantic fossils; and a single reconstructed landscape, showing the living *Dinotherium* in its inferred original conditions of life, browsing peacefully by the edge of a river, unconcerned at a lion's attack on a herd of horses in the background or by the volcano in full eruption in the far distance. This pair of vignettes neatly symbolized the act of reconstruction: the 'before' of careful excavation and the 'after' of imaginative reconstruction; or, conversely, the 'before' of the reality of the deep past, and the 'after' of its fragmentary survival into the present. That Klipstein and Kaup included a reconstructed scene in their otherwise conventional monograph is a sign of how they hoped

A lithograph of the *Dinotherium* in its natural habitat, drawn to decorate the title page of the monograph on its fossil remains by August von Klipstein and Johann Kaup (1836). (Reproduced by permission of the Syndics of Cambridge University Library.)

to transcend the goals of merely descriptive palaeontology; but that the scene was printed as a decoration to the title-page, and not on one of the full-page lithographed plates, is equally a sign of the tentative and experimental character of that act of reconstruction.[13]

Caricatures of the deep past

This scene by Klipstein and Kaup is, at least at present, the earliest known reconstruction of the prehuman past to be published, albeit marginally, as part of a standard scientific report.[14] But at least one other example of the genre must have been circulating informally in German scientific circles a few years earlier. In a letter written in 1831, the Oxford geologist William Buckland (1784–1856) told his fellow geologist Henry De la Beche (1796–1855) that 'a German parody of your *Duria Antiquior*' had just reached England.[15] The reference was to a recent lithograph in which De la Beche – an accomplished artist as well as a fine geologist – had portrayed a scene of 'a more ancient Dorset', namely the Dorset of the period when the famous ichthyosaurs, plesiosaurs and other Liassic fossils of Lyme Regis had been alive. De la Beche had reconstructed a scene which, although rather implausibly crowded with all the more spectacular denizens of Liassic Dorset, vividly brought those animals to life.[16] Almost every animal was shown eating, or being eaten by, another. Such habits had been inferred in the preceding years from a wealth of sober scientific evidence, notably by Buckland himself. The slightly vulgar showmanship of Buckland's lectures, in which the professor himself frequently acted the part of the extinct animals he was describing, had delighted most members of his audiences as much as it had been frowned upon by a minority of his more staid colleagues. Above all, however, Buckland's highly ecological approach to the study of fossils had made the deep past *anschaulich* as never before. De la Beche's half-humorous, half-serious portrayal of *Duria Antiquior* merely translated that vision on to lithographic stone and thence on to paper. In that form it circulated widely among the circle of gentlemanly geologists in England. At least in Buckland's opinion, a copy must then have been sent to Germany, to become the inspiration for a similar scene (as yet unlocated), perhaps based on the equally fine Liassic fossils of Württemburg. It may have been that German version that became the inspiration for Klipstein and Kaup's single reconstruction of a quite different period in the history of life, and hence, indirectly, for Unger's complete sequence of 'landscape representations'.

However that may be, Buckland's letter to De la Beche records more than that possible link from England to Germany. Like the happy accident of a Solnhofen slate, it preserves in unpublished form the earliest known 'fossil record' of the *idea* of constructing a whole sequence of scenes from the deep past.[17]

Buckland implored De la Beche not to stop with *Duria Antiquior*, but 'to put on the stocks in your best style 2 or 3 more restorations of scenes in the ancient world'. In fact he proposed three, which together with the existing one would have made a sequence of four scenes. His first suggestion ran as follows:

  I. The Period immediately preceding the formation of Diluvium

'A more ancient Dorset' (*Duria Antiquior*), the half-humorous lithograph drawn by Henry De la Beche in about 1830 to illustrate the life habits of the ichthyosaurs, pleisiosaurs and other animals found as fossils in the Liassic strata of Dorset. (Reproduced by permission of the Department of Geology, National Museum of Wales.)

> – a Land Piece – with only rivers plains & mountains – as in Pale-strina Pavement – exhibiting the gamboled Battles of Elephants & Rhinoceros & Mastodons – Hippopotami jumping into the Rivers – Megatherium sitting on his Haunches with one fore Paw against the trunk of a Tree and the other reaching down an enormous Branch – Horse Ox and Elk scampering before a Pack of Wolves and falling headlong into fissures – Hyaenas in their Den or dragging into its Mouth their Prey – Tigers crouching to spring on Deer.[18]

This lively verbal representation of the fauna of the late Tertiary – with the animals of the Old and New Worlds mixed up somewhat implausibly but to good dramatic effect – was followed by a scene from the early Tertiary, based on the celebrated Parisian research of Georges Cuvier (1769–1832) earlier in the century:

> II. A lake scene from the F[resh] water Period. Ponds full of Palaeotherium Anoplotherium and Chaeropotamus and all the Paris Pigs of those Days. Dogs and Sarigues at Montmartre. Birds and Reptiles, snakes and water Rats in Auvergne Tortoises Beavers Crocodiles – Volcanoes in the Distance.[19]

With *Duria Antiquior* already available to represent the much earlier

period of the Lias, Buckland then moved for his final scene still further
back in time:

> III. A sea scene – Sea with tropical Islands of Carboniferous and
> Transition Periods. Land very short of animals, but glorious
> hypertropical Vegetation – Sternberg's Lepidodendron, stems
> of gigantic Cactus. Sea full of Tropical islands – covered with
> Coal Plants – under water, encrinites, Corals, chain Corals –
> Orthoceratites and Nautili at Surface – Spirifer, Producta – Tri-
> lobites – and a few fish. The Trilobites wd caricature well.[20]

Buckland's sequence of verbal sketches was scribbled down with
scant regard for formal punctuation, and in the characteristic hand-
writing that led his friends to claim that he employed a trained spider
as amanuensis. But it expressed perfectly his vivid ecological imagination, and the lively reality with which in his mind's eye he transported
himself back into the remote prehuman past. His verbal sketches
could almost have been a rough draft of Unger's instructions to *his*
artistic collaborator, but for one striking difference. Unger arranged
his scenes in their sequence in real time, from ancient to recent; by
contrast, back near the origins of the genre, Buckland's were listed
in retrospective order. Like most geologists who summarized stratigraphical knowledge in the 1820s and 1830s – including Charles Lyell
(1797–1875), in the third volume (1833) of his *Principles of Geology*[21] –
Buckland considered it natural to probe backwards, from the relatively
well-known recent past towards the more obscure periods of the
deepest past. Nevertheless, despite that important difference in presentation, Buckland's letter to De la Beche marks the potential start
of the genre of a *sequence* of reconstructed scenes. De la Beche seemed
not to have acted on the proposal, and the idea may therefore have
remained for the time being in the private realm. Knowingly or not,
however, it was eventually to be realized in Unger's great series of
'landscape representations.'

Spy-holes into the past

How then did Buckland and his circle come to conceive of the idea of
depicting even a *single* scene from the deep past? A clue is provided
by the apparently casual comment at the end of Buckland's verbal
proposal; 'the Trilobites would caricature well'. The scenes he hoped
De la Beche would draw would be not only serious attempts to depict
the past, but also, at the same time, light-hearted *caricatures* of the
results of scientific research. De la Beche was the obvious man to approach, as he was the best-known geological caricaturist. In addition
to *Duria Antiquior* he had, for example, just lithographed a caricature

showing a 'Professor Ichthyosaurus' lecturing on a human skull: this was an imagined view into the post-human *future*, which ridiculed the implausible cyclicity in the history of life that Lyell had postulated in the first volume (1830) of his *Principles*.[22]

Caricatures such as these were privately – but widely – circulated among the gentlemanly geologists in the research triangle of London, Oxford and Cambridge. The earliest and most significant of them, however, dates from a decade before either *Duria Antiquior* or Professor Ichthyosaurus. It was drawn by Buckland's close friend and former Oxford colleague William Conybeare (1787–1857), and it reflects the immense impact of Buckland's celebrated analysis of the bone-bearing cave at Kirkdale in Yorkshire. This research won him the Royal Society's Copley Medal, and formed the centrepiece of his *Reliquiae Diluvianae* (1823). That title was a striking misnomer, though doubtless it helped the sales of the book.[23] The organic relics that Buckland described and analysed were not themselves relics of the Deluge at all, though he did argue that a transient diluvial event had sealed them into the cave and thus preserved them. They were relics of the period immediately *before* the Deluge, a period of tranquility and normality, or at least a period marked by ecological relations of predator and prey, scavengers and scavenged, closely analogous to those of wilder regions in the modern world.

The cave at Kirkdale, although far smaller and less spectacular than other bone-bearing caves already known on the continent, gave Buckland the evidence with which to reconstruct a den of prediluvial hyaenas, whose scavenging had concentrated and preserved a sample of an entire fauna. It gave him an opportunity to exercise his scientific imagination in a reconstruction not just of the bodies of single animal species, as Cuvier had done more authoritatively before him, but of an entire fauna interlocked in ecological relations with one another.[24] His analysis was so persuasive and his verbal reconstruction so vivid that it seemed to his contemporaries to disclose the reality of the prehuman past as never before.

This was the achievement that Conybeare translated into visual form. His caricature was not a sober reconstruction of the kind that came to fruition in Unger's work, but it was no less decisive historically. It depicted Buckland himself crawling into Kirkdale cave, candle in hand, and there encountering the hyaenas of the prediluvial period, very much alive among their scavenged bones. The geologist became in caricature a *participant* in the scene he had soberly reconstructed in words. The entrance of the cave became symbolically a passage through the epistemic barrier that separated the observable present from the prehuman past; the candle became symbolically the illumi-

'The Hyaena's Den at Kirkdale', the lithographed caricature drawn by William Conybeare to celebrate William Buckland's analysis (1822) of the bone-bearing cave in Yorkshire.

nation that the geologist could bring to that otherwise inaccessible past.

That point was made by Conybeare himself, at the end of some sixty lines of dreadful doggerel appended to the caricature:

Mystic Cavern, thy chasms sublime,
All the chasms of History supply;
What was done ere the birth-day of time,
Thro' one other such hole I could spy.[25]

The aspiration was of course exaggerated for poetic effect. Conybeare, who had just completed his great *Outlines of the Geology of England and Wales* (1822), knew perfectly well that the 'birth-day of time' was unimaginably remote, and that far more than *one* 'other such hole' into the deep past would be needed in order to understand the history that had filled it.[26] But the principle had been established. Buckland's cave research demonstrated, more vividly than ever before, the feasibility of penetrating the epistemic and imaginative barrier between present and past, and of rendering the past *anschaulich*. If Buckland could encounter the hyaenas through a spy-hole at Kirkdale, he and others could hope by careful research to construct other windows on to the past, and so to encounter the palaeotheria, the ichthyosaurs, the lepidodendrons, and so on back to the trilobites.

Back to the Ark

Conybeare's caricature of Buckland with the hyaenas seems to represent the very origin of the genre of reconstructed scenes from the prehuman past. Significantly, it is preceded only by representations that belong to a quite different tradition. A late example, dating from less than twenty years before Conybeare's caricature, formed the frontispiece to the first volume (1804) of the first substantial illustrated book on fossils to be published in England, the *Organic Remains of a Former World* by James Parkinson (1755–1824).[27] It depicted a rocky shoreline and a view out to sea, with the sun breaking through stormclouds. On the shore were the drifted shells of some of the fossils illustrated in the lavish hand-coloured engravings that embellished the book. But although the shells included ammonites, this was no view of ancient Oolitic or Liassic times. For the sky was marked by a rainbow; and at the rainbow's end, stranded on a rocky islet, was a distant but unmistakable Ark. Parkinson regarded *all* his fossils as the remains of a *single* former world; more precisely, it was 'the antediluvian world' that had been brought to an end by an event that was at least congruent with the Genesis narrative. Such a belief had long been abandoned by those who studied fossils on the continent – as Parkinson realized in time for his later volumes – but his original frontispiece can stand as a late example of an iconographical tradition that stretches back at least as far as the encyclopaedic *Noah's Ark* (1675) by the Jesuit polymath Athanasius Kircher (1601–80).[28]

The collective memory of that tradition may perhaps explain the hesitancy with which the newer genre came to be established. The scientific geologists may well have been concerned to distance their project from the earlier tradition, for theirs was self-consciously based on reasoning from natural evidence alone. Certainly, it is striking how 'marginal' the scientific genre remained in its early years. Born with an explicit caricature, lithographed for the entertainment of a close-knit circle of scientific friends, it only slowly shed its gentle humour and its semi-private circulation. The first fully public and fully realistic examples were still printed in a 'marginal' manner, separately from more conventional scientific illustrations. Even the first full series of scenes was presented somewhat apologetically, with the excuse that it had originally been devised for the benefit of

X

amateurs; and the most influential of later series was explicitly aimed at a juvenile readership.

Yet it may be no coincidence that this scientific genre of reconstructed scenes first arose within precisely the circle of geologists who were most concerned to maintain at least the *religious* authority of the biblical records, while conceding or even welcoming the interpretive insights of the new biblical criticism.[29] It is not fanciful to conjecture that the genuine piety of men like Conybeare enabled them to *imagine* the unitary biblical narrative with a greater sense of concrete reality than their less religious counterparts; and that that distinctive vision of the human past facilitated their first attempts to project a similar concrete reality into the deep time revealed by their geology.

On to Darwin

'He who can read Sir Charles Lyell's grand work on the Principles of Geology', wrote Charles Darwin in the *Origin of Species*, 'yet does not admit how incomprehensibly vast have been the past periods of time, may at once close this volume.'[30] Darwin did indeed need Lyell's extremely long time-scale – significantly longer, incidentally, than modern estimates based on radiometric methods – in order to validate his conception of extremely slow evolutionary change by means of natural selection.[31] But the fame of the intellectual lineage that runs through Lyell to Darwin should not blind us to its limitations. The evolutionary view of the natural world, which in its organic aspect we have come to associate so crucially with Darwin, needed far more than the mechanism of natural selection to lend it plausibility. It needed more than a Lyellian vision of vast *time*, within which natural selection could operate effectively. It needed equally, or perhaps even more, a concrete vision of an unimaginably lengthy prehuman *history*.

It is one of the ironies of the 'Darwinian Revolution', to the understanding of which John Greene has contributed with such distinction, that this sense of prehuman history was fed into the stream of nineteenth-century thought, not by the 'apostolic succession' of heroes that led through Lyell to Darwin, but by those who in an earlier historiography were cast as the villains, the opponents of 'progressive' evolutionary thinking. The reason, however, is not hard to find. Lyell and Darwin were ultimately concerned to discover the generalities of the natural causal *laws*, both inorganic and organic, by which the world maintained itself as a natural system. Others, like Conybeare and Buckland, who were formerly reviled as catastrophists or dismissed as diehard opponents of evolution, were ultimately more concerned to reconstruct the particularities of the *history* that had brought the world to its present state. In the jargon of historiographical theory, Lyell and Darwin set themselves objectives that were above all *nomothetic*; their opponents, by contrast, set themselves primarily *idiographic* objectives.

The two sets of goals were not, and are not, mutually exclusive; on the contrary, they are blended together in the modern evolutionary understanding of the earth and its biosphere. In our enthusiasm for Darwin's own role in his 'revolution', therefore, we should not overlook the historical importance of those who first made *some* kind of evolutionary theory seem a plausible possibility, whether intentionally or not, by displaying the vast scale and rich diversity of the *history* that needed causal explanation. In particular, we should recognize the achievements of those who made the long history of the earth and of its fauna and flora visually imaginable, *anschaulich*: first and tentatively to their fellow geologists, then more boldly to amateurs of the science, and ultimately to an even wider public – the public that read *The World before the Deluge* a century ago and that now throngs our museums of natural history.

Notes

This essay is based upon work supported by the National Science Foundation under grant no. SES-88-96206.

1. The main allusion is to John C. Greene, *The Death of Adam: Evolution and Its Impact on Western Thought* (Ames: Iowa State University Press, 1959), the first and perhaps most influential book in the *oeuvre* honoured by this volume. The other side of the allusion is a reminder – which may be needed by some historians of science, though not by John Greene – that the figure of Adam remains 'alive' in religious practice, because it is so rich in symbolic and therefore practical meaning; from this perspective, it is simply irrelevant that it is indeed 'dead' in modern natural-scientific practice.

2. The parody is adapted from Genesis 1:20–7 (RSV).

3. Visual reconstructions in books remain, of course, as important as those in museums. For some recent, artistically attractive, and scientifically authoritative examples, see Anthony J. Sutcliffe, *On the Tracks of Ice Age Mammals* (Cambridge, Mass.: Harvard University Press, 1985), especially the Welsh hyaena den on p. 119.

4. Constraints beyond my control limit severely the number of 'visual quotations' that can be included in this essay, although images, and not words, are its subject matter. The early history of other, more technical, visual genres in geology is explored in Martin J.S. Rudwick, 'The Emergence of a Visual Language for Geological Science, 1760–1840', *History of Science*, 14 (1976), 149–95. The felicitous phrase 'deep time' is borrowed from John McPhee,

Basin and Range (New York: Farrar, Strauss, Giroux, 1981).

5. Louis Figuier, *La terre avant le Déluge: ouvrage contenant 25 vues idéales de paysages de l'ancien monde* (Paris: Hachette, 1863 and later edns); *The World before the Deluge* (London; 1865 and later edns; also New York: D. Appleton, 1869). Other translations are listed in standard major library catalogues. Figuier's *La terre* was the first volume in his series entitled 'Tableau de la nature: ouvrage illustré à l'usage de la jeunesse'.

6. Alcide d'Orbigny, *Cours élémentaire de paléontologie et de géologie stratigraphique*, 2 vols. (Paris: Victor Masson, 1849–52).

7. *La terre*, 6th edn, p. 419n. Ironically, the authenticity of the Moulin-Quignon mandible was soon rejected by the experts; but a consensus on the main point, based on a much wider range of evidence, did indeed congeal at just this time: see Donald K. Grayson, *The Establishment of Human Antiquity* (London: Academic Press, 1983), ch. 9. Charles Lyell's book on *The Geological Evidences of the Antiquity of Man* (London: John Murray, 1863) was but one popularization of the new view.

8. *La terre*, 6th edn, fig. 322. The strikingly subtropical character of the vegetation reflects Figuier's claim that 'the appearance of Man' had taken place in its traditional location somewhere in Asia, *not* in the evidently harsher climate of Pleistocene Europe.

9. *La terre*, 1st edn, 1863, fig. 310. In some later editions this engraving was retained, incongruously, in addition to the one that was designed to supersede it; in the New York edition (1869) it was even elevated in position to form the frontispiece of the book!

10. F. Unger, *Die Urwelt in ihren verschiedenen Bildungsperioden: 14 landschaftliche Darstellungen mit erläuterndem Texte* (Vienna, 1847). I am preparing a modern edition of this important work.

11. In the second edition (Vienna, 1858), Unger added two new views, of the Silurian and Devonian periods, while adding the qualifier 'moderne' to his original view of the 'Epoque de Transition'.

12. Unger himself did believe in some kind of evolutionary explanation for the changing character of the organic world; but his visual representations did not require, let alone compel, such a conclusion.

13. A. von Klipstein and J.J. Kaup, *Beschreibung und Abbildung von dem in Rheinhessen aufgefundenen colossalen Schadel der Dinotherii gigantei, mit geognostischen Mittheilungen über die Knochenführenden Bildungen des mittelrheinischen Tertiärbeckens* (Darmstadt, 1836). The vignette was lithographed by R. Hoffmann and L. Becker, whom I have been unable to identify further.

14. Other reconstructions from the same period are all found in popular books, or in works that in some other way are not 'standard' scientific reports; e.g. John Martin's highly imaginative (and inaccurate) apocalyptic frontispieces for Gideon Mantell's *The Wonders of Geology; or, A Familiar Exposition of Geological Phenomena* (London: Rolfe and Fletcher, 1838), and for *The Book of the Great Sea-Dragons, Ichthyosauri and Plesiosauri, Gedolim Tanimim of Moses, Extinct Monsters of the Ancient Earth* (London: William Pickering, 1840), by the eccentric fossil collector Thomas Hawkins.

15. W. Buckland to H.T. De la Beche, 14 Oct. 1831, De la Beche papers,

National Museum of Wales, Cardiff. I am indebted to the Keeper of the Museum for permission to quote from this letter.

16. *Duria Antiquior* is analysed in Paul J. McCartney, *Henry De la Beche: Observations on an Observer* (Cardiff: Friends of the National Museum of Wales, 1977), pp. 44–7. See also James A. Secord, 'The Geological Survey of Great Britain as a Research School, 1839–1855', *History of Science*, 24 (1986), 223–75 (esp. 241–7). A fine, water-colour version, perhaps the drawing on which the lithograph was based, is in the National Museum of Wales, and is reproduced in colour on the cover of S.R. Howe, T. Sharpe and H.S. Torrens, *Ichthyosaurs: A History of Fossil 'Sea-Dragons'* (Cardiff: National Museum of Wales, 1981).

17. The Solnhofen 'slate' (in fact a fine-grained limestone), which was later to become famous for its *Archaeopteryx*, was already yielding exceptionally fine specimens of a wide variety of Jurassic organisms that would not have been preserved under more normal circumstances. Their discovery was due to the exploitation of the stone for use in lithography, a technique that came into general use only in the 1820s.

18. See n.15 above. The 'Palestrina pavement' was the Hellenistic 'Barberini mosaic' in the Italian town (the birthplace of the composer), which depicts the landscape of the Nile with its animal (and human) inhabitants.

19. The latest collection of Cuvier's work was in his *Recherches sur les ossemens fossiles, où l'on rétablit les caractères de plusieurs animaux dont les révolutions du globe ont détruit les espèces*, 3rd edn, 5 vols. (Paris and Amsterdam: G. Dufour and E. Ocagne, 1825). A *sarique* is an opossum. Buckland noted that De la Beche's own recently published *Geological Manual* (London: Treuttel and Würtz, Treuttel Jun. and Richter, 1831) would provide him with the data for sketching 'the appropriate vegetation' for that scene.

20. The reference in this last passage was to Kaspar, Graf von Sternberg, *Versuch einer geognostisch-botanische Darstellung der Flora der Vorwelt* (Leipzig, Prague and Regensburg, 1820–38).

21. Charles Lyell, *Principles of Geology, Being an Attempt to explain the Former Changes of the Earth's Surface, by reference to Causes now in Operation*, 3 vols. (London: John Murray, 1830–3).

22. This caricature, and the sketches that led to it, are analysed in Martin J.S. Rudwick, 'Caricature as a Source for the History of Science: De la Beche's Anti-Lyellian Sketches of 1831', *Isis*, 66 (1975), 534–60. See also McCartney, *Henry De la Beche*, pp. 50–3. Another of his caricatures, which circulated widely at the time although it was never even lithographed, is his telling visual comment (1834) on the nascent Devonian controversy: see Martin J.S. Rudwick, *The Great Devonian Controversy: The Shaping of Scientific Knowledge among Gentlemanly Specialists* (Chicago: University of Chicago Press, 1985), p. 104; also McCartney, *Henry De la Beche*, p. 31.

23. William Buckland, *Reliquiae Diluvianae; or, Observations on the Organic Remains contained in Caves, Fissures, and Diluvial Gravel, and on Other Geological Phenomena, attesting the Action of an Universal Deluge* (London: John Murray, 1823). That the title was a misnomer is noted by Nicolaas A. Rupke, *The Great Chain of History: William Buckland and the English School of Geology (1814–1849)*

(Oxford: Clarendon Press, 1983), p. 39. Buckland's book was enlarged from a long paper read to the Royal Society in 1822 and published in its *Philosophical Transactions*, vol. for 1822, pp. 171–236.

24. Cuvier seems never to have carried his reconstructions further than mere outlines of the bodies of extinct mammals: see his lively MS. sketch of a *Palaeotherium* reproduced in William Coleman, *Georges Cuvier, Zoologist: A Study in the History of Evolution Theory* (Cambridge, Mass.: Harvard University Press, 1964), p. 122; also the drawings published in Cuvier, *Recherches*, 3rd edn, III, pl. 66.

25. [W.D. Conybeare], 'The Hyaena's Den at Kirkdale, near Kirby Moorside in Yorkshire, discovered A.D. 1821', lithographed broadsheet, anonymous and undated. The poem is reprinted in C.G.B. Daubeny, *Fugitive Poems connected with Natural History and Physical Science* (Oxford: James Parker, 1869), pp. 92–4, where it is attributed to Conybeare and dated 1822.

26. Conybeare revised and greatly enlarged an earlier book by his nominal co-author (and publisher) into what became internationally a standard reference work on stratigraphical geology: William Daniel Conybeare and William Phillips, *Outlines of the Geology of England and Wales, with an Introductory Compendium of the General Principles of that Science, and Comparative Views of the Structure of Foreign Countries. Part I* [all issued] (London: William Phillips, 1822).

27. James Parkinson, *Organic Remains of a Former World: An Examination of the Mineralized Remains of the Vegetables and Animals of the Antediluvian World; Generally termed Extraneous Fossils*, 3 vols. (London: Sherwood, Neely and Jones, 1804–11). See J.C. Thackray, 'James Parkinson's *Organic Remains of a Former World* (1804–1811),' *Journal of the Society for the Bibliography of Natural History*, 7 (1976), 451–66.

28. Athanasius Kircher, *Arca Noë in Tres Libros Digesta, sive De Rebus Ante Diluvium, De Diluvio, et De Rebus Post Diluvium a Noemo Gestis* (Amsterdam: Johann Jansson, 1675).

29. See, for example, W.D. Conybeare, *An Elementary Course of Lectures on the Criticism, Interpretation, and Leading Doctrines of the Bible, delivered at Bristol College in the Years 1832, 1833* (London: John Murray, 1834).

30. Charles Darwin, *On the Origin of Species by Means of Natural Selection, or the Preservation of Favoured Races in the Struggle for Life* (London: John Murray, 1859), p. 282. The comment was somewhat overdone, because the reader would already have got through more than half the book before encountering it!

31. See Joe D. Burchfield, 'Darwin and the Dilemma of Geological Time', *Isis*, 65 (1974), 300–21.

XI

A year in the life of Adam Sedgwick and company, geologists[1]

INTRODUCTION

"Give you joy, Sir, give you joy! A fine boy, Sir, as like you, Sir, as one pea is to another". So, in the early hours of 22 March 1785, the midwife announced to the vicar of Dent the arrival of his son Adam Sedgwick. Almost eighty-eight years later, on 27 January 1873, that same life came peacefully to an end in rooms overlooking the Great Court of Trinity College, Cambridge.[2]

The custom of marking such dates allows us two chances per century to celebrate the great, but it has the odd result that often we overlook the centenaries of the achievements that lie between cradle and grave. In Sedgwick's case, however, there was no outstanding "Eureka!" moment, nor any memorable day of publication of an epoch-making work. So in this paper I shall exploit the very routine of his annual round of teaching and research, in order to portray a quite ordinary but characteristic year of his life. And what better way to choose a year than to go back a century and a half to 1835, the year in which Sedgwick reached the often traumatic milestone of a personal half-century, but a year in which, like Miss Brodie, he was fully in his prime (Figure 1)?[3]

However, this paper is not a slice of straight biography. Scientific biography remains a valuable genre, but historians of science are increasingly conscious of the distortions inherent in the older style of heroic life-story. It is not that we want to cut the great down to size, still less to claim unrealistically that egalitarianism rules in scientific achievement. It is rather that we are increasingly aware that even the greatest and most innovative scientists achieve whatever they do achieve, not in splendid isolation but in more or less intensive interaction with a network of colleagues, whether in collaboration or in rivalry or both. So this portrait of 1835 is not primarily of a year in Sedgwick's life, but rather of a year in the life of Sedgwick *and company*—the company of all his fellow geologists.[4]

The commercial allusion in the title has a further meaning. Companies produce goods and services by means of processes operated by personnel. It would be useless to try to understand the work of an industrial firm merely by analysing its finished and packaged products. Yet until recently, much of what passed for history of science was just such an analysis of polished publications, with little attention to the processes by which they were produced or to the practice of the scientists themselves. So this portrait of 1835 will focus on the practical *activities* of Sedgwick and ˙ company, as they contributed to one of the most lively and innovative periods in the history of the earth sciences.

Figure 1. A portrait of Adam Sedgwick, painted two years before he reached the age of fifty.

GEOLOGY AS A UNIVERSITY SUBJECT

Sedgwick's chief memorial is the museum in Cambridge that is named after him, which is now incorporated in the department that grew out of his university teaching. So it is appropriate to begin with those activities. But it would be highly misleading to regard Sedgwick's university work as functionally similar to what the current professors and curators now do in those buildings. In 1835 Sedgwick had not yet been allocated proper space for even a small museum; many of the acquisitions he had made since he was elected Woodwardian professor back in 1818 were piled in his college rooms, still packed in crates, unopened and unstudied (Figure 2).[5] His annual lecture course, however, like that of his Oxford counterpart William Buckland (51), was well established and deservedly popular, with dons as well as undergraduates.[6] But at both ancient universities such courses were strictly extra-curricular; neither led to any examination, nor did they carry any credit towards a degree. Sedgwick and Buckland shared the general opinion that if geology deserved a place at a university—and Buckland had to fight harder for that recognition than Sedgwick—it was on cultural, not vocational grounds. Their courses were not intended to train professional scientists—indeed the very word "scientist" in its modern anglophone sense had only been coined two years earlier, by Sedgwick's Trinity friend and colleague William Whewell (43) (Ross, 1962; Morrell & Thackray, 1981: 20). They were designed instead to be a cultural top-dressing for young men preparing themselves either to enter the traditional learned professions of Church, Law, and Medicine, or to play a worthy part in local or national affairs as town or country gentlemen.

CAMBRIDGE GEOLOGICAL MUSEUM, 1842.

Figure 2. Sedgwick's geological museum in Cambridge, opened six years after he reached the age of fifty. Until then, the collections were housed inadequately in a few small rooms, and could not be properly displayed.

In the academic year 1834–35 Sedgwick's lectures almost certainly followed the revised syllabus he had published only two years earlier, though with modifications to bring them right up-to-date with his own and his colleagues' latest research.[7] The balance and sequence of topics reflected his and their conception of what "geology" should properly contain. The terms commonly used for the science in the three main scientific languages—geology, *géologie,* and *Geognosie*—were all taken to represent the same agreed contents of established knowledge, current problems, and agenda for future research. Most of Sedgwick's lectures therefore dealt with the sequence of formations of strata and their fossils, surveyed in the customary manner from the more recent towards the more ancient. The reason for this apparently unnatural order was simple. Apart from the very puzzling superficial or "diluvial" deposits, which had formerly been attributed to Noah's Flood, the more recent strata were much better understood than the more ancient. Sedgwick's stratigraphy would be intelligible—and indeed broadly acceptable—to any modern geologist, down as far as the Old Red Sandstone. Only when his lectures reached the still older formations did Sedgwick have to point out that the correct sequence was still uncertain, and the fossils poorly understood. Were it not for those problems, the whole sequence would more properly be read the other way round, as a series of chapters in Nature's own records of the *history* of the earth and of life. The reconstruction of that stupendously lengthy pre-human history was regarded by Sedgwick and his contemporaries as a major goal of geological science; and it was a research programme that had already made progress of which they could be proud.[8]

That Sedgwick did indeed emphasise this historical element in geology is clear from a newspaper account of the excursion he led into the Cambridgeshire countryside, one day in the spring of 1835, to illustrate his lectures in the field.[9] At the head of "a class of 60 or 70 academic horsemen", the Woodwardian professor first showed his students the gravel pits at Barnwell, on the outskirts of Cambridge, which were well known for yielding bones of large extinct mammals. He emphasised that these ancient river deposits were far older than any human records, and immeasurably older than the nearby ruins of Barnwell abbey. Yet the gravels in turn were far younger than the underlying Chalk, Gault, Greensand and Kimmeridge Clay, all of which the party saw, before they "proceeded, for the sake of very necessary refreshment, to the Lamb, at Ely". Towards the end of the day, Sedgwick took his students to the top of Ely cathedral tower, and synthesised what they had seen into one grand historical perspective, right through to the epoch of the prehistoric river gravels, the building of the mediaeval abbeys, and finally the drainage of the Fens that stretched all round them. The professor and his students arrived back at Magdalene Bridge long after dark, with two horses and their riders covered in mud after failing to jump a fenland ditch, and "looking like fossil bodies become reanimate for the instruction of the party". Their 40-mile ride had evidently been as much fun—and as gentlemanly—as a day out with hounds. They had had fossils to chase instead of foxes, but above all they had gained a new and vivid perspective on the vast history of the natural world.

That perspective was indirectly what made geology seem a proper academic study at the ancient universities. The previous year Sedgwick's work had been publicly attacked in a pamphlet entitled "Popular geology subversive of divine revelation!" But Sedgwick dismissed all such works—the indirect antecedents of modern creationism—as ignorant and ill-conceived.[10] Like Buckland and a host of lesser reverend geologists in England, he saw nothing in his research to impede the sincere and conscientious fulfilment of his duties as an Anglican clergyman. But while he firmly excluded so-called "Mosaic" or "scriptural" geology, Sedgwick did allow and indeed welcome another area of theological concern within the tacit boundaries of scientific or "philosophical" geology. He had emphasised this in his famous Trinity sermon of 1832, which, published as *A Discourse on the Studies of the University,* was already into a fourth edition. Sedgwick's views were paralleled by Buckland in Oxford, not only in his lectures but also in the Bridgewater Treatise he was expected to publish before the end of the year.[11] Both geologists claimed that geological science reinforced a traditional approach to *natural* theology, by proving that living organisms had been the products of beneficent design, even in the unimaginably remote eras before the creation of human beings. For both of them this was perhaps the supreme justification for including geology in the studies of a university.

Elsewhere in Britain there was little formal teaching of geology. At King's College London, which was only in its fourth academic year, a short course was given in 1835 by the new part-time professor of geology, a visiting provincial lecturer from York. This was John Phillips (35), the nephew of William Smith and an up-and-coming specialist on fossils. But Phillips's course was only notable for the way he reversed the usual order and attempted to give a truly chronological description of the strata (Phillips, 1834). Apart from Phillips's course, the only geology that was taught was included in courses on natural history and other sciences, for example the course given in Edinburgh by Robert Jameson (61). On the Continent, by contrast, geology had by 1835 won well established recognition in universities and similar institutions, not only as a major branch of natural history but also as the scientific basis for the mining industry. In Paris, for example, Cuvier's early collaborator Alexandre Brongniart (65) was still lecturing in mineralogy at the Muséum royale d'Histoire naturelle, and his son Adolphe (34) in fossil botany; Léonce Élie de Beaumont (37) was giving courses on geology for the Collège de France and at the École des Mines, and Constant Prévost (48) at the Faculté des Sciences.

In fact, most of the states of continental Europe, and those of Scandinavia and Russia, supported university professors who were paid at least modest salaries for teaching natural history or mining technology, and many of them included geological subjects in their courses.[12] But neither Sedgwick nor Buckland, for example, could have begun to support even a modest gentlemanly life-style, without major sources to supplement their meagre professorial income. Buckland had benefitted from Tory patronage back in 1825, by being appointed a Canon of Christ Church; this had given him a handsome assured income of £1,000 as well as enabling him to get married. Sedgwick, as a Whig, had to wait another decade for a similar piece of ecclesiastical patronage. But soon after the start of the academic year 1834-35 Lord Brougham, in one of his last acts as Lord Chancellor in Melbourne's administration, offered Sedgwick a prebendal stall at Norwich. Sedgwick accepted with alacrity, suspended his lecture course until the Lent Term, and immediately began his two-month stint of duties at the cathedral. Though he did not much enjoy it, his new post at once tripled his income, adding some £600 to the meagre £200 a year he received as Woodwardian professor (and perhaps another £150 from his dividend as a Fellow of Trinity).[13] Better still, it was an annual ecclesiastical chore that could be combined with his university Chair and his college Fellowship, so he could continue doing his science in the academic world he had come to value above the attractions of matrimony.

THE LURE OF GEOLOGICAL FIELDWORK

Sedgwick was evidently an eloquent and persuasive lecturer, but he himself got even greater pleasure from his fieldwork. Almost every summer, he and most other leading geologists exchanged town for country, fashionable clothes for weatherproof ones, and the company of gentlemanly colleagues for that of innkeepers, farmers and quarrymen (Figure 3).[14] Once in the field, Sedgwick's Cambridge headaches, gout, dyspepsia and general hypochondria all vanished miraculously, to be replaced by a healthy appetite and boundless energy. Sedgwick was well known for his feats of outdoor endurance; but geologists anywhere regarded it as normal to be in the field from dawn to dusk and to cover anything up to twenty miles a day on foot, and even more if on horseback. Their equipment was simple, and would still be familiar to modern students on a first-year field course. A hammer was of course the most distinctive item, together with a collecting bag and a notebook, a topographical map and a magnetic compass, and perhaps a clinometer for measuring the dip of strata and an acid bottle for detecting limestones.[15] With even the leaders of the science requiring so little for their research in the field, it is no wonder that so many amateurs also felt the lure of geological fieldwork.

Figure 3. Henry De la Beche's caricature of the contrast between a field geologist (himself!) and a *salon* geologist or theoretician (probably Charles Lyell), drawn in criticism of the first volume of Lyell's *Principles of Geology* (1830).

By 1835 Sedgwick had already made substantial progress in tackling the British end of a major research problem of international dimensions. This was the daunting task of unravelling the oldest known strata, the so-called "Greywacke" or "Trans-ition" rocks lying below the Carboniferous formations, which included the econom-ically vital Coal Measures. (In the 1830s the "Carboniferous Group" was generally extended downwards to include the Old Red Sandstone.) The fieldwork needed for this project was almost everywhere exceptionally difficult, owing to the strong folding of the strata, their confusion by slaty cleavage, the scarcity and poor preservation of the fossils, and—in Britain—the frequently atrocious weather. A sample of the field notes that Sedgwick made in Cumbria in 1835 gives a fair impression of the painstaking detailed work on which any grander conclusions had to depend.[16]

> Go to *Beck Side:* ¼ m. up the stream (Black Cwm Gill) is a mill, just behind which is a bed [which] strikes N.E. by N., underlies or dips N.W. by W. at 65°—It is very hard, quartzose, with greenish spots, specks of pyrites. 30 f. thick . . . [etc.]—return and detained at Broughton by the rain.

These ancient rocks in Cumbria were similar to those Sedgwick had been studying in Wales for the past four years. The previous summer he and his collaborator the London geologist Roderick Murchison (43) had met in the Berwyn Hills to link up their respective research areas. They had agreed that Sedgwick's strata were definitely lower and therefore older than Murchison's less confusing and more fossiliferous ones on the borders of Wales. Murchison had already established that those strata in turn were older than the Old Red Sandstone. So between them they seemed well on the way to unravelling a reliable sequence in the Greywacke strata of Britain. Murchison, unlike Sedgwick, could afford to spend as much time as he liked on geology, having a wife who was a wealthy heiress as well as a skilful geological artist. In the summer of 1835 he successfully traced his formations, particularly by their distinctive fossils, all the way from Shropshire through south Wales to the tip of Pembrokeshire. Ever anxious to secure proper credit for his research, he took the precaution of publishing a quick note in the July issue of the *Philosophical Magazine*—the functional equivalent of *Nature* or *Science* today—to stake his claim to priority by terming his rocks the "Silurian System" (Murchison, 1835).[17]

A few other examples of fieldwork in 1835 will begin to bring its international dimension into clearer view. The London geologist and author Charles Lyell (38) spent part of the summer in the Swiss Alps, primarily to check at first hand some of the evidence for his interpretation of mountain chains as the products of the long-continued operation of ordinary geological processes. Before leaving England, he sent his publisher John Murray the fourth edition of his controversial *Principles of Geology*, which embodied that approach. The four-volume work was already bringing him such satisfactory royalties that he had felt able to resign from the professorship at King's College London and to devote himself full-time to geology. Passing through Paris, he first had a chance to argue his case with two of his most formidable opponents (Lyell, 1881, i: 450-2; Wilson, 1972: 413-5). Leopold von Buch (61), the doyen of Prussian geologists, was also in Paris, on his way to a summer season of fieldwork among the celebrated extinct volcanos of Auvergne, intending to evaluate his own more paroxysmal ideas of crustal movement. Élie de Beaumont, on the other hand, stayed in Paris that summer, but only because he and his colleague Pierre Dufrénoy (43) were putting the finishing touches to their great geological map of France, which had taken them into the field every summer for the past ten years.[18] By contrast with their generous support from the French state, the geological map that was being completed at the same time for the Ordnance Trigonometrical Survey cost the British taxpayer a mere £300. That was the paltry fee paid to Henry De la Beche (39) for three years' mapping in Devonshire. He had been forced to ask for the commission, because his income from the Jamaican plantation he had inherited had slumped disastrously to an ungentlemanly level; of the leading British geologists, only he was short of cash (McCartney, 1977: 28). He was therefore in the field for far more of 1835 than any of his colleagues, but not by choice.

For all the other leading geologists, however, fieldwork was a welcome change from the routine of gentlemanly social life; for Sedgwick, it was also an escape from often tiresome duties in college and cathedral, as well as from the more pleasurable duties of the lecture room. But fieldwork was also, of course, the primary locus of encounter between geologists and the empirical phenomena they sought to understand and explain. Whether their attitude to theorising was cautious and sceptical, like Sedgwick and De la Beche, or enthusiastic and confident, like Lyell and Élie de Beaumont, they all deferred to what they called the "Baconian" foundations of their science, at least by insisting rhetorically on the primary significance of first-hand observation in the field.[19]

GEOLOGICAL LETTER-WRITING

In reality, however, the centre of their geological lives lay not in fieldwork but in discussion and argument with others. While giving lip-service to the primacy of fact-collecting they followed in practice a quite different conception of science. Their actual behaviour showed they believed the route to reliable knowledge lay not only through fieldwork, but also and more decisively through the exchange of information and ideas with others, an exchange often marked by heated argument and intense controversy.

It was not necessarily easy, however, for an individual geologist to engage face-to-face in this collective process of lively exchange. Even with a comfortable income assured, Sedgwick's participation was, for example, severely limited by demands on his time, by his college and university duties and now by his ecclesiastical duties too. (Buckland had the demands of a large family as well.) Except during the long vacation, neither could often afford the time for the long and tiring day's coach ride that would take them to London. Yet the capital city was unquestionably the scientific centre of the nation, and one of the three or four greatest centres in the world. Still, although Sedgwick and his contemporaries lacked the ambivalent benefits of the telephone, they did enjoy the unmixed blessing of a postal system more efficient than that of the 1980s.[20] Though his visits to London were infrequent, by letter Sedgwick could communicate swiftly and reliably—though not cheaply—with his colleagues in London and beyond. A letter sent by the evening mail-coach from Cambridge could be read in London, say by Murchison, the next morning, and Murchison's response could be back on Sedgwick's breakfast table the morning after that. A letter from Élie de Beaumont in Paris would reach him in only two or three days, thanks to the recent introduction of reliable steamships across the Channel; and even a letter from von Buch in Berlin, the third major centre of world science, would arrive within a week, via a steamer from Hamburg to London.

Such speed and reliability, combined with the relative discomfort and high cost of travel, gave letter-writing an importance and centrality in scientific life that it had never had before this period and—arguably—was never to have again. Sedgwick certainly felt isolated from intellectual company while he was in Norwich. Even in Cambridge, where he had helped found the Philosophical Society to bring together the small band of dons with scientific interests, he felt cut off from discussions with geologists of his own calibre, apart from the omnicompetent Whewell (Clark & Hughes, 1890, i: 205-7; Hall, 1969). But the postal system kept him tied into a tight network of scientific communication. Indeed, since fluent letter-writing was then as routine an accomplishment as fluent telephone conversations are now, correspondence was a highly effective substitute for face-to-face discussion.

ARENAS OF GEOLOGICAL DEBATE

Of all occasions for such face-to-face discussions, the Royal Society might have been expected to take first place. But it did not. All the leading English geologists were Fellows of the Royal Society, almost as a matter of course, but they rarely chose to have their papers read there.[21] In the 1834–35 season, Lyell's Bakerian lecture, on the evidence for the continuing slow elevation of the land around the Baltic, was the only major piece of geological research presented there (Lyell, 1835), and then only because otherwise the Society could hardly have awarded him its Royal Medal.[22] The senior scientific body occupied only a marginal position in the lives of English geologists. One reason for this is that the Royal Society covered in principle the whole range of the natural sciences, whereas another body in London provided a more specialised arena for geology. But another reason is quite as important. Papers read at the Royal Society were greeted with a dignified silence. Discussion was prohibited, at least until the Fellows had adjourned for coffee and cakes; for it was considered that discussion or argument in the formal setting would imperil the public status of science as a body of authoritative knowledge. In any case, the lay-out of the meeting room, dominated by the president and secretaries on a dais, with the Fellows facing them like a passive audience, would surely have inhibited any lively exchange.[23]

The more specialised arena, in which such prohibitions and inhibitions had been overcome, was the Geological Society. On three of the occasions in 1835 when Sedgwick did make the journey to London, it was to hear his own papers read at the Society and to attend its anniversary meeting.[24] It is hardly possible to over-estimate the importance of the Society in the life of English geologists in this period, whether they lived in or outside London. Any foreign geologist visiting London would as a matter of course be taken as a guest to its meetings: in 1835, such foreigners included the palaeontologists Louis Agassiz (28) from Neuchâtel and Edouard de Verneuil (30) from Paris. Conversely, any British geologist visiting Paris would be taken to the meetings of the Société Géologique de France, as for example Lyell was in the same year. Although smaller and much younger, this was closely modelled on the society in London.[25]

The leading members of each society—"the great men", as Charles Darwin (26) was to call them when he joined the one in London a year later (Burkhardt & Smith, 1985: 512-4)—were a quite small group. Many were almost continuously on the elected council or *bureau* that officially ruled each society; but they wielded great power behind the scenes even when they were not. In London they were also leading members of the Geological Society *Club,* an inner circle of the most clubbable members, who would fortify themselves with a convivial dinner—and settle much business informally—before facing the formal reading of papers at the meeting of the Society itself. In London at the start of 1835 George Greenough (57), co-founder of the Society in 1807 and its original president, was nearing the end of his third term in that office; at the Anniversary meeting in February, after giving the customary review of the previous year's research, he handed over to Lyell. Until then Lyell had been the Society's foreign secretary, so De la Beche took over from him. Murchison was one of the four vice-presidents, and Buckland and Sedgwick were ordinary members of the Council, all three having already served as president. Likewise in Paris the elections at the start of 1835 put Ami Boué (41), a medical man of independent means and a leading figure in the foundation of the Société in 1830, into the presidential chair; the onerous task of writing the annual review of French geology was assigned to one of the secretaries, the *ingénieur-géographe* Claude Rozet (37); and other members of the *bureau* included Élie de Beaumont, the elder Brongniart and Prévost.

The only known picture of a meeting in progress at the Geological Society during its first decades (Figure 4) almost certainly dates from several years earlier than 1835,

Figure 4. The Geological Society of London in session, probably in the 1820s, drawn from behind the president's chair. The more prominent geologists are seated on the front benches; the lesser figures behind them are scarcely sketched in at all. Note the items under discussion: an ichthyosaur's head on the table, and a geological section on the far wall.

before it moved into more spacious premises in Somerset House.[26] But it does show a striking contrast to the Royal Society. The characteristic parliamentary arrangement (which survived until the 1970s) put the Fellows on to benches that faced each other, with the president in the Speaker's position flanked by the two secretaries. There was also a parliamentary contrast between the front benches and the back. The "great men" on the front benches were depicted in a highly individual manner—De la Beche is easily recognised by his spectacles—whereas those on the back benches were scarcely sketched in at all. Leading geologists such as Buckland and Sedgwick, Greenough and De la Beche, Murchison and Lyell, habitually took their places on the front benches, whereas lesser mortals crowded into the rows behind them. This reflected their respective roles in the social drama that was enacted in the meeting room: like their equivalents in Parliament, the front benches contributed far more than their number would suggest.

   Out of thirty-seven papers read in London during the 1834–35 session, Murchison contributed four, all of them offshoots from his stratigraphical work on the borders of Wales. Sedgwick contributed two, one of them the celebrated paper that first clearly distinguished between bedding, jointing and slaty cleavage, and thereby allowed the Greywacke regions to be reliably unravelled. Greenough delivered the massive Anniversary Address, summarising and commenting on the whole of the previous year's work by members of the Society. De la Beche put an observational cat among Murchison's and Lyell's theoretical pigeons, by reporting fossil plants of Coal Measures species among the ancient-looking Greywacke rocks of Devonshire. Lyell reported on his work the previous summer on the Chalk strata of Denmark, which were relevant to his controversial claim that there was a vast unrecorded break between them and the Tertiary strata. Buckland enlarged his zoo of extinct vertebrates by reporting a gigantic new reptile from the Oolitic strata of Buckinghamshire.[27]

Likewise in Paris, out of forty-two papers read during the same session, Dufrénoy contributed two: one summarised his recent mapping in Brittany and Normandy, and made an important distinction between older and younger parts of the Greywacke or *terrain de transition;* the other dealt with the correlation of the Tertiary strata of the Midi. Prévost twice discussed the vexed question of the stratigraphical relations of some of the Tertiary strata of the Paris basin, an apparently local matter that was in fact crucial to the interpretation of the alternating marine and non-marine conditions under which several Tertiary sequences seemed to have accumulated. Rozet paralleled Greenough's address with an equally exhaustive review of the previous year's work by members of the French society.[28]

The number of papers just mentioned, however, does not adequately reflect the importance of the contributions of the small inner circle of each society, because it does not take into account the intensive discussions that followed the reading of the papers. It had been an audacious innovation when, a few years earlier, under the presidency of the wealthy physician William Fitton (55), the Geological Society had first allowed discussions at all. "I will not call them discussions, still less debates," Fitton (1828: 61) had said soon after the "experiment" began, fearful that the arguing would become rancorous and divisive. But outsiders were more forthright. "Though I don't much care for geology," commented John Lockhart, the editor of the influential Tory *Quarterly Review,* "I do like to see the fellows fight" (Allen, 1976: 70). Their fighting was doubtless facilitated and even encouraged by the parliamentary seating: it was easier to have a good argument with a man facing you than with one in a row behind your back. But the fighting was not unproductive. On the contrary, the development of many of the problems that were debated at the Geological Society shows how opinions on both sides were repeatedly modified by the experience of discussion. This can be seen particularly clearly in the prolonged and often heated arguments—soon to be dubbed the "great Devonian controversy"— which broke out after the reading of De la Beche's paper on Devonshire at the end of 1834.[29] But under the implicitly neutral chairmanship of a president in the position of Mr Speaker, it was only rarely that the argument transgressed the bounds of gentlemanly behaviour. In any case, when the Société Géologique was founded, not long after discussions began to be permitted in London, the same custom was adopted almost at once.[30] In 1835, for example, the debates following the papers by Dufrénoy and Prévost on Tertiary correlations seem to have been as lively, if not quite so vehement, as some of those in London.[31]

After papers had been read at either society, summaries would be printed in the *Proceedings* and the *Bulletin* published every few weeks in London and Paris respectively. These gave prompt and wide circulation to whatever conclusions had been put forward in the two meeting rooms. Each society would also appoint referees, asking them to report whether the papers deserved to be printed and illustrated in full in the societies' more prestigious publications, the *Transactions* and the *Mémoires*. The refereeing was generally kept within the small inner circle of each society. During the 1834-35 season in London, for example, Buckland reported on De la Beche's paper and Lyell on one of Sedgwick's; Fitton, Sedgwick, Murchison and De la Beche all reported on papers by lesser authors; while at least five referees were chosen from among other present or former members of the Council, for the sake of their more specialised expertise.[32]

Members of the two societies who purchased the *Transactions* or the *Mémoires* in 1835 received respectively two or one hundred pages of handsome quarto text, with hand-coloured plates of geological maps and sections, and fine illustrations of fossils. But the papers that were given this first-class treatment had often been so long in the pipeline that their value in current debate was much reduced. With only one exception, the ten papers published in the *Transactions* during 1835, and the five in the *Mémoires,* had all been read between 1829 and 1833, while those that were read during 1835 were to remain unpublished—with the same single exception—for several years. The exception was Sedgwick's paper on jointing and slaty cleavage: that dilatory author had to be asked not to withdraw it after it had been read, for it was considered so important that it was allowed to jump the queue and be published with almost unprecedented speed (Sedgwick, 1835).

Within this usually lengthy procedure, from the initial reading of papers to their eventual publication, contemporary records reveal that face-to-face discussions were far more important to the leading geologists than their polished publications. But such discussions were witnessed by larger audiences than the small inner core of each society. It is at this point that the full membership of the two societies must be brought into the picture.

CLUBS FOR GEOLOGICAL GENTLEMEN

Scientific London and scientific Paris were both quite small in geographical scale. When Sedgwick, for example, stayed in his usual lodgings near Trafalgar Square, he was hardly more than half a mile from the Geological Society in Somerset House on the Strand, and even closer to the Athenaeum at the bottom of Regent Street.[33] At the Athenaeum, as Lyell had noted, one could get "a genteel elegantly served dinner for 2s. 6d. with all the newspapers, etc." (Wilson, 1972:318). For members more sociable than Lyell, the "etc." could include not only all the latest books and periodicals, but also the conversation of most of the leading London intelligentsia. In fact the Geological Society itself, with its library and other public rooms, also functioned somewhat like a social club. It was routinely used as a rendezvous for those with geological business to discuss; as in other gentlemen's clubs, members who lived in London were required to pay more for their heavier use of its facilities. Provincial members of either society, coming to London or Paris to one of the fortnightly meetings, would arrange to stay in the metropolis at least a couple of nights, so as to have time to inspect the newly acquired books and specimens in the library and museum, or to talk informally with others, before the meeting began in the evening. Many of the leading metropolitan members, in turn, lived within walking distance, or at most within a short cab ride, of each other and of the premises of the geological and other scientific institutions. Their letters and diaries refer to a round of working breakfasts, teas and dinner parties. In London, the grandest of all such occasions were the regular scientific soirées held by the wealthy bachelor Greenough in his Italianate villa on the edge of the Regent's Park, not far from the recently established Gardens of the Zoological Society.

Two general points may already be apparent from this verbal sketch. First, although the two geological societies were supremely important arenas for specifically geological discussion, and although many of their members were certainly geological specialists, they were at the same time participants in a much wider intellectual culture—*wissenschaftlich* in its scope rather than just "scientific" in the then recently narrowed anglophone sense of the word. In London, for example, the interpenetration of the various scientific and learned societies is strikingly revealed by some of the other affiliations of the Fellows of the Geological Society. In 1835, no fewer than 25% (159 out of 644), including Sedgwick and all the other leading members, were also Fellows of the Royal Society; 17% (107), including Lyell, Buckland, Murchison and others interested in fossil zoology, were Fellows of the Linnean; 9% (55) were Fellows of the Society of Antiquaries, which included interests that were later to be termed prehistory; and 6% (38) were members of the Royal Astronomical.[34] It is no wonder that at every Anniversary dinner of the Geological Society, a long series of toasts and speeches celebrated those links, as well as those with the universities, in gentlemanly style. After the 1835 dinner, for example, Lyell reported how his first occasion as president had gone (K. Lyell, 1881, i: 447):

I got [the marquis of Lansdowne] to give [i.e. propose a toast to] Oxford and Buckland. Fitton gave Cambridge, answered by Sedgwick; Sedgwick the Royal Society, answered by Lubbock; Buckland the Linnean; I, the Astronomical, answered by Baily; Greenough the Geographical, answered by Murchison. [etc.]

The second point is that both geological societies drew their members predominantly from strata of society that were gentlemanly, relatively wealthy, and decidedly "establishment" in character. Entrance fees of six guineas and twenty francs respectively, and annual fees of three guineas and thirty francs, would have been enough to deter the lower orders. But in London they were also excluded by more subtle means, since candidates for membership had to be personally known to existing members in order to guarantee their social respectability. The professional artist and engraver James de Carle Sowerby (47), for example, was not a member, although his great expertise with fossils was invaluable to those who were, and Lyell paid him by the hour to teach him conchology. Most strikingly of all, the veteran land surveyor William Smith (66) was likewise not a member, even though Sedgwick, as president, had awarded him the Society's first Wollaston Medal only four years earlier. Conversely, however, the leaders of both societies evidently prized the social prestige that came from having among their members substantial numbers of the titled nobility: in 1835 nearly 6% of the members of each society represented the aristocracy (37 in the English, 13 in the French), ranging downwards from four English dukes (Bedford, Buckingham and Chandos, Devonshire and Northumberland) and two French marquises (de Dalmatie and de Drée).[35] In addition to those of the English nobility who sat in the House of Lords, the Geological Society could draw on no fewer than 30 M.P.s (5% of the membership) to watch over its interests in the House of Commons, two of them being currently on its Council. In Paris there were at least a couple of peers of France (one of them the veteran Auvergnat geologist the comte de Montlosier) and a couple of members of the Chamber of Deputies (one of them the distinguished physicist François Arago).

The backbone of each society, but particularly the British, lay however in the learned professions. In 1835, 8% (49) of the London society and 10% (23) of the Parisian held medical qualifications, though they did not necessarily practise: for example, Fitton in London and Ami Boué in Paris. Lawyers of one kind or another made up 8% (18) of the French society and probably a similar proportion of the English: for example the Bow Street magistrate William Broderip (46) was prized for his zoological expertise, Lyell had worked as a barrister before turning full-time to geology, and de Verneuil was an advocate attached to the Ministry of Justice in Paris. But in the third of the traditional learned professions there was a stark contrast. In the London society no fewer than 13% (85) were clergymen, ranging from two deans (of Carlisle and Windsor) and "Reverend Professors" such as Sedgwick and Buckland down to humble country parsons; the great majority of them Anglicans, but also including a few non-conformists (e.g. the Unitarian James Yates) and even one Roman Catholic (John McEnery). But the relatively anti-clerical atmosphere of science in France at this period—and the anti-scientific temper of its Catholicism—is reflected in the fact that the clergy were represented in the Société Géologique by just one Catholic priest, the *abbé* Croizet (48), *curé* of a rural parish in the Massif Central and co-author of a local palaeontological monograph in the style of Cuvier.[36]

It is difficult to estimate the proportions of those for whom geological questions and geological knowledge impinged more directly on their livelihood. In the French society, 9% (21) can be identified as academics and museum personnel concerned with the natural sciences, ranging from *professeurs* such as the Brongniarts and Élie de Beaumont in the elite Parisian institutions, down to freelance lecturers and provincial teachers and *conservateurs*. In the English society only 6% (37) were academics; and even this figure includes not only teachers of the natural and medical sciences at the universities, such as Buckland and Sedgwick, but also other Fellows of Oxford and Cambridge colleges for whom geology was no more than a sparetime hobby. The 9% (22) of the French society and 5% (35) of the English who held military rank included several members employed on topographical surveying work for the Corps-Royal des Ingénieurs-géographes and the Ordnance Trigonometrical Survey respectively. But the far greater involvement of the French government in technical matters related to geology only emerges in the figure of no less than 17% (40) of the Société Géologique who were employed in the state mining service, *ponts et chaussées* and similar bodies. This included Dufrénoy and Élie de Beaumont, together with many of the provincial *ingénieurs des mines* who had assisted them locally with their great geological map of France. By contrast, the original inspiration for that map, namely the one of England and Wales that Greenough had published fifteen years earlier, was being revised by him as a characteristically British piece of private enterprise. At the end of 1835 De la Beche had only just managed—with the official support of Buckland, Sedgwick and Lyell—to have his *ad hoc* mapping of Devonshire transformed into a tiny geological section of the Ordnance Survey, and to set up a small and makeshift Museum of Economic Geology in London—the forerunners of the modern British Geological Survey and Geological Museum respectively.[37]

Finally, several members of the French society were noted in the membership list as being *propriétaires des mines*, or at least involved in the management of mines, ironworks and the like. In contrast, the more gentlemanly character of the English society is underlined by the fact that only one member was officially noted as "civil engineer," and only two gave industrial addresses (namely, ironworks at Merthyr Tydfil in the South Wales coalfield). However, it should be remembered that the agricultural applications of geology, and for example the prediction of sites for artesian wells, were regarded as quite as important as those connected with mining and heavy industry; and the perceived relevance of the science to the many country landowners in both membership lists should not be underestimated.[38]

PROVINCIALS AND FOREIGNERS

All these figures have been based on the quite impressive total membership of the two societies: excluding foreign members (of whom more later), there were no fewer than 644 in the Geological Society in 1835, and 234 in the Société Géologique. But were those large numbers of any real scientific significance? The answer to that question depends, of course, on one's definition of science. Certainly no special qualifications or attainments in geology were required for membership of either society: only an expressed interest in the science and an appropriately attested social respectability. At the very least, however, the majority of the two memberships constituted an important audience for the much smaller number of those who performed with papers at the meetings. They were the faceless ranks ranged on the back benches in the Geological Society's meeting room (Figure 4), waiting to be instructed and entertained by the "fighting" among the front-benchers; they constituted the bulk of the readers of the *Proceedings* and the *Bulletin*; and some were also purchasers of the *Transactions* and the *Mémoires*. At the very least, such publications added tone and even beauty to the library table in any town or country house with aspirations to scientific culture.

In fact, however, the dramaturgical metaphor of performers and audience is highly misleading if it is taken to imply a sharp dichotomy between the producers and consumers of science. In geology, even more than in other natural history sciences such as botany, the intrinsic spatial dimension of the subject-matter ensured that even the most local of observations were potentially valuable, provided they were made with a basic minimum of scientific rigour: provided, for example, that the localities of specimens were accurately and reliably recorded. It was therefore of great importance that the members were so widely distributed geographically. In both societies, about half the total indigenous membership lived outside the capital: 52% (332) of the English society, 45% (107) of the French. In the less centralised elite culture of England, of course, the figure included men like Sedgwick and Buckland, who cannot properly be termed provincials. But even discounting such leading geologists, it is clear that the two societies could if necessary call on the strictly local expertise of members in all but the most remote parts of the British Isles and in most of the French *départements*; certainly there was an excellent scatter in both countries (Figure 5 and Figure 6).[39]

Futhermore, the proceedings of the two societies in 1835 reveal the active involvement of what might otherwise be thought a *lumpen* "audience" of provincial members. Even their donations of specimens from their local areas could be of great interest to the metropolitan geologists. But the provincials were also quite well represented among those whose papers were read in the two capital cities. Meetings in London heard, for example, papers by Robert Austen (27), a young country gentleman in Devonshire, and by David Williams (43), a country parson in Somerset. Likewise, in Paris in the same season the French geologists heard papers by—among others—a land surveyor in Caen, a military engineer in Grenoble, and a legal official at Châlons-sur-Marne.[40] Nor were all these papers of merely marginal significance: Austen's for example, on a raised beach near his home, impinged directly on the major problem of geologically recent crustal elevation—the problem that Lyell had dealt with in his Bakerian lecture at the Royal Society.

Provincials with such interests were linked into the wider network of exchange not only by their donation of specimens and presentation of occasional papers to the societies in London and Paris, but also conversely by their ability to help visitors from the metropolis. Geologists making an initial study of a provincial area could scarcely hope to find what might *vaut le voyage* or even *mérite un détour*, except under the guidance of local geologists with an intimate knowledge of the best rock exposures and fossil localities, the fruit of long and patient local fieldwork. So for example when in 1835 Rozet reviewed the previous year's research by the French society, he made a point of mentioning for the benefit of the Parisians that three provincial members in Épinal (Vosges)—one a local doctor, another an *ingénieur des ponts-et-chaussées*—were eager to show visitors the geology of their region.[41] Likewise towards the end of 1835 Sedgwick, feeling uneasy about De la Beche's interpretation of Devonshire but still cautious about Murchison's rival one, began to make his own enquiries from Major William Harding (43), a country gentleman at Ilfracombe and a keen local fossil collector; and the following year he and Murchison were to make use of Harding's local expertise in the field.[42] The scatter of valuable local geologists could be exploited across national boundaries too. On his way to the Alps in the summer of 1835, for example, Lyell was able to tap the local knowledge of two provincial members of the Société Géologique—a mining engineer at Vesoul (Haute-Saône) and a landowner just over the Swiss frontier at Porrentruy—while two local geologists from Neuchâtel showed him over another part of the Jura.[43]

Like such provincials, but on a broader scale, was the scatter of members of the two premier societies living or working beyond their respective national frontiers. In 1835 the London society had among its members British diplomats in Frankfurt and Naples and a consul in Brazil; a total of 13 Britons in the Americas and 11 in India (including Major Everest of the Trigonometrical Survey there). Sir John Herschel was observing the southern stars at the Cape of Good Hope, and collecting fossils in his spare time; and there were even members on St Helena and in Australia (Figure 7). The Paris society likewise counted among its members an engineer in Algiers, a doctor in New York and a scientific traveller in Chile (Figure 6). Like members in the provinces, such overseas residents and travellers were no mere passive audience for metropolitan geology, but contributed both specimens and scientific papers. In 1835 the London society heard, for example, a paper on Naples by the traveller Captain Basil Hall (47), the son of Hutton's friend Sir James Hall; reports by other naval officers about the coasts of South America and West Africa; a paper on New South Wales and one from a parson on Ascension Island; and, last but not

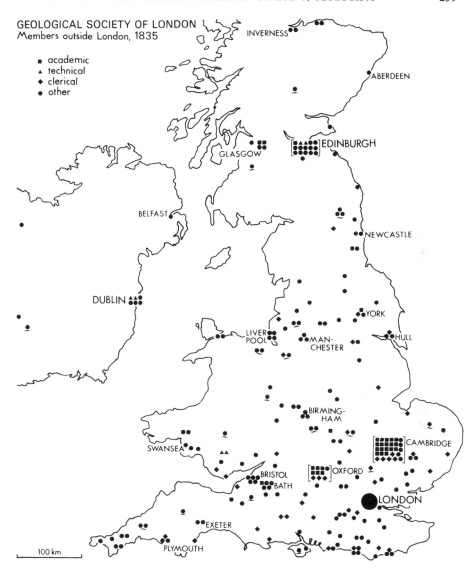

Figure 5. Distribution map of members ("non-resident Fellows") of the Geological Society of London who lived outside London in 1835 (for those who lived outside the British Isles, see Figure 7). Note the concentrations in the university towns or cities, and in the traditional centres of provincial culture (e.g. Bristol and Bath). The only notable cluster related to a mining or industrial area is that in Cornwall, in the far south-west of England.

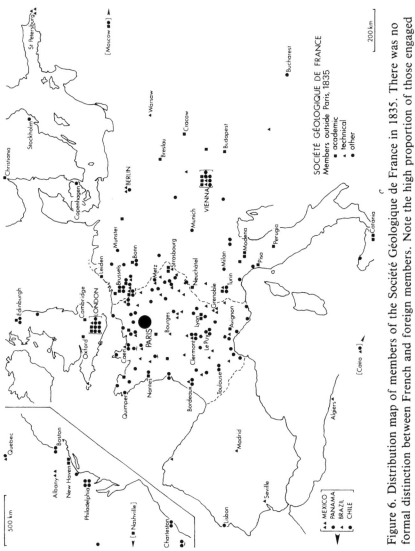

Figure 6. Distribution map of members of the Société Géologique de France in 1835. There was no formal distinction between French and foreign members. Note the high proportion of those engaged in technical activities—mostly in mining—and the broad geographical scatter throughout Europe and even beyond. The member in Cambridge was Sedgwick.

least, a paper from the young naturalist Charles Darwin, who was informally attached to a naval vessel surveying in South American waters.[44] As with papers by provincial members, these were of no merely marginal significance for current debate. For example, Hall's paper on the famous "Temple of Serapis" (used by Lyell for the frontispiece of his *Principles*), and both of those on South America, impinged directly on the hotly-debated problem of recent movements of the earth's crust.

Also important for international exchange were the foreign members of the two societies. The Geological Society chose its "Foreign Members" and a trio of foreign "Members of Royal Blood" primarily to enhance its own social standing and as a mark of scientific recognition, and it charged them no fee for the honour. The Société Géologique, in a contrasting gesture of *égalité*, was open to foreigners on exactly the same terms as it was to Frenchmen. Many of the 57 foreign members of the London society would have been obvious choices in the eyes of any knowledgeable English geologist, simply for their unquestioned distinction: for example Élie de Beaumont and the elder Brongniart in Paris, von Buch and the great Alexander von Humboldt (66) in Berlin. But in general the distribution of the foreign members looks like the result of a calculated policy of placing useful international contacts in strategic positions: a modest scatter throughout the French-, German- and Italian-speaking states of Europe, and up into Scandinavia and Russia, and a few in the United States (Figure 7). Since membership of the Paris society was by contrast a matter of application rather than invitation, the scatter here must reflect more directly the usefulness of the society as perceived by the foreigners themselves. So it is striking that in 1835 they constituted almost 35% (124 out of 357) of the total membership of the Société Géologique. Of those foreigners, 17 were British; Sedgwick had been one of the first two to join, soon after the society was founded, and by 1835 all the leading members of the Geological Society had joined too. The Paris society had an even better scatter through Europe than the London one, while 17% (21) of its foreigners were in North America (Figure 6).

These international links, like those between metropolis and provinces, were far from being merely nugatory or symbolic. In the course of 1835 the Paris society received from London, for example, authors' presentation copies of De la Beche's *Researches on Theoretical Geology* and the latest edition of Lyell's *Principles*; from Liège came the important monograph on cave bones by Philippe Schmerling (44), and from Leiden a paper on artesian wells by Jacob van Breda (47); and other papers arrived from as far afield as Moscow and Philadelphia. De la Beche sent a note about his controversial discoveries in Devonshire, which Élie de Beaumont promptly read in Paris. Similar gifts of books and papers were received by the London society from foreign members such as Dufrénoy and von Buch, and Agassiz gave a progress report on his work on fossil fish at a meeting there.[45] And all such gifts were of course in addition to the periodicals that flowed regularly into the libraries of the two societies, many of them in exchange for the societies' own publications.

CONCLUSION

In this paper I have used Sedgwick's fiftieth year as a convenient benchmark for surveying the whole geological community in the period when he, and indeed the

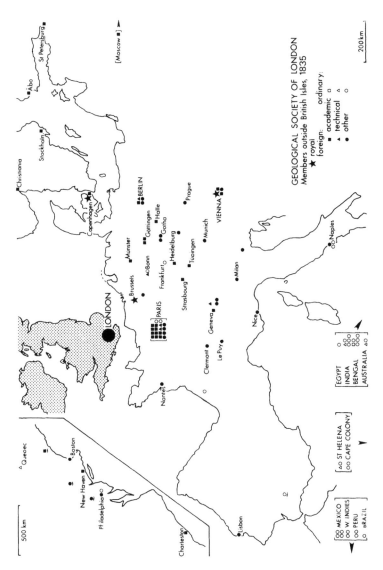

Figure 7. Distribution map of members of the Geological Society of London who lived outside the British Isles in 1835. In addition to the scatter of "Foreign Members" (solid symbols), the distribution of overseas Britons (open symbols) is clearly related to British colonies, and commercial and mining interests elsewhere.

"early classical" geology he did so much to consolidate, were in their prime. I have perhaps mentioned a bewildering number of individual names and apparently minor events, but it is only through such detailed *histoire événementielle* that any broader features can be discerned. I want in conclusion to draw out three main points.

First, the shape and composition of the geological community one hundred and fifty years ago. The many individuals mentioned by name exemplify what I have elsewhere (Rudwick, 1985: 418–426, Figure 2.3) called the "socio-cognitive topography" of the science in that period. The relative competence of individuals, as perceived and judged by themselves and their contemporaries, ranged them along a tacit gradient of evaluation. At the upper end were unquestioned "elite" geologists such as Sedgwick and Lyell, Élie de Beaumont and von Buch, who were generally acknowledged to be competent to pronounce on the most fundamental issues in the science—but certainly not necessarily to agree about them among themselves. Lower down was a middle zone of those recognised as "accomplished" in some more limited sphere, as for example Dufrénoy on the older strata of Brittany, Prévost on the Tertiary strata around Paris, and the young Darwin in South America. Still lower were mere "amateurs" such as Williams in Somerset and the trio of local enthusiasts in Épinal, who were judged competent to give valuable local information on fossil localities and the like, and to amass reliably documented collections, but not to pronounce on their significance. Beyond them lay the general public, who were not reckoned competent to pronounce on anything at all, but only to sit and listen to their betters. The two major geological societies were in effect superimposed on this tacit topography of evaluation, while the scriptural geologists were kept out at the cognitive margins of the science.

Second, the intensity of discussion and the sheer *pace* at which opinions were exchanged, nationally and internationally, in geological practice a century and a half ago. On what I have elsewhere (Rudwick, 1982b; 1985: 429–435) termed "the continuum of relative privacy," the central zones of discussion and argument were more decisive in the resolution of controversy and in the shaping of new consensual knowledge than either the relatively private zone of individual fieldwork or the relatively public realm of formal scientific papers and other publications.

Third, the heterogeneity of geological debate, during this sample year from another age. There was unquestionably a "quiet background" of steady cumulative effort in the science, represented by the many straight-forward papers that were read at the geological meetings in London and Paris. But there were also several distinct "hot spots" of international argument and controversy, centred on refractory "focal problems" of particular theoretical importance. These problems included for example the interpretation of the ancient Greywacke Group, *terrains de transition*, or *Übergangsgebirge*, and of the oldest fossils then known; the explanation of anomalous fossils giving contradictory evidence of relative age, such as those already giving rise to the "great Devonian controversy"; the reality or otherwise of an overall progression in the history of life, and the antiquity of the human race; the correlation of the Tertiary strata and their relation to the present world of geological processes; and the character and causation of geologically recent movements of the earth's crust. Each of these focal problems had its own "core-set" of leading geologists who were decisively involved in its attempted resolution; but each core-set depended heavily on evidence provided by other individuals, less intensely involved or less competent, including local amateurs. The resolution of these focal problems, which constituted the landmarks in the development of the science, therefore depended in different degrees on the whole range of competence and ability represented within the international geological community.[46]

In 1835 a geologist like Sedgwick, even when his duties confined him to Cambridge or Norwich, could legitimately feel himself part of an extensive network of richly varied scientific exchange, spanning all the countries of Europe and, more sparsely, far beyond. I hope my sketch of geological activities in the year of Sedgwick's half-century will have suggested the futility of regarding the history of geology either as a story of the exploits of a few great heroes or as an achievement bounded by any political or linguistic frontiers.[47] Sedgwick was indeed a fine geologist; but his memory would be best honoured by emulating his own generosity towards his colleagues. It was he and they together—Sedgwick *and company*—in intensive interaction, whether friendly or acrimonious, who created the "early classical" geology that forms the foundation of our modern science of the earth.

NOTES

[1] An abbreviated version of this paper was given in Cambridge on 22 March 1985, as the sixth Ramsbottom lecture to the Society for the History of Natural History, during a meeting to celebrate the bicentenary of the birth of Adam Sedgwick. The research for this paper was assisted by a grant from the Royal Society of London, for which I am very grateful.

[2] Clark & Hughes (1890 i: 46; ii: 477). This Victorian 'Life and Letters' remains an indispensable source; but see Speakman (1982) for valuable material on Sedgwick's Yorkshire background.

[3] Figure 1 is reproduced from an engraving after a portrait painted in 1833 by Thomas Phillips; it is the nearest in date to 1835.

[4] In this paper the calendar year 1835 will be extended back by two or three months into 1834, in order to include one complete academic year, and its equivalent in the annual cycle of the major geological societies.

[5] The collection had long outgrown the single room containing John Woodward's original cabinets, but little further space was made available until a small public museum was opened in 1841 (Clark & Hughes, 1890, i: 197, 233, 262, 428; ii: 18, 26, 350). Figure 2 is reproduced (by kind permission of the Curator of the Sedgwick Museum) from an undated engraving, probably published to mark the opening of that museum.

[6] Four years later, in 1839, Sedgwick described his audience as "two Heads of Colleges—many Masters of Arts—and about 60 undergraduates" (Rudwick, 1985: 331). Relative to the small size of the whole university, the attendance was impressive. On Buckland's lectures, see Rupke (1983). Here and in the rest of this paper, figures in brackets after the first mention of individuals denote their ages in 1835, and are given in order to convey some sense of the age-range within the community of active geologists.

[7] Sedgwick (1832), a revision of what he had first published in 1821, soon after beginning his annual course. The accelerating pace of progress in the science is reflected in the much shorter interval that elapsed before he produced a third edition, in 1837.

[8] On the role of 'historicism' in geology, see Oldroyd (1979); on the 'structural' approach that was more prominent in the everyday *practice* of geologists, see Rudwick (1982a). Other branches of geology, such as mineralogy and palaeontology—the latter term only just coming into use—were regarded as ancillary disciplines, linking geology to other natural sciences such as chemistry and zoology.

[9] *Cambridge Chronicle*, 10 April 1835, reprinted in Clark & Hughes (1890) ii: 491–4. This was apparently the first instance of what became a celebrated annual event.

[10] Cole (1834). For a classic assessment of this literature, see Gillispie (1951); also Millhauser (1954).

[11] Sedgwick (1833); in the event, Buckland's book (1836) was delayed until the following summer. Their treatment of the "argument from design" was not identical: see Sedgwick's criticisms of Buckland's book, in Clark & Hughes (1890) i: 469-471.

[12] There is still no adequate comparative study of European geological institutions in this period, but Zittel (1899) remains a valuable source of factual information (the English translation of 1901 is abridged and lacks the references).

[13] Clark & Hughes (1890) i: 432-5. The Trinity dividend fluctuated from year to year; Sedgwick's net income from a small living in rural Cambridgeshire had shrunk to nothing: Clark & Hughes (1890) i: 265, 384.

[14] Figure 3 is reproduced (by kind permission of the Director of the British Geological Survey) from a field notebook of De la Beche. It is one of a series of anti-Lyellian caricatures, which are reproduced and analysed in detail in Rudwick, 1975b.

[15] See for example the detailed instructions for the field geologist in Boué (1835-36).

[16] Sedgwick, MS notebook 28, notes dated 1 October 1835 (Sedgwick Museum, Cambridge: quoted by kind permission of the Curator).

[17] Secord (1986) describes and analyses the collaboration and subsequent conflict between Sedgwick and Murchison, over the definitions of their "Cambrian" and "Silurian" systems.

[18] Brochant de Villiers (1835), their superior, summarised the progress of their survey to the Académie des Sciences at the end of the year; the map was then complete, but problems with the engraving delayed its publication for another six years (Dufrénoy & Élie de Beaumont 1841).

[19] "Baconian" was the term favoured at the time, though its relation to the ideas of Francis Bacon himself is questionable. On its rhetorical functions, see Yeo (1985).

[20] Robinson (1948) describes how the system of horse-drawn mail-coaches was at its most efficient in the 1830s, just before it began to be superseded by the new railways.

[21] M.B. Hall (1984) describes the nineteenth century reforms at the Royal Society, as a result of which Fellowship was transformed from a badge of social respectability into a coveted honour for professional scientists.

[22] The medal was given primarily for the *Principles* (Lyell, 1881, i: 443; Greenough, 1835: 169-170); but there would have been some disapproval if he had never presented *any* of his research at the Society.

[23] See the contemporary illustration of a meeting, reproduced in Needham & Webster (1905: 235). See also Forgan's (1986) analysis of the lay-out of scientific meeting-rooms.

[24] This section draws on the Society's archives, and particularly its Ordinary and Council minutes for 1834-35, as well as its printed records.

[25] No comparable society was to be founded in the fragmented German-speaking world for another thirteen years, and then only as a reflection of the growing political dominance of Prussia: the Deutsche Geologische Gesellschaft was founded in Berlin in 1848 (see its *Zeitschrift*, i, pp. 1-40, 1849).

[26] The sketch reproduced as Figure 4 (by kind permission of the Council of the Society) is undated. The ichthyosaur's head was probably the one described in Conybeare & De la Beche (1821); the section has not been identified. By 1838 Lyell was describing a meeting at Somerset House as gathered round a "great horseshoe table" (K. Lyell, 1881, ii:37), evidently much larger than this one. For a comparable lay-out at Somerset House, see the contemporary illustration of a meeting of the Society of Antiquaries, reproduced in Needham & Webster (1905:239).

[27] Summaries were printed in the Society's *Proceedings*, ii: Murchison's papers are at pp. 114-122, 193-5; Sedgwick's, at pp. 181-5, 198-200; De la Beche's, at pp. 106-7; Lyell's, at pp. 191-2; Buckland's, at p. 190. For the presidential address, which was printed in full, see Greenough (1835).

[28] Summaries were printed in the *Bulletin* of the Société, vi: Dufrénoy's papers are at pp. 239-247, 250-1; Prévost's, at pp. 114-5, 292-4. Rozet's 91-page address is printed in full in an appendix. The *Bulletin* reported on a much wider range of material than the *Proceedings*, ranging from long abstracts of formal papers to brief extracts from letters received; the figure of 42 items represents those substantial enough to be printed in a distinctive typeface, and it is comparable to that given for the London society. Geological papers were also read at the Académie des Sciences, much more frequently than at the Royal Society; but the greater prestige of a presentation at the Académie was offset by the inhibiting formality of its weekly meetings.

XI

[29] Rudwick (1985); the argument that precipitated the controversy is described on pp. 99–103. No accounts of the discussions were ever officially recorded, still less published, and they can only be reconstructed from the correspondence of those present.

[30] Discussion was first recorded in the *Bulletin* (i, p. 55) in the report of the fourth meeting (not counting some early business meetings); thereafter, discussions were reported frequently and fully in print, in contrast to the London society.

[31] *Bulletin*, vi, pp. 93–5, 252–3.

[32] Referees' reports for 17 of the papers read during the session 1834–35 are preserved in the Society's archives (COM.P4/2), pp. 2, 13, 15, 30, 42–4, 59–60, 82–3, 111, 125, 174, 205, 210, 212, 248). The 'specialist' referees were Broderip, Clift, Conybeare, Stokes and Turner.

[33] See the map of London in Rudwick (1985 Fig. 2.6).

[34] "List of the Geological Society of London, March 1, 1835" (copy in the Society's archives). The other society whose Fellows were recorded in the list was the Horticultural (85, or 13%), but its prestige was social rather than intellectual. Members of the Zoological (founded 1828) were *not* listed, probably because it was still at this time of dubious political and social status in the eyes of those who ran the Geological: see Desmond (1985).

[35] This and subsequent figures are based on an analysis of the 1835 membership list of the Geological Society (note 34), and the similar list of the Société Géologique dated October 1835 and printed in the *Mémoires*, ii, pp. v–xv. At this point in the present essay, non-British and non-French members (respectively) are excluded. The Geological Society figures also exclude the "Honorary Members", a closed and dwindling category (43 in 1835) of early provincial members who had declined its invitation to become Fellows (and to pay the appropriate fee!) after it was awarded a royal charter in 1828.

[36] Croizet & Jobert (1826–8). The French list is more explicit than the English about the professions of members: the English list gives no indication, for example, of legal qualifications.

[37] On the work on the French map, see Brochant de Villiers (1835) and the memoir accompanying the map (Dufrénoy & Élie de Beaumont 1841); on the parallel developments in England, see De la Beche (1839) and McCartney (1977), also Greenough (1820, 1840).

[38] Many of the French landowners—but probably not all—are identified in the membership list as "propriétaires"; many of those in the English society can be identified—at a first approximation—from the form of their rural addresses. Some landowners, of course, may have owned industrial as well as agricultural property.

[39] Figures 5 and 6 are based on addresses given in the 1835 membership lists of the two societies; symbols with a wavy underlining denote individuals who have been located only to the level of county or *département*. The geographical spread was even wider than the maps suggest, because many members listed with metropolitan addresses, particularly the aristocracy and landed gentry, also had residences in the provinces.

[40] *Proceedings*, ii, pp. 102–3, 111–2; *Bulletin*, vi, pp. 132–3 (Castel), 255–9 (Breton), 294–300 (Drouet).

[41] *Bulletin*, vi, pp. lxxxviii–ix. The three were Mougeot, Hogard and Jacquiné.

[42] Rudwick (1985:149, 152). Harding had not yet been elected to the Geological Society.

[43] K. Lyell (1881, i: 453) the local geologists were, respectively, Thirria, Thurmann, Montmollin and Coulin.

[44] *Proceedings*, ii, pp. 114 (Hall), 179–180 (Freyer), 188–9 (Belcher), 109–111 (Cunningham), 189 (the younger Hennah), 210–2 (Darwin). Darwin was elected to the Society soon after his return to England in 1836.

[45] De la Beche (1834); Lyell (1834–5); Schmerling (1833–4); *Bulletin*, vi, pp. 83, 90, 95, 231; *Proceedings*, ii, pp. 99–102, 133–4.

[46] For a fuller discussion see Rudwick (1985:426–8). On "core-sets", see Collins (1981).

[47] In this as in other ways Karl von Zittel led the way, when he gave his contribution to the multi-volume *Geschichte der Wissenschaften in Deutschland* an international dimension, pointing out that "in hervorragenden Mass sind Geologie und Paläontologie Disciplinen für welche es keine politischen und sprachlichen Grenzen gibt" (Zittel, 1899:1).

REFERENCES

ALLEN, D. E., 1976 *The naturalist in Britain: a social history.* London.

BOUÉ, A., 1835-36 *Guide du géologue voyageur, sur le modèle de l'agenda geognostica de M. de Leonhard.* Paris.

BROCHANT de VILLIERS, A., 1835 Notice sur la carte géologique de la France. *Comptes-Rendus, Séances de l'Académie des Sciences* 1835: 423-429.

BUCKLAND, W., 1836 *Geology and mineralogy considered with reference to natural theology.* London.

BURKHARDT, F. & SMITH, S. (eds.), 1985 *The correspondence of Charles Darwin.* 1. Cambridge.

CLARK, J. W. & HUGHES, T. McK. (eds.), 1890 *The life and letters of Adam Sedgwick.* Cambridge.

COLE, H., 1834 *Popular geology subversive of divine revelation!* London.

COLLINS, H. M., 1981 The place of the "core-set" in modern science: social contingency with methodological propriety in science. *History of Science* 19:6-19.

CONYBEARE, W. D. & DE LA BECHE, H. T., 1821 Notice of a discovery of a new fossil animal, forming a link between the ichthyosaurus and the crocodile; together with general remarks on the osteology of the ichthyosaurus. *Transactions of the Geological Society of London* 5:558-594.

CROIZET, J. B. & JOBERT, A. C. G., 1826-28 *Recherches sur les ossemens fossiles du Département du Puy-de-Dôme.* Paris.

DE LA BECHE, H. T., 1839 *Report on the geology of Cornwall, Devon and West Somerset.* London.

DESMOND, A., 1985 The making of institutional zoology in London, 1822-1836. *History of Science* 23: 153-185, 223-250.

DUFRÉNOY, P. & ÉLIE de BEAUMONT, L., 1841 *Carte géologique de la France.* Paris.

FITTON, W. H., 1828, Address delivered on the anniversary, February 1828. *Proceedings of the Geological Society of London* 1 (6): 50-62.

FORGAN, S., 1986 Context, image and function: a preliminary enquiry into the architecture of scientific societies. *British Journal for the History of Science* 19: 89-113.

GILLISPIE, C. C., 1951 *Genesis and geology. A study in the relations of scientific thought, natural theology, and social opinion in Great Britain, 1790-1850.* Cambridge, Mass.

GREENOUGH, G. B., 1820 *A geological map of England and Wales.* London.

GREENOUGH, G. B., 1835 An address delivered at the anniversary meeting of the Geological Society of London, on the 20th of February, 1835. *Proceedings of the Geological Society of London* 2(39): 145-175.

GREENOUGH, G. B., 1840 *A physical and geological map of England and Wales.* London.

HALL, A. R., 1969 *The Cambridge Philosophical Society. A history, 1819-1969.* Cambridge.

HALL, M. B., 1984 *All scientists now: the Royal Society in the nineteenth century.* Cambridge.

LYELL, C., 1835 On the proofs of the gradual rising of the land in certain parts of Sweden. *Philosophical Transactions of the Royal Society of London* 1835: 1-35.

LYELL, K. (ed.), 1881 *Life, letters and journals of Sir Charles Lyell, Bart.* London.

McCARTNEY, P.J., 1977 *Henry De la Beche: observations on an observer.* Cardiff.

MILLHAUSER, M., 1954 The scriptural geologists. An episode in the history of opinion. *Osiris* 11: 65-86.

MORRELL, J. B. & THACKRAY, A., 1981 *Gentlemen of science. Early years of the British Association for the Advancement of Science.* Oxford.

MURCHISON, R. I., 1835 On the Silurian system of rocks. *Philosophical Magazine and Journal of Science* ser. 3, 7: 46-52.

NEEDHAM, R. & WEBSTER, A., 1905 *Somerset House, past and present.* London.

OLDROYD, D. R., 1979 Historicism and the rise of historical geology. *History of Science* 17: 191-213, 227-257.

PHILLIPS, J., 1834 *Syllabus of a course of eight lectures on geology.* London.

ROBINSON, H., 1948 *The British Post Office. A history.* Princeton.

ROSS, S., 1962 Scientist: the story of a word. *Annals of Science* 18: 65-85.

RUDWICK, M. J. S., 1975 Caricature as a source for the history of science: De la Beche's anti-Lyellian sketches of 1831. *Isis* **66**: 534-560.

RUDWICK, M. J. S., 1982a Cognitive styles in geology. *In* DOUGLAS, M. (ed.) *Essays in the sociology of perception.* London.

RUDWICK, M. J. S., 1982b Charles Darwin in London: the integration of public and private science. *Isis* **73**: 186-206.

RUDWICK, M. J. S., 1985 *The great Devonian controversy. The shaping of scientific knowledge among gentlemanly specialists.* Chicago.

RUPKE, N. A., 1983 *The great chain of history. William Buckland and the English school of geology (1814-1849).* Oxford.

SECORD, J. A., 1986 *Controversy in Victorian geology. The Cambrian-Silurian dispute.* Princeton.

SEDGWICK, A., 1832 *A syllabus of a course of lectures on geology.* Cambridge.

SEDGWICK, A., 1833 *A discourse on the studies of the university.* Cambridge.

SEDGWICK, A., 1835 Remarks on the structure of large mineral masses, and especially on the chemical changes produced in the aggregation of stratified rocks during the different periods after their deposition. *Transactions of the Geological Society of London,* ser. 2, **3**: 461-486.

SPEAKMAN, C., 1982 *Adam Sedgwick. Geologist and dalesman, 1785-1873. A biography in twelve themes.* Heathfield.

WILSON, L. G., 1972 *Charles Lyell. The years to 1841. The revolution in geology.* New Haven.

YEO, R. 1985 An idol of the market-place: Baconianism in nineteenth century Britain. *History of Science* **23**: 251-198.

ZITTEL, K. A. von, 1899 *Geschichte der Geologie und Paläontologie bis Ende des 19. Jahrhunderts.* Munich and Leipzig.

XII

Travel, Travel, Travel:
Geological Fieldwork in the 1830s

'We must preach up travelling, as Demosthenes did 'delivery', as the first, second, and third requisites for a modern geologist, in the present adolescent state of the science.'

'We want nothing short of a radical reform in geology & we shall have one soon if honest men will travel and write and travel again.'

Lyell to Murchison, Naples, February 1829

Geology in the 1830s

Appropriately for a conference on British culture in the 1830s, I take preaching as a serious genre, and start with texts. They come from a remarkable sequence of letters that Charles Lyell wrote to his erstwhile geological companion Roderick Murchison, immediately after his fieldwork in Sicily on the eve of the 1830s. This was the culminating segment of a nine-month tour through France and the future Italy, which, as he wrote later, 'made me what I am in theoretical geology'. That was no mere expression of hindsight: Lyell's notebooks from his tour are packed with rehearsals of arguments he was clearly planning to use in a major theoretical treatise, and as soon as he returned to Britain he began rewriting earlier drafts into what became the *Principles of Geology* (1830–33). In this paper I shall use Lyell's tour of 1828–29 – with apologies for stretching the 1830s as historians routinely stretch their centuries – to exemplify the character of geological fieldwork in the British scientific culture of our chosen decade. More specifically, I want to examine the relation between the practice of fieldwork, the epitome of what many of our 'men of science' regarded as the collection of 'facts', and the practice of theory-building on the highest level of generality and explanatory scope.

The prominent place of geology in the pecking order of the sciences in the Britain of the 1830s is well known. The sequence of symbols used by the British Association for the Advancement of Science for its specialized Sections was clearly no accident, even though they began as merely conventional labels for the doors of lecture-rooms. What is striking to modern eyes about the order chosen is

XII

the high place given to geology, as 'Section C', immediately after the physical and chemical sciences, and before any kind of biology, let alone the social or human sciences. But in the eyes of the 1830s, and certainly in those of the metropolitan gentlemen who were busy hijacking the provincial Association for their own purposes, geology clearly deserved its distinguished place, by the imputed sublimity of its subject-matter, the acknowledged eminence of its leading exponents, and the broad popularity of its basic practices.

The practices of geology included, more prominently than any other, the practice of outdoor fieldwork. It was this that enjoyed a popularity unmatched before or since. Fieldwork appealed to a range of devotees that transcended the many boundaries and barriers of British society: to young and old, male and female, aristocrats and working men. What they expected from their fieldwork, and what they found in it, were of course as varied as their stations in life. But there was enough in common to constitute a distinctive culture. Its outward badge was a miner's hammer, which, in the hands of many who were very obviously not miners, became a geological hammer. Following an iconographic tradition that stretched back into the previous century, portraits of geologists almost invariably featured a hammer, as a prop that was as distinctive as St Peter's keys. No one dared lay claim to the title of geologist without having experience of fieldwork to point to as a source of authority.

The practice of geological fieldwork in the early nineteenth century invites deconstruction and other forms of cultural analysis, and the opportunity has not been missed. But the resultant debate has, I believe, missed the main point. It seems undeniable, for example, that fieldwork was regarded by some as a Romantic activity. Certainly some of the most spectacular geological sites, such as the Giants' Causeway and Fingal's Cave, were also considered highly Romantic sights. But those who visited such places with ambitions to inform themselves or to investigate, rather than merely to admire, did so in a decidedly un-Romantic manner. They wanted to see the rocks in a hard and clear light, not wreathed in swirling clouds. If they could not visit in person, the proxies or pictorial representations they valued were the painstakingly accurate and hard-edged engravings of topographical artists such as the Daniells, not the Romantic impressions of Turner. Geologists who made their own visual records of dramatic scenery, or who paid someone else to do so for them, preferred a drawing made with the camera lucida, in order to avoid Romantic exaggeration of the vertical scale. Romantic evocations were indeed exploited by geologists on occasion, as when Gideon Mantell got John Martin to depict the reconstructed 'country of the Iguanodon'; but significantly that commission was designed to help sell a work aimed at the general book-reading public (*Wonders of Geology*, 1834), not to illustrate a scientific memoir.

Likewise it is undeniable that some geologists regarded their feats of energetic outdoor fieldwork as expressions of macho virility; Murchison was not the only one who treated it as surrogate fox-hunting or as a quasi-military campaign. But the countryside also teemed with women fossil collectors, who clearly regarded fossilizing as an activity no less ladylike than botanizing, both to be pursued outdoors in the field. Fieldwork did not by its nature require great feats of heroic endurance, and the rhetoric that stressed its manliness was directed at effete urban life rather than at femininity.

Finally, geological fieldwork unquestionably acquired part of its macho image from the workaday activities of lower-class practical men or mineral surveyors such as William Smith and John Farey. But at the same time many of their gentlemanly rivals in the Geological Society went out of their way – in more senses than one – to distance their fieldwork from any merely utilitarian goals, regarding for example the Lake District rather than the Newcastle coalfield as an appropriate site for their research. They may have enjoyed the sartorial disguise of fieldwork clothes, and relished the social inversion of being on occasion mistaken for working men; but the joke would have turned sour if they had not been able to re-establish their status at the end of the day, whether in a local inn or at the country seat of a nearby amateur geologist.

All such cultural interpretations miss the point, in part because they are set up as mutually exclusive alternatives. On the contrary, it seems to me to be unsurprising that geological fieldwork was motivated and legitimated in such diverse ways, given the social variety of its practitioners. What all such interpretations miss, however, is what the actors themselves – even the most amateurish of them – would have stressed the most: that they hoped their fieldwork would contribute, in however humble a way, to scientific knowledge. There are times, and I believe this is one of them, when we analysts of science should pay attention to the actors, and not treat them merely as objects ripe for deconstruction. Unless we take seriously the claims made for fieldwork as the main epistemic input for geology, our interpretations are likely to be highly defective.

The Practice of Fieldwork

It would be salutary to begin with a recognition of our shocking ignorance, as analysts, of two of the three main places of scientific knowledge. We now have many insightful analyses of the experimental practices that take place in laboratories, past and present. For the 1830s, David Gooding's work, for example, has enabled us to imagine what it would have been like to peer over Michael

Faraday's shoulder while he tinkered in his lab in the basement of the Royal Institution, or to be in his audience upstairs when he performed a well-rehearsed experiment. But not all scientific knowledge is constructed in laboratories, although physicists and historians of physics often imply that it is. Much of science is shaped – and was in the past – in museums and in the field. We need far more historical attention to both those sites. Here I am only concerned with the field, and specifically with the role of travel and fieldwork in the geology of the 1830s.

To constrain the topic into a manageable size, I restrict myself further, and without apology, to the geological elite. However unfashionable it may be to point it out, it was the opinions of the Lyells and Murchisons, the Bucklands and Sedgwicks, the Whewells and Herschels, and their Continental cousins, that counted in the debates; it was their opinions that were shaped into consensual geological knowledge. Those conclusions were of course dependent on an infrastructure built by lesser men, and many women, some of them known by name, but many almost invisible to history. But the relation was asymmetrical: local amateurs and even regional experts simply provided the materials; the elite geologists made global sense of them. The elite not only assumed authority to do so; except at times of controversial crisis, that authority was generally conceded by others. It was conceded willingly, above all because the elite geologists were *well-travelled.*

Travel, considered here as a specific form of fieldwork, was respected because it enabled the well-travelled to compare and contrast in a way that even the most expert in one region could not. Again unfashionably, I would argue that this appreciation of well-travelled fieldwork was grounded in the intrinsic character of the science's empirical materials (which would explain the continuing appeal of Lyell's prescription, even to modern geologists). In its roots in eighteenth-century mineralogy, the future science of geology – like the botany and zoology of the time – had been a science primarily of specimens: specimens collected during fieldwork in provincial or exotic places, and then assembled for collation and comparison at central or metropolitan sites. Such collecting could be, and often was, carried out by students, assistants and other underlings. But for what would now be called the earth sciences, naturalists such as Saussure had argued eloquently that first-hand field observation by the savant himself was imperative, if the meaning of mineral specimens was to be understood in their proper context of mountains, volcanoes, rock formations and other large-scale features. That emphasis on fieldwork – which by the 1830s was taken for granted among geologists – was no 'mere' rhetoric, for it was based on the recognition that those large-scale features, as much as or even more than portable specimens, were the primary materials for understanding the earth.

There was no substitute for travelling to see such features with one's own eyes, and for two distinct reasons. First, the features themselves were by their nature immobile; Mahomet had to go to the mountain, or Lyell to Etna. Seeing the mountain with one's own eyes and climbing to the top was not just an exhilarating experience, though it could be that too; it was also a scientific necessity, if multiple impressions and observations were to be integrated into a whole. As a poor and pale analogy, consider the difference between seeing a Henry Moore in a landscape, walking all round it and watching it in different lights as the sun moves round the sky, feeling it all over to appreciate the texture of the stone or bronze – and inspecting even the best set of reproductions of the same sculpture in a coffee-table book.

That analogy points to the second reason for the importance of geological travel. Without the observational experience that only travel could bring, the geologist had to rely on visual or verbal representations that were mobile; or, to cut the trendy jargon, on pictures and texts. In the 1830s, these were in most cases poor proxies for the real thing. Most engravings – mobiles that were immutable at least within a given edition – were degraded pictorially by comparison with the original drawings; and even the latter, if the geologist had access to them, were usually limited by the artist's lack of understanding of what made the view scientifically significant. Hence the least unsatisfying pictures, or the best proxies for a first-hand visit to the locality, were those drawn either by a geologically informed professional artist or by a geologist with passable artistic competence and a well-defined interpretation to convey. Examples would be, respectively, the views of the spectacular strata in the cliffs around the Isle of Wight (1816), by the Geological Society's functionary Thomas Webster; and the overtly theory-laden panoramas of the extinct volcanoes of Auvergne (1827), by Lyell's friend 'Pamphlet Scrope', the geologist, political economist and Member of Parliament. As for textual descriptions of such features, their limitations as proxies for first-hand visual experience were regarded as self-evident; at best, they had some limited value as verbal commentaries on the pictures.

Given the importance of seeing immobile large-scale features at first hand, it is not surprising that certain sites or regions had acquired an almost canonical status among geologists. Ever since the pioneer work of naturalists such as Desmarest, Saussure and Hamilton in the previous century, regions such as Auvergne and the Alps, and specific features such as Vesuvius and Etna, had come to constitute an almost stereotypical Grand Tour for all geologists with pretensions to be regarded as well-travelled. Such travellers did not mind that these places were far from virgin territory; on the contrary, what put them into the three-star *vaut le voyage* category was precisely that they were *well* known. Geologists travelled in order to see such features through their predecessors' eyes; and having done so, they hoped to add some further insight or to improve on the traditional interpretations. They had no illusions about being 'mere' Baconian fact-collectors, any more than their laboratory-based colleagues who tried repeating classic experiments.

A Geological Grand Tour

In this way the practice of geological fieldwork, not least in its elite form of extensive inter-regional and international travel, was embedded in the social practices of geological debate and discussion, and also in the reading practices of a coherent scientific tradition. When Lyell and Murchison planned their joint tour to the Continent in 1828, it was – at least for Lyell – embedded in a long-term theoretical research project. In writing an essay-review of Scrope's memoir on central France for the *Quarterly Review* (1827), Lyell had already been taken on what might be termed a 'virtual' field-trip of major importance. Scrope had in effect guided him through the classic country of Auvergne and Vivarais. Lyell had 'virtually' seen the celebrated extinct volcanoes of the Massif Central through Scrope's eyes, aided by the author's interpretative panoramic sketches. (Scrope's own fieldwork, in turn, had not been in virgin territory; on the contrary, he had drawn on a rich tradition of earlier fieldwork, particularly – and with scant acknowledgement – from Desmarest's mapping half a century earlier.)

Noting Scrope's conclusion that the French volcanoes proclaimed Nature's message of 'Time! Time! Time!', Lyell had realised that his own broader project would be well served by converting the virtual field-trip into an actual one. So Auvergne became the first major goal on his and Murchison's itinerary. Far from relying on Scrope alone, Lyell tapped into the network of informal knowledge of the region by consulting other English geologists such as Buckland and Daubeny, who had also been there. Likewise, when he crossed the Channel, he went first to Paris, not only as the hub of all French travel, but primarily to ask geologists such as Prévost and Élie de Beaumont for their similar tips on where exactly to go and what to look out for. So the way he saw the volcanoes and other features, when at last they arrived in Auvergne, was a seeing disciplined by a whole community of contemporaries and predecessors. In confronting Nature they were surrounded by a cloud of witnesses.

Other witnesses were closer at hand, and even accompanied them in person. Although Lyell was inclined to disparage them in his letters back home, he and Murchison were in fact heavily dependent on the detailed advice of local experts. For example, the comte de Montlosier, Auvergnat landowner, historian and geologist, could distil several decades of detailed fieldwork into a concentrate that his visitors could imbibe in a few days. When Lyell tried to find on his own a feature he ardently desired on theoretical grounds – an intrusive contact between granite and Tertiary sediments – he failed miserably. It just wasn't there, and Montlosier would surely have found it if it had been. On the other hand, Lyell did find one pleasing locality unnoticed by the locals – a volcanic cone and lava flow that he regarded as refuting Buckland's diluvial interpretation of the area with particular force – so that he was not merely following in the footsteps of either locals or predecessors. What they led him to, what he found for himself, and what he failed to find, all had discernible differentiating effects on his on-the-spot theorising.

In other cases, it was the local experts who, intentionally or inadvertently, put theoretical constructs into Lyell's mind. For example the accomplished conchologist Risso, the professor of natural history at the lycée in Nice, showed him the fossil shells of some local Tertiary sediments, listed their species, and calculated that 18% were still living in the Mediterranean nearby. Only from that point onwards did Lyell's notebooks and letters begin to mention percentages, as he began to interpret the differences between various Tertiary faunas as differences in relative *age*, rather than as differences due to biogeographical factors. By the time he reached Sicily, he was theorising in terms of a quite general pattern of piecemeal faunal change, on which a quantitative dating method could be based. He had begun to be able to guess the approximate relative age of any sample of Tertiary fossils, just by looking at them; but such a rough-and-ready method was not good enough. For accurate estimates, he was totally dependent on the taxonomic expertise of conchologists: Risso in Nice, Bonelli in Turin, Guidotti in Parma, Costa in Naples, and above all – on his way back – Deshayes in Paris. For Lyell, the technique of conchological dating, although of his own devising, was frustratingly an impenetrable black box, just as radiometric dating is for most modern geologists. He could not escape from dependence on his scientific community; whether he liked it or not, his fieldwork had to be keyed into museum work done by others, if it was to yield its fullest meaning.

His theorizing, on the other hand, did not need to wait for work by others; it could be, and was, done right in the middle of his fieldwork. Far more than Murchison's notebooks (both sets have survived), Lyell's show that detailed observation and even the highest-level theorizing were intimately entwined. Again and again, seeing some specific feature evidently triggered a set of theoretical reflections of great generality. For example, seeing specific Auvergnat villages perched on hilltops, and ancient basalts likewise capping specific plateaux, triggered a mini-essay entitled 'Analogy of Geology & History'. Lyell argued that in both cases only a truly historical interpretation could explain the anomalous positions: mediaeval villages had to be built in defensible positions, not in the fertile valleys, because of the turbulent times; lavas always flowed naturally down valleys, but subsequent erosion had left ancient lava-filled valley-floors high and dry above the modern valleys. Whether Lyell intended such essays as dry runs for the treatise he was planning is uncertain (their polished prose suggests he did); what matters is that he was clearly seeing geological features in the light of the debate he was already carrying on with himself (and probably with Murchison), and that he planned in due course to continue on a wider stage. Fieldwork was not divorced from the agonistic context of his broader project, which he expressed – appropriately for the times, back in England – in terms of bringing Reform itself to his chosen science.

The differentiating effects of the empirical inputs that Lyell experienced during his fieldwork show themselves even more clearly in the way his itinerary was modified in the light of what he saw. While still in the Massif Central, he concluded that the volcanoes of Auvergne and Vivarais not only refuted Buckland's diluvialism but also showed that non-marine Tertiary sediments were far more significant – both in extent and in theoretical implications – than the Parisians appreciated. Above all, the region seemed to Lyell to demonstrate a correlation between volcanic activity and the uplift of whole areas of land. This hinted at a general causal explanation for the elevation of mountains and continents. After he and Murchison reached the Mediterranean, Lyell decided on a radical change of plan, to test this global hypothesis by seeing for himself the more recent, indeed active, volcanoes of Italy and Sicily. (Murchison and his wife later split off and followed the original plan by returning across the Alps and through the German states.)

From that point on, Lyell's tour took on some of the liminoid character that Victor Turner discerned in the structure of pilgimage. Vesuvius, and even more the greater cone of Etna, became the quasi-sacred goals of his quest for a geological theophany. On finding what he took to be fossil evidence of geologically recent elevation in the vicinity of the volcanoes, first on Ischia and then on the flanks of Etna itself, his mounting excitement is palpable, both in his private notebooks and in the letters he wrote to Murchison and to his own family. But as with the pilgrimage experience (at least in the view of pilgrims), exalted emotions led not to a dulling of the critical faculties but rather to a heightening of sensual and cognitive receptivity. The weeks that Lyell spent on Sicily, exploring Etna and its regional context, were a period in which he accumulated vivid visual experiences that became the key to his view of the globe as a dynamic system operating over a vast timescale. Sicily was not virgin territory, of course, any more than Auvergne:

Etna had been a three-star attraction for naturalists ever since Borelli's time, and Lyell was conducted to the summit by one of the Gemmellaro family, just as they conducted many other visiting geologists around the same time. But Lyell was able to stamp his own specific interpretation on Sicilian geology, because he brought to it an equally specific set of other visual experiences from other regions: the modern limestone-*in-statu-nascendi* near his Scottish family home, the analogous Tertiary sequences of Hampshire and the Isle of Wight near his English family home, and now the Tertiaries and volcanoes of Auvergne and those of Italy. It is no wonder that he urged the preaching of travel as the supreme requisite for a geologist.

Fieldwork and Debate

That prescription was modified significantly, however, in the second of the texts I quoted at the start of this paper: not just travel, but – in effect – 'Travel! Write! Travel!' Once again, this indicates how self-consciously Lyell was integrating his practice of largely solo fieldwork with that of communal debate. In fact, he had begun practising what he preached, long before he returned from Sicily to Naples and wrote that letter to Murchison. Not only was he using the privacy of his notebooks to jot down what could be used later as drafts of key passages for publication; not only did he use Murchison as a sounding-board for his ideas (in letters after they parted, and doubtless face-to-face earlier on). More formally, as soon as they left the Massif Central and reached the Mediterranean, Lyell began composing papers that would announce some of their conclusions to the wider geological world. (The papers were nominally by both geologists, but style and content betray Lyell's dominant hand; joint names greatly strengthened the authority of the reports, by indicating that they represented the consensual conclusions of two competent witnesses, not the possibly idiosyncratic ideas of just one.)

These papers show careful strategic planning, whether the strategy was fully conscious or not. One was directed at Lyell's home community, the Geological Society in London, the other to its main – and generally friendly – rival, the Parisian geologists (soon to be organized in parallel manner as the Société Géologique de France). The first, on the excavation of valleys by their present streams, used specific examples from Auvergne and Vivarais to put forward an interpretation of general scope; it was duly read at a meeting in London soon after Lyell's return, and predictably generated a heated debate with the diluvialists. The other paper, on the great Etna-sized ancient volcano of Cantal, likewise had general implications embedded in its regional subject-matter; it was written and published in French for the *Annales des Sciences Naturelles*, the most lively Parisian journal of the time for geological debate. Even before Lyell completed his tour and returned to Britain, some of the results of his fieldwork were thus already being fed into the stream of international discussion. Almost as soon as he returned, he started the major work of revising earlier drafts of what might have been mere *Conversations on Geology*, incorporating a massive amount of material from his Continental fieldwork, and shaping it instead into what became his *Principles of Geology*.

It may be objected, in conclusion, that this brief sketch of Lyell's great field trip of 1828–29 tells us nothing of general significance about the culture of geological fieldwork in Britain in the 1830s: not because it fell outside the 1830s by a few months, but because most geologists were not Lyell. The premise is of course true, but I believe the conclusion is mistaken. Not only do scientific elites, like social elites, deserve their social and cultural histories as much as plebs. In the case of scientific elites, it is manifestly the case that those who earn a place in the elite – and they are not born to it – are those whose opinions come to count in the shaping of what is taken to be reliable knowledge. In the case of geology, in the 1830s as much as today, an important qualification for being taken seriously in this respect was the recognition of being well-travelled. In this paper I have narrowed my scope still further, to Lyell alone, because I follow Howard Gruber's prescription of many years ago, when he chose to study scientific creativity by focussing on one outstanding individual (Darwin) rather than the generality of less creative people. But my analysis of Lyell's fieldwork has shown, I believe, that the outstanding individual never ceases to be embedded in a social matrix. Fieldwork brought the geologist of the 1830s into direct contact with the natural world – the hammer striking the rock symbolized as much – but even in solitude the geologist remained tied into the social world of discussion and argument, collaboration and controversy. Lyell for one would not have wanted it otherwise.

XIII

The Group Construction of Scientific Knowledge: Gentlemen-Specialists and the Devonian Controversy

Introduction

Several years ago Sir Peter Medawar urged analysts of science to study in detail "what scientists *do*" in the course of their research. As a distinguished practicing scientist, Medawar himself was well aware that what scientists *say* about their activity can never be taken at face value. Instead he urged that "only unstudied evidence will do — and that means listening at a keyhole."[1] In fact, however, the few analysts of science who have followed this prescriptive suggestion effectively have preferred to establish themselves as participant observers on the inner side of the keyhole, to penetrate into the laboratory itself and to watch the activities of scientists from the perspective of the ethnographer or anthropologist.[2] Valuable and provocative as such studies are, however, I think their authors have underestimated the extent to which historical studies can also contribute to a better understanding of scientific practice.[3] Clearly it is desirable in any case that we should trace how the practice of research has changed over the *longue durée* of the history of science, in conjunction with changing social and cognitive circumstances. But quite apart from that, there are also ways in which much more 'fine-grained' historical studies may give us better access than research on modern science can, to the "unstudied evidence" that Medawar saw was needed; and

XIII

194

they may avoid the observational or interpretative 'distance' that was implied in his metaphor of the keyhole.

Even for contemporary science, there is an ever-present danger of interpreting the activities of scientists and the processes of research practice with the benefit of hindsight. Realizing this, some cognitive sociologists of science have deliberately chosen to study problems or controversies that are as yet unresolved, in order to avoid any possible use of hindsight, by the scientists they interview or by themselves.[4] But there is another way to avoid the same danger, and it is a way that allows the historian to make a non-retrospective analysis even of problems that have long since been resolved. The method is simply to reconstruct the chosen episode with the strictest attention to that scorned and neglected component of historical practice — precise chronology. One must think oneself back into the lives of the historical actors *as their research proceeded*: not just in a general sense of getting inside the skin of their 'world-view' or even the contemporary state of their particular discipline, but in the far more specific sense of reliving what they did and said and wrote and argued about, week by week and month by month.

At present, most of the best examples of such 'fine-grained' reconstructions are focused on the work of some particular outstanding individual whose *Nachlass* is rich and complete enough to allow a detailed reconstruction of the development of his or her research. An outstanding example would be the fine body of recent research on Charles Darwin's early work on a theory of evolution.[5] Yet even here, it is only recently that historians have become aware of the dangers of retrospective analysis; only recently have some of them begun to treat each phase of Darwin's work as a cognitive entity in its own right, and not just as a building-block added to a cumulative structure, or as an almost inevitable step on Darwin's path to a finally successful theory.[6] It is at least arguable, however, that such examples of 'fine-grained' reconstruction, however brilliantly carried out, are seriously atypical, precisely because they are focused on individual scientists who were so outstandingly original that their work proceeded in relative isolation from other scientists.

Without falling into the opposite and currently fashionable trap of scientific egalitarianism, which assumes that Nobel prize-winners are no more worthy of attention than run-of-the-mill Ph.D.'s, we can surely allow that much, if not most research in science, at least in the past two centuries, involved an interactive *group* or network of specialists.[7] Yet we still lack any substantial body of examples to show precisely how, in 'fine-grained' chronological detail, new knowledge is built up out of these group

XIII

interactions. Material for such examples may not be as rare as historians of science tend to assume. The contingencies of certain places and periods in science, and certain social configurations of scientists, may have preserved unexpectedly rich evidence of the social processes by which particular pieces of claimed scientific knowledge were constructed and validated. These episodes may or may not turn out to be typical of science as a whole. The very circumstances which occasioned the exceptional preservation of documentary evidence may bear witness to their atypicality. But unless we try to assemble some examples of this kind, we shall not even discover the parameters of possible variation in the historical practice of scientific research.

Background and Setting

In this paper I shall give a brief progress report on my current work of writing a 'fine-grained' and nonretrospective narrative and analysis of one such episode.[8] The period is that extraordinarily fertile time for many of the natural sciences, the second quarter of the nineteenth century. The place is mainly England and particularly London, but with important extensions to Paris, to the rest of Europe, and, eventually, to the rest of the world. Combining the place and the period, it is not surprising that the social setting is that of 'gentlemen of science'.[9] The main historical actors were *gentlemen-specialists*: gentlemen both by their social class, and by their possession of resources that enabled them to carry out substantial scientific research without the need of paid employment for it; specialists by virtue of the fact that even the most polymathic of them concentrated their main efforts in one or a few specialized branches of natural knowledge.[10] The science chosen in this instance was geology, which was then experiencing its first and greatest boom in conceptual innovation, empirical expansion, and public approval and interest.[11] The episode is one that was widely known and discussed at the time in scientific circles, though it has since dropped into obscurity; but this is a positive advantage for the historian, because the story can be told without the readers knowing beforehand who were the goodies and who the baddies, or how the plot was to end.

I first became aware of the importance of this episode many years ago, while I was sorting through the scientific correspondence — at that time still privately owned — of one prominent geologist of the period.[12] For I came across a bundle of letters, apart from the rest, tied up with red tape and labeled 'Great Devonian Controversy'. Reading these letters made me aware that here were the traces of a controversy that had evidently been central to geology for almost a decade. It had raised questions at the heart of contemporary geological practice, and it had had far-reaching implications for the careers of some of those involved. But above all, these letters revealed at once the kind of vigorous informal argument, combative and persuasive by turns, that I was then experiencing at first hand as a practicing scientist. Yet I knew also from that experience that in the mid-twentieth century such informal arguments vanish almost entirely into thin air above the coffee cups at scientific conferences and similar occasions, leaving only the most distorted and misleading traces in formal papers and the unreliable memories of participants. It became clear to me, therefore, that the Devonian controversy might be an unusually favorable strategic site for an analysis of "what scientists *do*," or at least what they *did* a century and a half ago in one specific social setting. So in the interstices of other research I began tracking down the other sides of this correspondence in other archives. By an informal process, analogous to the sociologists' technique of snowball sampling, the network of those who turned out to have been involved in the Devonian controversy grew slowly outwards from where I had first happened to find it, until it came to comprise — in minor if not major roles — most of the leading geologists of the time in Britain and many of those on the Continent too.

It was not only leading geologists, however, who played roles of various kinds in the Devonian controversy. Any adequate narrative and analysis of this episode must also attend to the roles of many lesser figures too. In place of the customary dichotomy between active scientific performers and their relatively passive audience, with its implicit definition of a strong-boundaried scientific community, we need a mental image of the social and cognitive 'topography' of scientists that allows for much weaker boundaries and for many kinds of contingent variation in different sciences at various periods. For geology at the time of the Devonian controversy, a topography of concentric weak-boundaried *zones of competence* is appropriate (Figure 1).[13] Such zones, and the individuals who populated them, can be recovered from the historical record by analyzing the ways in which specific individuals can be seen to have treated the work of others. This analysis recovers a tacit pecking order or gradient of competence in geology — the competence being of course that ascribed by the geologist at the time to themselves and others. Even where they argued vehemently with each other, they implicitly acknowledged that their opponents' arguments needed to be taken seriously. So there was a high degree of tacit consensus about the form of this invisible topography: those in zones of lesser competence generally accepted that their proper place was there, even if they had ambitions to climb higher.[14]

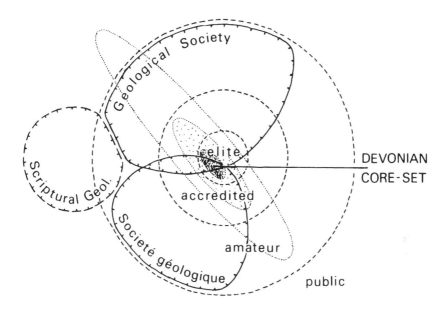

Figure 1. The social and cognitive topography of geology in the 1830s, and of the Devonian controversy, drawn as a Venn diagram. This shows (a) the three weak-boundaried concentric zones of ascribed competence in geology (elite, accredited, amateur) surrounded by the general public; (b) the analogous zones of relative involvement in the Devonian controversy (the 'core-set' of major involvement is densely stippled; the surrounding 'matrix' of moderate or slight involvement is stippled more lightly or left blank); and (c) the strong-boundaried major geological societies (Geological Society of London, Societé géologique de la France) and the marginal "scriptural geologists."

Geologists in the zone I have termed the 'elite' deemed themselves competent to put forward theoretical interpretations of the Earth and its history at the highest level of generality. 'Accredited' geologists, by contrast, were deemed competent to give reliable accounts of the geology of the regions or strata or other phenomena of which they had first-hand knowledge, but not to give high-level interpretations beyond those limits. 'Amateurs', in the sense I use the term here, were deemed competent to make reliable local observations and to assemble reliable collections of specimens, but not to offer even local interpretations of their significance. Members of the general public were not even deemed competent to make reliable observations: their reports were treated with scepticism, unless and until they had been checked by someone with at least 'amateur' status.[15]

An episode like the Devonian controversy was a *focal problem* or temporary 'hot-spot' that disturbed the otherwise even tenor of much routine research in the science.[16] As such, it was of central importance to a certain subset of the elite geologists, and it also engaged the attention of subsets of the accredited and amateur geologists in major or minor ways. To represent this, a second-order 'topography' must be superimposed on the first, making it a complex Venn diagram (Figure 1). Those elite geologists who were involved in the controversy as a major part of their current work, and who were familiar with all the relevant arguments as it proceeded, were also those who — by virtue of their elite status — were deemed competent to pronounce on its widest theoretical implications. These men constituted the 'core-set' for this particular focal problem in science; they were the small set of those through whose changing opinions the controversy was ultimately deemed to have been settled.[17] Beyond that core-set, however, was a wider circle or matrix of those who contributed to the debate in important but lesser ways. Some of them were elite geologists who might simultaneously be in the core-set that was debating some other focal problem in the science.[18] Others were accredited geologists with highly regarded local or specialized expertise that was deemed relevant to the Devonian problem. Others again were merely amateur geologists who assembled reliable local collections of specimens, which were then offered to those in higher zones of competence for evaluation and interpretation. Finally there were some who did not even enjoy amateur status — notably quarrymen and miners from far below the gentlemanly social class — whose collections, made for payment, were nonetheless accepted as important once their authenticity had been checked.

These two graded social topographies — the one for the Devonian focal problem being in a complex manner a subset of the one for geology as a whole — must also both be regarded as cognitive in character, since they were tacitly defined by the ways in which claimed knowledge or information was treated according to its point of origin and in its subsequent transmission across the topography. Of course, no visual representation of such topographies can adequately depict the subtleties of the real social and cognitive interactions that sustained them. But the value of such a diagram is that it encourages the historian to attend to the structure of the *whole* network of information-flow through which new knowledge was constructed.[19] Certainly in the case of the Devonian controversy, any analysis would be seriously defective if it ignored the effects on the debate of empirical material derived from even the lowest zones of ascribed competence. Specimens collected by amateurs or quarrymen with little or no understanding of the higher theoretical issues were nonetheless a vitally important empirical input into the debate.

My diagram, however inadequate, does at least do rough justice to the social shape of the whole of geology as it was practiced internationally at the time of the Devonian controversy, since there was substantial communication and mutual evaluation between the main centers of research in different countries. The formal institutions serving geology at this time — the sister societies in London and Paris were by far the most important — can be mapped as a further element onto the informal tacit topography, making it a still more complex Venn diagram (Figure 1).[20] If this topography represents the entire social 'field' of geology, then the Geological Society of London can be identified as the most important arena for the social dramas that formed the high points of the Devonian controversy.[21] It was the oldest society anywhere to be devoted to geology; but it was also by common consent one of the liveliest arenas for debate in *any* science at that period. Alone among learned societies, it allowed and even encouraged argumentative discussion after the reading of scientific papers. Even quite acrimonious argument was in practice tolerated, because it took place among gentlemanly social equals, and no account of these private discussions was allowed to be published.

The importance of this social convention became apparent when, at one crucial phase of the Devonian controversy, similar arguments were aired in a very different arena. At the annual meetings of the newly founded British Association for the Advancement of Science, the 'Section' devoted to geology was organized by much the same group of men who ran the London society. But meetings of the Association were open to a much wider social range of participants (including, grudgingly, the female sex).[22] So there was great concern among geologists when divergent opinions about the correct geological structure of Devonshire were discussed in this socially less exclusive arena, because one leading geologist (Roderick Murchison) publicly impugned the scientific competence of another (Henry De la Beche) and thereby imperiled the latter's career prospects.[23] But this was in fact almost the only point at which macrosocial factors impinged on the Devonian controversy. These two prominent geologists epitomized two alternative futures for their science within British society: the one (De la Beche) seeking financial support from the government and anticipating a more professionalized future; the other (Murchison) looking to aristocratic circles for social approval and a more diffuse sense of legitimacy for the life of science.

Undoubtedly, this contrast contributed to the acrimony of their argument in the Devonian controversy. Yet there is no adequate evidence to support an interpretation of the controversy as a whole in terms of the divergent social interests they represented. There was no discernible pattern of social alignment within the controversy that would correlate with such interests; indeed the very fluidity of the cognitive alignments within the controversy, when it is traced in detail, makes any such interpretation highly implausible. While social interests, along with ideological conflicts, probably did operate constitutively in certain other controversies in nineteenth-century science, in the Devonian controversy their cognitive effect was marginal. So was that of the occasional charges of scandalous conduct or attempted fraud. But this does not mean that the cognitive processes involved in the Devonian controversy were therefore nonsocial. On the contrary, a 'fine-grained' analysis shows that the normal processes of scientific practice were *social* processes through and through, in the Devonian controversy as in any other example of the construction of scientific knowledge.[24]

Private and Public Science

The historian of science who studies any period before the twentieth century is deprived of the sociologist's classic form of evidence, the interview. But historians who have had to deal with the written reminiscences of earlier generations of scientists may well feel that this deprivation is less than disastrous. For the recollections of participants, years or decades after the events being recalled, are notoriously and systematically unreliable.[25] The historian of science is on safer ground if forced by the nature of the record to focus attention on material that is strictly contemporary with the events being reconstructed and analyzed. In fact, to describe the forms of documentary evidence is, in effect, to outline simultaneously the shape of scientific practice. The historian may have no unmediated access to the thoughts of the scientists being studied; but the documents certainly give mediated access to the activities that constituted their practice.

Instead of the customary sharp dichotomy between published and unpublished material, it is more useful to review the range of documentary evidence in terms of a continuum stretching from the most private to the most public forms of activity and their corresponding records.[26] Taking the Devonian controversy as a concrete example, and beginning at the private end of the continuum, the historian has access firstly to field notebooks, which constitute a trace of the day-by-day activities of at least the more active individuals, during their periods of most intensive '*intra*-personal' interaction with the natural objects that were accepted as the empirical basis for all theoretical arguments. Geology in the early nineteenth century was above all a field science; the empirical focus of geological practice was the field excursion, far more than the laboratory or even the museum. Few geologists kept daily records of their research, except during periods of field work. But what field notebooks lack in continuity, in comparison with laboratory notebooks in the experimental sciences, they may amply compensate through vivid immediacy.[27] They were often compiled by a solitary geologist or simultaneously by a pair of colleagues, during prolonged periods of study in romantic rural scenery. Such individuals were far from home, and temporarily suspended in 'liminoid' isolation away from the ordinary pressures of the larger disciplinary community.[28] In these notebooks, intended only for the compiler's own eyes, the mundane details of observation are found inseparably linked to their spontaneous provisional interpretation. And while deciphering and reconstructing the daily course of field work, the historian's perseverance is occasionally rewarded by reading an interpretative note which gives a sudden insight of startling transparency into the line of reasoning that the individual was following on the most private level at that moment.

But notebooks give no specially privileged access to the thought processes of the individual. They record one relatively private aspect of the whole structure of scientific practice, but they also reveal the integration of solitary reasoning into a larger fabric of social interaction. Interpretative notes are often explicit rehearsals of arguments to be used against real or imagined critics. Even the most innocently 'factual' of observations can often be seen from their context to have been guided by theoretical objectives, such as the need to search for new empirical evidence to reinforce the compiler's own position or to undermine that of his critics.

Closely linked to the evidence of notebooks, but opening out away from the private end of the continuum, is the evidence of correspondence, the trace of generally dyadic or *inter*-personal interactions. In the case of the Devonian controversy, this evidence is quite exceptionally rich. This may be due partly to the perceived importance of the controversy at the time, which may have led the main participants to preserve the relevant letters with particular care.[29] Partly it is also the result of various contingent circumstances which caused those participants to be geographically separated for much of the time, although they knew each other well enough to correspond on easy and informal terms, even when vehemently criticizing each other's opinions.[30] More generally, they were not cursed with the telephone; and the postal service — particularly within Britain — was much more efficient than it is today.[31] And they belonged to a society — or more accurately, to a social stratum within that society — in which a spontaneous and vigorous style of letter-writing was a routine accomplishment. All these circumstances combine to provide, in the case of the Devonian controversy, an exceptionally well-preserved record of a network of interactions. This network linked individuals situated in all parts of the social-cognitive topography of the science, from elite to amateur geologists; and it can be reconstituted at a level of temporal resolution, which at times becomes as fine-grained as several times a week.

Like field or laboratory notebooks, scientific correspondence gives the historian no straightforward or unproblematic trace of individual thoughts or opinions. Even more clearly than notebooks, letters were at this period an indispensable medium for the continuing process of argument and persuasion underlying the construction of knowledge. In these exchanges of letters — sometimes dashed off in haste, sometimes evidently composed with deliberate care — are preserved the traces of complex moves that were at one and the same time both social and cognitive. If read in a strictly contemporary context, even the most apparently straightforward factual reports can be seen to have been directed by the writer toward altering the opinions of the recipient: consolidating one alliance, detaching the recipient from another, enhancing the writer's own credibility, undermining the credibility of one of his critics. Nothing is quite what it appears on the surface: the most significant remarks may be introduced with an apparently casual "By the way..." or relegated to a "P.S." Needless to say, such 'readings' by the historian are themselves never uniquely definitive; but they can and do receive cumulative corroboration, the more fully the individual letter is embedded in its immediate context of contemporary argument.[32] It is here, perhaps more than at any other point in the task of reconstruction, that precise chronology becomes essential; for even a single letter, misdated perhaps by careless deciphering of a smudged postmark, can throw the meaning of many other letters into confusion.

The informality of these exchanges by letter — even when they were acrimonious — was closely related to the character of the main arena in which the drama of the Devonian controversy was played out. As already mentioned, the Geological Society of London was renowned at the time for the vigor of its argumentative discussions. These discussions, and the formally presented papers that occasioned them, constitute a further stage along the continuum from private to public science; but, as noted earlier, they were significantly less than fully public events. Formal summaries of the papers were published promptly in the Society's *Proceedings*, circulated among the membership, and quite widely reprinted or abstracted in general scientific periodicals in Britain and abroad. But, like modern scientific papers, they are often highly misleading in that they systematically omitted or concealed much that was speculative or controversial. Reports of the reading of such papers, as recorded in letters to absent members of the Society, are often more reliable sources, at least to supplement the printed record, even if the historian has to make allowance for the known biases of the writer. And such private and informal reports are the *only* source from which the course of the famous discussions themselves can be reconstructed. Nonetheless, at least for some occasions the discussions can be reconstructed with some confidence, particularly when independent accounts of what was said can be checked against each other. Though often faint and obscure, such accounts of discussions are the traces not only of face-to-face interactions across the floor of the Society's meeting room, but also of those more subtle '*trans*-personal' interactions by which it seemed occasionally that the meeting as a whole had reached some kind of provisional collective opinion transcending those of the individual geologists present.[33]

Finally, at the most fully public end of the continuum, relevant papers provided the occasion for almost annual discussions of the controversy at the meetings of the British Association, open to all who could pay the fee, and fully reported in the general cultural weeklies (particularly the *Athenaeum*) and the scientific periodicals.[34] On other occasions, when priority had to be established in a hurry, or when the controversy became more personally acrimonious than the guardians of the Geological Society's gentlemanly norms would tolerate, the usual procedures were simply bypassed, and 'letters to the editor' were given rapid publication in the scientific press (particularly the *Philosophical Magazine*). But generally, papers presented to the Society would, if substantial enough, eventually receive full publication in its lavishly illustrated *Transactions*; but the publication delays were so long that by the time they were published they had generally lost their novelty value in the debate, becoming little more than monuments to their authors. As in present-day science, such finished products of research were in any case highly stylized and artfully 'objective'. Yet when read in the context of the debates to which they were designed to contribute, it is not difficult for the historian to read 'between the lines', as knowledgeable contemporaries certainly did, and to discern there the continuing trace of the process of argumentative persuasion.

Thus throughout the continuum, from the most private to the most public forms of documentation, the historian has ample resources from which to reconstruct the intrinsically social processes by which a new piece of claimed knowledge, in this case the 'Devonian system' of strata and 'Devonian period' of earth history, came to be constructed and consensually validated.

XIII

A 'Coarse-grained' Analysis

The Devonian controversy arose in the course of development of 'normal science'. The dominant cognitive enterprise within geology in the 1830s was an attempt to order the sequence of strata in the Earth's crust. The increasingly explicit goal of this collective effort was to reconstruct the history of the Earth and of life on Earth before the advent of human beings.[35] A variety of techniques had been developed for this purpose, based above all on the careful detection of the correct sequence of the strata in particular regions, and the attempt to 'correlate' such sequences in different regions by matching their characteristic rock types and fossils. How in practice such techniques were to be employed was very much a matter of the shared tacit knowledge of geologists. Indeed, geologists' mutual evaluations of competence, and hence their ascribed places on the tacit social topography of the science, were largely determined by their degree of perceived success in applying this kind of unformalizable craftsmanship.

The collective application of these geological techniques had led to the compilation of a sequence of major groups or *'systems'* of strata.[36] Much of this sequence was accepted consensually as valid throughout the well-surveyed regions of Western Europe, and there were many who believed its validity might eventually extend worldwide. This collective enterprise had been less successful, however, in dealing with the lowest and therefore oldest strata. These typically rose to the surface in upland or mountain regions; they were usually crumpled, fractured or distorted; and they were generally lacking in easily distinguished rock types or well-preserved fossils. Some geologists feared that these oldest strata would remain forever a realm of 'chaos' — the epithet was their own. But to others such rocks presented a challenge to their skill and perseverance; for if they could produce order out of chaos *here* — if they could unravel the correct sequence and demonstrate its validity over a wide area — they would gain the credit for deciphering the earliest 'chapters' of the history of the Earth and of life itself.

In the early 1830s the London geologist Roderick Murchison was prominent among those pursuing this risky but valuable prize. In the Borderland between Wales and England he detected a group of these so-called 'Grauwacke' or 'Transition' strata lying in an unusually clear sequence below the well-known 'Carboniferous' system with its valuable coal deposits; and he found that the older strata contained distinctive animal fossils but no land-plant fossils at all. In the visual language of geology (then and now), his conclusions can be represented as a columnar 'section' or diagrammatic pile of the strata involved (Figure 2, column 1). He termed

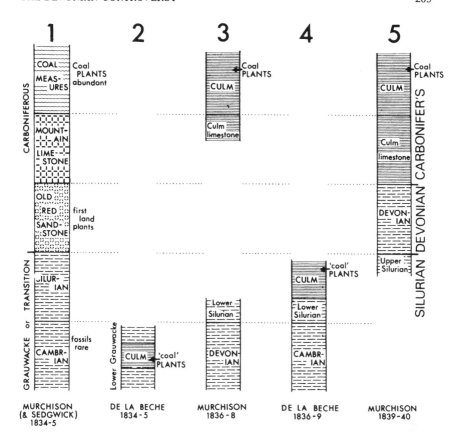

Figure 2. Diagrammatic representation of four successive interpretations of the sequence of strata in Devonshire (columns 2 to 5), matched or "correlated" with the "standard" succession in central England as extended downwards by Murchison (and Sedgwick) in Wales and the Welsh Borderland (column 1). The thicknesses of strata in these columnar ("vertical") sections are not to scale: the shadings represent various predominant types of rock.

'his' strata *Silurian*, after the tribe of Silures who had lived in the Borderland in Roman times.[37] His growing confidence in the general validity of the Silurian as a major period in the Earth's history was marred only by persistent reports from several other countries that similar ancient or pre-Carboniferous rocks contained abundant plant fossils and even coal seams. But in his project to establish the Silurian system, he did not feel inhibited by such apparent anomalies, until one was reported close to his own territory, and in a particularly serious form.

In southwest England, Henry De la Beche discovered plant fossils in ancient-looking strata in the course of his geological survey for the government, and these fossils were identified by a competent specialist as being species well known from the coal strata of the Carboniferous system. This report, presented late in 1834 at the Geological Society, precipitated the Devonian controversy. De la Beche insisted that the fossils came from well within an unbroken sequence of strata that had long been regarded as ancient in date, and certainly pre-Carboniferous (Figure 2, column 2). Murchison, on the other hand, felt so confident about his own emerging theoretical scheme, with the plantless Silurian system below the Carboniferous, that he flatly contradicted De la Beche, although he had never yet studied the area in question. The fossil plants, he claimed, must have come from a hitherto unrecognized patch of coal strata of Carboniferous age. This implied a huge time gap, and hence a structural discontinuity or 'unconformity', separating them from the genuinely ancient strata of Devonshire (Figure 2, column 3). But De la Beche insisted, and continued to insist, that there was no sign whatever of any such gap or discontinuity.

When eventually, in 1836, Murchison did go to Devonshire to look for himself, he concluded that De la Beche had indeed been seriously mistaken in the way he had deciphered the structure and therefore the local sequence there. According to Murchison's new interpretation, the plant-bearing strata were uppermost, as he had earlier guessed, so that it was plausible to regard them as truly Carboniferous. This eliminated the anomaly that had become so damaging to his Silurian scheme; but simultaneously it created another, for he had to gloss over his failure to find any trace of the gap or 'unconformity' that his interpretation required (Figure 2, column 3).

At the 1836 meeting of the British Association, Murchison gleefully announced De la Beche's gross 'mistake' in deciphering the structure of Devonshire. De la Beche was furious because, when made in such an inappropriately public arena, this claim was bound to lead the politicians to doubt his competence, and it thereby jeopardized his future employment. But in the cognitive dimension De la Beche soon accepted his critics' interpretation of the *sequence*, though not of its *dating*. Since he was convinced there was no perceptible gap in the sequence, and the plant-bearing strata were now to be placed at the top, he inferred that they might be Silurian (Figure 2, column 4). But of course this put them back just where they most seriously compromised Murchison's conception of his Silurian system.

Murchison wrestled unsuccessfully with this recalcitrant anomaly for two or three years more, until, in 1839, he suddenly announced a radically new interpretation (Figure 2, column 5). Tacitly accepting what De la Beche had maintained all along, namely that there was no gap in the sequence in Devonshire, Murchison proposed that the older strata were much less ancient than all the relevant geologists — himself included — had hitherto inferred. Instead, he now assigned most of them to a position (and age) equivalent to a distinctive group of strata — the 'Old Red Sandstone' — found at the base of the Carboniferous system in other parts of Britain. In the eyes of other geologists, however, this seemed no more than an ad hoc device to get Murchison off his own theoretical hook. For what he now termed 'Devonian' strata (incidentally changing the meaning of that term from an earlier usage: compare Figure 2, column 3) were radically different from the Old Red Sandstone in both rock types and fossils, and therefore made a highly implausible 'correlation'. Murchison now argued that it was the Old Red Sandstone strata, not the Devonian rocks, that were atypical of the period of their formation. The Devonian strata, he argued, *must* be of that intermediate age because the relevant specialists had pronounced their fossils to be intermediate in character between those of his Silurian system below and those of the rest of the Carboniferous system above.[38]

However, this implausible interpretation did embody an explicit set of predictions; and in the following two field seasons Murchison undertook extensive field work in Belgium and Germany, and then in Russia, to search for less ambiguous sequences of strata. He hoped these would establish the validity of what he came to regard as a full-fledged Devonian 'system', sandwiched between his Silurian system below and the now redefined and narrowed Carboniferous system above. Despite many setbacks, local anomalies and periods of doubt about this scheme, he eventually produced evidence that more and more of his geological colleagues found convincing. The 'Devonian system' of strata came to be consensually accepted as globally valid, and as representing a corresponding 'Devonian period' in the history of the Earth and of life. So after about a decade of vehement argument, the Devonian controversy subsided in the later 1840s, with only a handful of marginal men refusing to accept the new interpretation.[39] The Devonian system (and period) became a piece of increasingly solid scientific knowledge. Its solidity has been greatly enhanced in the subsequent century and a half by the consistent filling in of its detailed contents on a worldwide scale, in a way that would surely have impressed, but not surprised or perplexed, those who first proposed and supported the Devonian interpretation.

This brief summary of the Devonian controversy bears scarcely any trace of my earlier emphasis on the varied sources of evidence for analyzing the construction of the 'Devonian' as a piece of new knowledge, nor of my emphasis on that construction as a result of social processes spread across the diversified topography of geologists. It is a 'coarse-grained' summary that could have been written almost entirely from an analysis of formal published papers; and it describes the controversy almost entirely in terms of an argument between just two protagonists, Murchison and De la Beche.

Even in this crude form, however, the summary already shows one important feature which rarely appears in the idealized and formalized accounts of science that philosophers generally use for their own purposes. The interpretative scheme for Devonshire geology that was ultimately judged 'successful' in terms of its increasingly fruitful use worldwide, namely Murchison's proposal of a Devonian system (Figure 2, column 5), did not simply triumph over an earlier scheme with lesser explanatory power. It was a scheme that had not been foreseen by anyone, certainly not by Murchison himself, when the controversy started. Yet it was no simple compromise between De la Beche's scheme that precipitated the controversy (Figure 2, column 2) and the incompatible rival scheme that Murchison devised in response (Figure 2, column 3). De la Beche modified his first scheme to incorporate explicit concessions to Murchison's arguments (Figure 2, column 4), and Murchison modified *his* first scheme to embody similar — but only tacit — concessions to De la Beche's arguments (Figure 2, column 3 is a composite representation of Murchison's earlier and later schemes). The finally 'successful' scheme (Figure 2, column 5) was significantly different from *either* of the two earlier rival schemes, even in their modified forms. It tacitly absorbed De la Beche's continuing insistence that there was no evidence for any discontinuity in the sequence. But it also embodied Murchison's continuing insistence that fossil plants so similar to those of the coal deposits elsewhere must be of truly Carboniferous age, and not Silurian or even older. Over and above those elements inherited from the earlier rivals, however, the finally 'successful' scheme also introduced a radically novel element. This was the claim — initially a highly implausible one — that the older strata of Devonshire were much younger than all the geologists concerned had until then supposed. The perceived plausibility of this 'Devonian' scheme was strengthened only as the predictions it embodied were successively fulfilled elsewhere in Europe.

In crude or 'coarse-grained' outline, therefore, the Devonian controversy illustrates a dynamic pattern of theory construction that may be much more common than either philosophers or historians of science have generally recognized. The 'successful' scheme emerged from a process of dialectical or mutual modification of incompatible rival schemes, yet it transcended them in a way that made it far more novel than any compromise, and in a way that was quite unexpected at the outset and highly implausible when first proposed.

Toward a 'Fine-grained' Analysis

The crude simplicity of my 'coarse-grained' summary of the Devonian controversy gives no hint of the astonishing complexity that emerges once the full range of documentary evidence is brought to bear on the reconstruction. The formal published papers of the chief participants then appear merely as the deceptively isolated tops of far larger icebergs. To change the metaphor to a more appropriate one, the published texts are the occasional press releases that emerge from a long and complex process of diplomatic negotiation behind closed doors. This process of negotiation can only be reconstructed by a detailed analysis of all its documentary traces: above all in the network of correspondence that passed between all the participants; but supplemented by the records of their field work, by reports of their arguments in the semi-public arena of the Geological Society and elsewhere, and — with appropriate reading 'between the lines' — by the texts of their formal papers. Only such a 'fine-grained' reconstruction will ever give an adequate picture of the dynamic processes of argument and persuasion that lay behind the construction of the Devonian system as a new piece of solid and reliable scientific knowledge.

Pending such a detailed narrative and its analysis, I can at least mention three examples of the ways in which the full documentation enlarges and transforms the coarse-grained summary. My examples are themselves of increasingly fine 'grain' or degree of temporal resolution. First, we can trace the formation of a consensus on the finally 'successful' interpretative scheme among six British members of the small core set of highly competent geologists. When the controversy broke out, they were divided on the issue into two unequal parties. De la Beche's scheme (Figure 2, column 2) was found plausible by three other influential elite geologists: Greenough, virtually the founder of the Geological Society; Buckland, who taught geology at Oxford; and Sedgwick, his counterpart at Cambridge. Murchison's objection to De la Beche was upheld by only one comparably important elite figure, namely Lyell, the self-appointed chief theoretician of British geology. The two parties were divided in effect by the confrontation of two apparently irreconcilable propositions. These can be summarized crudely in slogan form as, respectively, 'No gap in the sequence of Devonshire strata!' and 'No Carboniferous plants in any ancient strata!'

XIII

210

The subsequent course of negotiations, as revealed by the mass of precisely dated documents available, shows the importance of Sedgwick's gradual defection to Murchison's side, firmly established by their joint field work in Devonshire in 1836. It shows the complex modifications of the schemes proposed by some of these individuals, embodying certain concessions to their opponents in the light of empirical evidence that they had to admit as compelling. It also shows the quite rapid consensus of the core-set around the 'Devonian' interpretation, within about a year of its first public proposal by Murchison. But this process of negotiation and consensus formation, if conceived in terms of the cognitive trajectories of individuals, fails to give an adequate impression of the interactions between them.

It is also difficult to bear in mind that the 'Devonian' scheme was *not* a compromise between the initial rivals or even their later modifications, unless another aspect of the social process of negotiation is also emphasized. This is that a new 'conceptual space' opened up in the course of time between the positions that had earlier seemed irreconcilable. The two slogans in which I summarized those embattled positions were not of course logical opposites; but the possibility of a noncontradictory 'space' between them only became apparent through the intervention of individuals who were *not* members of the core-set of elite geologists, but men whose tacit status in the social topography of the science was only that of accredited geologists. More specifically, it was the young Devonshire geologist Robert Austen, and the fossil specialist William Lonsdale, the Geological Society's curator, who first produced evidence and opinions that implicitly opened up the new conceptual space that made the 'Devonian' scheme a conceivable possibility.

It was Murchison who eventually did conceive that possibility; and judging from the published record alone, it might appear to be a classic example of a sudden 'Eureka-experience' or *Aha-Erlebnis*. But — and this is my second example — a finer-grained reconstruction of the negotiations that led to his public proposal of the 'Devonian' scheme reveals a far more complex process of argument and persuasion. The cognitive pathway from Murchison's earlier scheme (Figure 2, column 3) to the new one two and a half years later (Figure 2, column 5) was far from straight. It involved a complex sequence of moves — proposals, counterproposals, new empirical reports, arguments about their significance, and so on — made by a varied cluster of individuals. These ranged from Murchison himself, with Sedgwick, his fellow-member of the elite core-set, acting as a kind of friendly but critical sparring partner, through accredited geologists like Austen and Lonsdale with local or specialized expertise, to merely amateur geologists like the Devonshire country gentleman William Harding, who discovered significant new fossil plants at a crucial moment in the intricate process of argumentative negotiation. If one follows Murchison's 'Devonian' scheme with strict attention to chronology and abstains from the dubious benefits of hindsight, it can be seen to have emerged in a quite unpredictable way from a sequence of incomplete and inconclusive arguments involving many others besides him. It is this social process of complex interactive negotiation that makes it legitimate to term the ultimately 'successful' Devonian interpretation a social *construction.*

Turning a still higher power of historical magnification onto this process of negotiation, for my third and last example, almost any substantive letter between members of the core-set in the Devonian controversy would demonstrate the argumentative, persuasive and rhetorical character of their interactions. A particularly vivid example is one where the message was conveyed in visual as well as verbal form. When De la Beche sent the Geological Society specimens of the fossil plants he had found in strata of ancient ('Grauwacke') appearance in Devonshire, the reliability of his report was immediately queried by Murchison, backed up by Lyell, although these critics had not seen for themselves the rocks and the locality De la Beche had described. As already mentioned, it was this incident that precipitated the Devonian controversy. De la Beche heard reports of the argument after the meeting, and was understandably incensed that his competence had been impugned in his absence. He wrote at once to both Greenough and Sedgwick, enclosing a caricatured version of what had occurred (Figure 3). He depicted himself standing before the other members of the Geological Society, and declaring "This, Gentlemen, is my *Nose.*" They, however, answered him, "My dear fellow! — your account of yourself generally may be very well, but as we have classed you, *before we saw you,* among men *without noses,* you *cannot possibly have a nose.*"[40]

This superficially innocent joke was pervaded with serious rhetorical purpose: it was a carefully devised move in the larger field of persuasive negotiation. De la Beche entitled his caricature "Preconceived Opinions *v[ersus]* Facts," in order to appeal to the well-known 'Baconian' ideals of his two correspondents. By contrast, he depicted his critics viewing the evidence of their senses through the colored spectacles of illegitimate theorizing. Furthermore, he showed himself dressed in the rough topcoat of a field geologist, ready to brave all weathers to study the raw facts of nature, whereas his critics were dressed in elegant tailcoats more suitable for purely speculative debate in a fashionable London *salon.* He sent this iconographically loaded caricature to Greenough, already his patron and champion in London, knowing full well that Greenough would show it

XIII

212

Figure 3.   Caricature by De la Beche, criticizing Murchison's and Lyell's reaction to his report (see Figure 2, column 2) of fossil plants of Coal Measures species in the 'Grauwacke' strata of Devonshire.
Caption: Preconceived opinions vs. facts
De la Beche: This, Gentlemen, is my *Nose*.
His critics: My dear fellow!— your account of yourself generally may be very well, but as we have classed you, *before we saw you*, among men *without noses*, you *cannot possibly have a nose*.

round among Geological Society members, and hoping that it would strengthen collective opinion in his favor before the controversy went any further, as he knew it was bound to do. He sent a separate copy to Sedgwick in Cambridge, hoping to undermine the working alliance that had grown up in other areas of research between Sedgwick and Murchison, and knowing that Sedgwick would also read the message as supporting implicitly his own criticism of Murchison's other ally, the theoretically overambitious Lyell. In this way, read in context, De la Beche's joke can be seen to have had the serious rhetorical purpose of altering the balance of forces in the incipient controversy in his own favor, particularly by isolating his critics Murchison and Lyell from their most influential potential ally, Sedgwick. But what De la Beche exceptionally put into visual form is no different from what he and all other geologists involved in the controversy put into verbal forms whenever they wrote to each other, and no doubt also in their face-to-face conversations of which the historian has no direct trace.

   The entire web of their interactions constitutes what Latour and Woolgar, in their 'anthropological' study of *Laboratory Life* in modern neuroendocrinology, have termed the 'agonistic field' of persuasive forces: forces which continually push in different directions the perceived plausibility of rival theoretical suggestions.[41] The main schemes of interpretation that I outlined in my coarse-grained summary of the Devonian controversy acted like locally 'gelled' portions of the agonistic field; they were relatively solid and secure pieces of plausible knowledge, candidates for further solidification, yet with enough defects for their dissolution into implausibility to be a conceivable outcome too. But when the 'Devonian' scheme gained the consensual support of all the elite geologists involved, with only a few marginal men holding out against it, this did *not* constitute a unique and irreversible 'inversion' into a spuriously solid scientific 'fact', as Latour and Woolgar suggest for their own case study.[42] The Devonian system and period became a more solid or 'fact'-like piece of knowledge only through proving itself valid *in use* over a progressively wider range of regions and in progressively greater and more consistent detail. It was entirely conceivable at the time that it might have failed in this test of use; it might have declined in plausibility and fallen into oblivion, as other comparable schemes had done, by proving inapplicable or inconsistent beyond the regions where it was first constructed.

That the Devonian system and period did prove so successful in descriptive and explanatory use must surely raise grave doubts about the extreme epistemological scepticism which is adopted by many of those currently working in the cognitive sociology of science.[43] The Devonian controversy is a particularly cogent example in this debate, because one can readily agree that in this case the entity produced by the social processes I have described was itself an *artificial* or human construct.

The Devonian system and period were not simply 'discovered' as natural objects. The total sequence of strata in the Earth's crust and the continuum of the Earth's entire history could in principle have been divided for descriptive purposes in many other, equally useful and valid ways. It is clearly a matter of historical contingency — a complex result of the European location of the relevant pioneer research — that the strata (and the history they represent) have come to be divided in the way they are, with the Devonian system (and period) defined and delineated as a particular portion. Nonetheless, that portion has been (and is) claimed to represent a part of the unique and real history of the Earth, however imperfectly described and understood.

I conclude that a historical example such as the Devonian controversy, if reconstructed and analyzed in fine-grained detail, fully confirms the view that scientific knowledge is *constructed* through a complex *social* process of

XIII

interactive negotiation, among individuals of differing competences holding widely divergent initial opinions. But within this process of social construction, the common currency in the negotiations is that of the empirical phenomena that are consensually accepted as relevant and ultimately binding, however ambiguous their interpretation may initially seem to be. In other words, the natural phenomena constitute an input that has a constraining and differentiating effect on the outcome of the social processes; and the entity that is socially constructed is thereby a mediated representation of an external reality that exists independently of those who construct it. But if philosophical realism is to be revived in the analysis of scientific knowledge, as I believe it should be, then it must be in a form that acknowledges that the human learning processes that generate reliable natural knowledge are *social* in character through and through. The individual scientist confronts the natural world only through the mediation of group interactions in the social world.

Notes

Earlier presentations of similar material were given as a Special University Lecture in London in January 1981 and at the ERISS conference ("Epistemologically Relevant Internalist Sociology of Science") organized by Syracuse University in June 1981. For valuable comments and criticism on these and other occasions I am especially indebted to David Bloor, Donald T. Campbell, Gerald L. Geison, Patrick de Maré, J.B. Morrell and Sylvan S. Schweber.

1 Peter Medawar, *The Art of the Soluble* (London: Methuen, 1967), p. 151.

2 An important and characteristic example is Bruno Latour and Steve Woolgar, *Laboratory Life: The Social Construction of Scientific Facts* (Beverly Hills and London: Sage, 1979). Latour, who had earlier done anthropological field work in the Ivory Coast, was a participant observer at the Salk Institute, La Jolla, California, in 1975–77.

3 For example, Latour and Woolgar (*op. cit.,* note 2, pp. 106–7) disclaimed any intention of writing a *history* of the "Construction of a Fact" of neuroendocrinology; yet a history is precisely what they were obliged to write, in order to make sense of developments stretching back years before the period of participant observation.

4 For example, H.M. Collins's studies of claims to the observation of high-flux gravitational radiation: "The Seven Sexes: A Study in the Sociology of a Phenomenon, or the Replication of Experiments in Physics," *Sociology* 9 (1975) 205–224; and its sequel, "Son of Seven Sexes: The Social Destruction of a Physical Phenomenon," *Soc. Stud. Sci.* 11 (1981): 33–62. See also Trevor Pinch's study of the continuing debate on the observation of solar neutrinos: "Theoreticians and the Production of Experimental Anomaly: The Case of Solar Neutrinos," in: *The Social Process of Scientific Investigation*, ed. Karin D. Knorr et al. (Dordrecht, Boston and London: D. Reidel, 1981), pp. 77–106; and "The Sun-Set: the Presentation of Certainty in Scientific Life," *Soc. Stud. Sci.* 11 (1981): 131–158.

5 For example, David Kohn, "Theories to Work By: Rejected Theories, Reproduction, and Darwin's Path to Natural Selection," *Stud. Hist. Biol.* 4 (1980): 67–170; Sandra Herbert,

"The Place of Man in the Development of Darwin's Theory of Transmutation," *J. Hist. Biol.* 7 (1974): 217–258, and 10 (1977): 155–227; Silvan S. Schweber, "The Origin of the *Origin* Revisited," *ibid.* 10 (1977): 229–316.

6 This has been most explicit in studies by Howard E. Gruber: see for example his "The Evolving Systems Approach to Creative Scientific Work: Charles Darwin's Early Thought," in: *Scientific Discovery: Case Studies*, ed. Thomas Nickles (Dordrecht and Boston: D. Reidel, 1980), pp. 113–130; and *Darwin on Man. A Psychological Study of Scientific Creativity*, second ed. (Chicago: Chicago University Press, 1981), esp. chap. 8.

7 This claim must be distinguished from the assumptions that underlie most research using citation analysis: see David O. Edge, "Quantitative Measures of Communication in Science: a Critic Review," *Hist. Sci.* 17 (1979): 102–134.

8 In this article the focus is on the implications of the case for the historical sociology of scientific knowledge. It is designed to complement one that I wrote for an audience of geologists, which is more "technical" in content: M.J.S. Rudwick, "The Devonian: A System Born from Conflict," in: *The Devonian System*, ed. M.R. House et al. [London: Palaeontological Association (*Spec. Pap. Palaeont.* 23), 1979], pp. 9–21.

9 The phrase is contemporary, but is aptly used in J.B. Morrell and Arnold Thackray, *Gentlemen of Science. The Early Years of the British Association for the Advancement of Science* (Oxford: Oxford University Press, 1981).

10 "Gentlemen-specialists" is my own phrase: see Martin J.S. Rudwick, "Charles Darwin in London: The Integration of Public and Private Science," *Isis* 73 (1982): 186–206. See also Roy Porter, "Gentlemen and Geology: the Emergence of a Scientific Career, 1660–1920," *Hist. Jl.* 21 (1978): 809–836.

11 There is no adequate general history of geology in this period. For the immediately preceding period, see, however, Roy Porter, *The Making of Geology. Earth Science in Britain, 1660–1815* (Cambridge: Cambridge University Press, 1977).

12 George Bellas Greenough (1778–1855), cofounder and first president of the Geological Society of London, the first specialist society in the world to be devoted to geology.

13 The following passage is summarized from Rudwick, "Darwin in London" (note 10).

14 Only the socially marginal "scriptural geologists" — the distant intellectual ancestors of modern creationists — disturbed this general consensus, in that they claimed that they, and not the mainstream geologists, were the true fount of authentic knowledge about the Earth and its history. Leaving them aside, however, the otherwise consensual form of this graded topography can be described quite simply. The scriptural geologists still await their historian: for a preliminary survey, see Milton Millhauser, "The Scriptural Geologists. An Episode in the History of Opinion," *Osiris* 11 (1954): 65–86.

15 Individuals could of course move quite rapidly across this topography — Darwin's early career is a good example: see Rudwick, "Darwin in London" (note 10), Figure 1. Furthermore, on a longer time-scale the form of the topography itself changed gradually, for example with the increasing specialization and esoteric character of the subject matter of the science. Any such topography is therefore an attempt to represent in visual form a set of contingent circumstances that were specific to a particular branch of science at a particular period.

16 "Focal problem" is my own term: see Rudwick, "Darwin in London" (note 10). The "hot spot" metaphor is used by H.M. Collins, "The 'Core-set' in Modern Science: Social Contingency with Methodological Propriety in Science," *Hist. Sci.* 19 (1981): 6–19.

17 See Collins, "Core-set" (note 16).

18 For example, Darwin was undoubtedly in the core-set that was simultaneously debating the reality or otherwise of the widespread elevation of continental land masses within recorded human history; but he played only a marginal role in the Devonian controversy. See Martin J.S. Rudwick, "Darwin and the World of Geology," in: *The Darwinian Heritage*, ed. David

Kohn (Chicago: Chicago University Press, forthcoming) and "Darwin in London" (note 10).

19 "Information flow' is used here in an information-theory sense; it does *not* imply that what flowed were theory-free empirical observations.

20 On the two societies, see Rudwick, "Darwin in London" (note 10), pp. 192–3. The scriptural geologists were almost equally strong-boundaried, although not organized formally; they had a mutually hostile frontier against — particularly — the Geological Society of London.

21 The terms "arena" and "social drama" are particularly apposite to convey the dramaturgical and (even) gladiatorial character of some of these discussions: see Rudwick, "Darwin in London" (note 10), and James A. Secord, *Cambria/Siluria: the Anatomy of a Victorian Geological Debate*, Princeton University Ph.D. dissertation, 1981.

22 See Morrell and Thackray, *Gentlemen of Science* (note 9), chap. 3.

23 *Ibid.*, pp. 462–5; also Rudwick, "Devonian" (note 8).

24 This point is rightly emphasized in — and is indeed central to — recent "anthropological" research on modern scientific practice (see note 2); but I do not follow such writers in their radical epistemological scepticism (see below).

25 For an analysis of a modern example, attributing the systematic remodeling to the needs of public presentation, see Steve Woolgar, "Discovery: Logic and Sequence in a Scientific Text," in: *Scientific Investigation* (note 4), pp. 239–268. A famous historical example is Darwin's recollection, late in life, that he had worked on "Baconian principles" and "without any theory" [*The Autobiography of Charles Darwin, 1809–1882*, ed. Nora Barlow (London: Collins, 1958), p. 119], when compared with the evidence of his 'species notebooks' that he was theorizing in a wide-ranging — and of course highly creative — way at the relevant period (see references in notes 5, 6).

26 The following passage amplifies my brief description of such a continuum, in Rudwick, "Darwin in London" (note 10).

27 For the importance of analyzing laboratory notebooks, see F.L. Holmes, "The Fine Structure of Scientific Creativity," *Hist. Sci.* 19 (1981): 60–70. His forthcoming major study of the work of Hans Krebs is represented in preliminary form in "Hans Krebs and the Discovery of the Ornithine Cycle," *Proc. Fed. Amer. Soc. Exper. Biol.* 39 (1980): 216–225. As far as I know, there are as yet no comparably detailed studies based on geological field notebooks.

28 Victor Turner extends and generalizes van Lennep's classic concept of the liminal to include a wide range of 'liminoid' situations: see *Dramas, Fields and Metaphors. Symbolic Action in Human Society* (Ithaca and London: Cornell University Press, 1974). One of his major examples, that of pilgrimage, illuminates in my opinion the phenomena of geological field expeditions.

29 Some of the most tantalizing gaps in the extant correspondence are due to the disappearance of certain series of letters *since* the compilation of the relevant late Victorian "Lives and Letters," which only printed incomplete and unreliable extracts.

30 For example, of the major actors in the Devonian controversy, Adam Sedgwick (1785–1873) was tied for much of the time by academic and ecclesiastical duties to Cambridge and Norwich; Henry De la Beche (1796–1855), by his governmental surveying duties, to Cornwall and South Wales; William Buckland (1781–1856), by academic duties, to Oxford; whereas Roderick Murchison (1792–1871), Charles Lyell (1797–1875) and Greenough (note 12) were based in London. They all met fairly frequently in London, however, particularly at meetings of the Geological Society and over the convivial dinners of its Club.

31 For example, Sedgwick and Murchison sometimes exchanged letters every day of the week by the overnight mail between London and Cambridge (or Norwich). Letters between London and the furthest end of Cornwall were delivered in two days, and those between London and Paris or Bonn took little longer. The high cost of postage (until the introduction

of the penny post in 1840, while the Devonian controversy was still raging) did not greatly deter these gentlemen of science.

32 Some sociologists of science have recently argued that, because all statements contained in such documents are necessarily indexical, i.e., dependent on their context for their meaning, no interpretative reading is ultimately more reliable than any other. My argument is that this conclusion does not follow from its premise, and that specific readings can become more reliable than others, the more fully their original context is reconstituted.

33 I use "trans-personal" in the sense in which it was originally developed to describe the dynamics of human groups in therapeutic contexts: see for example P.B. de Maré, *Perspectives in Group Psychotherapy* (London: Allen and Unwin, 1972). Some such term is undoubtedly needed to do justice to the historical reports of a "sense of consensus" on certain occasions.

34 See Morrell and Thackray, *Gentlemen of Science* (note 9), pp. 139–148. Several major actors in the Devonian controversy used these annual meetings regularly to rehearse arguments they planned to present later to a more expert audience at the Geological Society.

35 The originally *structural* and therefore *non*-historical goal of what is now termed stratigraphy is reflected in its conventional forms of visual representation: see Martin J.S. Rudwick, "The Emergence of a Visual Language for Geology, 1760–1840," *Hist. Sci.* 14 (1976): 149–195. This structural style is characterized in Rudwick, "Cognitive Styles in Geology," in: *Essays in the Sociology of Perception,* ed. Mary Douglas (London: Routledge and Kegan Paul, 1982), pp. 219–241.

36 The modern sense of the term "System," as a major assemblage of strata representing a specific period of time and therefore recognizable (in principle) anywhere in the world, came into use partly as a consequence of the outcome of the Devonian controversy. During and before that time, the term was used in many different senses, each with its own theory-loading.

37 His colleague Adam Sedgwick, professor of geology at Cambridge, termed still older but much more obscure strata "Cambrian," after the Roman name for Wales. Their famous long quarrel over their respective "Systems" was only indirectly related to the Devonian controversy; it is described in detail in Secord, *Cambria/Siluria* (note 21).

38 The fossils of these two Systems only became well known, and therefore available for this kind of comparison, *during* the Devonian controversy itself, with the publication of John Phillips's *Illustrations of the Geology of Yorkshire, Part II* (1836) and Murchison's own *Silurian System* (1839). The latter relied on fossil identifications by specialists such as William Lonsdale (1794–1871) and James de Carle Sowerby (1787–1871), who, with Phillips, were also responsible for evaluating the fossils from the disputed Devonshire strata.

39 The most vocal of the residual sceptics was the Somerset clergyman David Williams (1792–1850), who had contributed actively to the controversy on the basis of detailed, but only local, geological knowledge of southwest England. De la Beche and Greenough conceded the validity of the "Devonian" interpretation in substance, while saving face by avoiding using the term "Devonian System" itself.

40 De la Beche's scientific caricatures, some of them lithographed by himself to give them wider circulation, were well known at the time. Several are reproduced in Paul J. McCartney, *Henry De la Beche: Observations on an Observer* (Cardiff: Friends of the National Museum of Wales, 1977). The successive drafts that led to one famous example are analyzed in Martin J.S. Rudwick, "Caricature as a Source for the History of Science: Henry De la Beche's Anti-Lyellian Sketches of 1831," *Isis* 66 (1975): 534–560.

41 Latour and Woolgar, *Laboratory Life* (note 2), chap. 6.

42 *Ibid.,* chap. 4.

43 *Ibid.;* see also most of the essays in *Scientific Investigation* (note 4) and *Knowledge and Controversy, Studies of Modern Natural Science,* ed. H.M. Collins, special issue of *Soc. Stud. Sci.* 11 (1981): 1–158.

XIV

THE GLACIAL THEORY

Studies on Glaciers, preceded by the *Discourse of Neuchâtel*. Louis Agassiz, translated and edited by Albert V. Carozzi (Hafner, New York, 1967). $27.50.

The publication of this magnificent volume makes available in English one of the central documents of the glacial theory.[1] This was a theory that played a crucial role in the conceptual development of the earth sciences in the nineteenth century, but its importance has not yet been fully recognised. This must be attributed in part to the primitive state of the historiography of this branch of science. Historians of geology are in general, perhaps, still too brainwashed by Charles Lyell's persuasive polarisation between himself and his opponents to be able to give full weight to a theory that does not fit neatly into either 'uniformitarianism' or 'catastrophism'.[2] That Victorian polymath William Whewell can hardly be blamed for the use or misuse of the many scientific terms he coined, but it is becoming increasingly clear that the two labels with which he characterised the Lyellian debate have become a pair of historiographical mill-stones around the neck of nineteenth-century geology. They may have been useful originally to catch some of the striking features of that debate, but they left the meaning of 'uniformity' and 'catastrophe' confused and ambiguous, and more seriously their use implied too uncritical an acceptance of the battle-lines that Lyell had drawn. Moreover, there has since been a tendency for them to degenerate into Whiggishly applied terms of praise and condemnation respectively; and it might be best for historians of nineteenth-century science to agree now to abandon them altogether as tools for historical analysis, retaining them only for descriptive purposes within the strict historical context in which they were first applied.

These preliminary remarks are a necessary introduction to any re-evaluation of the glacial theory. For its history poses an important problem, which is in fact briefly mentioned on the first page of Professor Carozzi's introduction to the volume under review. "It may seem strange", he writes, "that a question such as the action of glaciers, which is an admirable illustration of the principle of uniformitarianism, with visible and undisputed proofs within easy reach of any observer, should have encounted such a powerful opposition from first-rate geologists like Alexander von Humboldt, Leopold von Buch and Élie de Beaumont."[3] It is indeed strange, at first sight, that virtually all the best geologists of the period found it difficult to believe that present glaciers are the shrunken remnants of far more widespread glacial conditions. In retrospect, the theory of an 'Ice Age' in the geologically recent past can be seen to have solved many of the most puzzling problems in early nineteenth-century geology; yet the full acceptance of such a theory took two or three decades even after the publication in 1840 of Agassiz's *Études sur les glaciers*. Moreover, there is another related historical problem: if, as Professor Carozzi suggests, the proofs of glacial extension should have been clear to any Alpine traveller, why was the formulation of a glacial theory so long delayed? The full solution of these problems will require much more detailed research on the geological sciences of this period; but meanwhile I think it is worth considering some of the areas to which that research could usefully be directed, and some of the possible solutions that merit further testing.

The first point that needs some conceptual sharpening concerns Professor Carozzi's use of one of Whewell's labels: what precisely was the 'principle of uniformitarianism' that Agassiz's work should have illustrated so persuasively? It is to Professor Hooykaas that we are indebted for a pioneer attempt to 'unpack' the notion of geological 'uniformity' into its logically distinct components; and though his analysis could be taken even further, it is sufficient here to draw his distinction between actualism, as a method of geological research based on attempted comparison with processes observable at the present day, and uniformitarianism in its proper and original sense, as a scientific theory that events in earth-history have maintained an essentially steady-state pattern with only minor fluctuations around a stable mean.[4] In the light of this distinction, Professor Carozzi's comment on Agassiz's theory clearly refers to its actualistic component: most of the *Études* is devoted to the description of present-day glaciers, explicitly in order to provide a key for interpreting the traces of their former extension.[5] This is indeed an excellent illustration of actualism: in the terms of Lyell's *Principles of geology*, it is "an attempt to explain the former changes of the earth's surface, by reference to causes now in operation".[6] Therefore it is not surprising to find Lyell among the early converts to this actualistic part of Agassiz's work. Yet Lyell, like other geologists of quite different theoretical persuasions, was extremely reluctant to go the whole way with Agassiz. They were all eager to use Agassiz's studies to detect the traces of former glaciers, but extremely unwilling to accept his conclusion that there had been a severe 'Ice Age' in the recent past. Agassiz's use of actualistic method was clearly not all that was needed for the acceptance of his theory.

One possible factor in the delayed formulation and ambivalent reception of a glacial theory is mentioned by Professor Carozzi. He follows his statement of the problem with the suggestion that "the old belief in the transportation of great boulders by huge water and mud currents, in relation to the universal deluge of the Mosaic tradition, was so deeply implanted in the mind of laymen as well as scientists" that it took the joint efforts of Agassiz and his glacialist predecessors to displace it. This usefully draws attention to the fact that some of the most striking evidence for an Ice Age was already explained in terms of another hypothesis, so that Agassiz's interpretation had to make its way against an established alternative.

But the nature of this alternative, the *diluvial* theory, urgently needs historical study, to get behind the obviously biassed description of it which Lyell, for his own polemical purposes, generally gave. It is particularly important that such a study should avoid the English provincialism that makes William Buckland into the 'typical' diluvialist: the theory needs to be examined in an international context. How typical, in fact, was Buckland's identification of the geological 'diluvial' episode with the

XIV

biblical 'Flood'?[7] The recognition of so-called 'extra-scientific' factors in the history of nineteenth-century geology has been on the whole a beneficial development; but there is now a danger that every historical puzzle will be regarded as 'solved' by prescribing a dose of the influence (subconscious? subliminal?) of "the Mosaic tradition". Buckland was widely criticised, even by those who believed there had been *some* kind of diluvial episode, for straining the evidence in order to assert the unique and universal character of the 'deluge', and also for straining the interpretation of Genesis in order to make the Flood into an extremely violent event. John Fleming was not alone in finding "The Geological Deluge, as interpreted by . . . Professor Buckland, inconsistent with the testimony of Moses *and* the Phenomena of Nature" (my italics).[8]

Furthermore the diluvial theory needs to be seen historically, not as a static idea embodied for all time in the *Reliquiae diluvianae* or any other work. Though the process deserves to be studied in detail, it seems that during the 1820s the character of the theory was transformed, as it gradually became clearer that there had been more than one 'diluvial' episode, and that none of them had been universal. This may have weakened still further the residual links with the Mosaic flood, but it by no means dissolved the problem of the 'diluvium'.

Indeed, in assessing the diluvial theory it must never be forgotten that nature played a harsh trick on the science of geology. Early nineteenth-century naturalists were right to conclude that there had been an extremely peculiar episode in the recent history of the earth. Glacial periods are very rare events, and the fact that the earth emerged from the most recent glacial episode a mere ten thousand years ago was bound to make its traces all the more striking. Such a retrospective comment may be unfashionable, but it is impossible to understand the scientific power of the diluvial theory without recognising the strength of the evidence that seemed to demand some such drastic explanation.

The "great boulders" mentioned by Professor Carozzi were perhaps the most striking phenomena of all. These 'erratic blocks' of rock could be traced from their source-areas across tens or hundreds of miles, and were often perched on hill-tops or otherwise unrelated to the present drainag system. Some were the size of small houses, and could not have been moved by the "causes now in operation" around them, no matter how lengthy the time allowed. Those that were scattered over the north German plain were separated from their sources in Scandinavia by the obstacle of the Baltic; others were perched on the slopes of the Jura, on the far side of the Swiss plain from the Alps where they had originated. The erratics themselves were associated with other peculiar deposits; 'boulder clay', for example, was a chaotic mass of materials quite different from either the deposits of present rivers or the regular strata of more ancient epochs of earth-history; 'diluvial' gravels were spread in great sheets, often on plateaus and unrelated to present rivers. Moreover, the organic world also seemed to have been affected drastically, for the 'diluvial' gravels contained the bones of huge mammals of extinct or climatically alien species.

Given such peculiar and striking phenomena, any satisfactory explanation had to incorporate some equally dramatic causal agency. In retrospect, of course, the puzzle is that an aqueous agency was preferred to a glacial. Here "the Mosaic tradition" may well have exercised an *imaginative* influence on the retention of a diluvial theory; but the historical problem is that an aqueous agency continued to be favoured even after the early link with the Flood had been severed, and after the potential efficacy of the glacial agency had been recognised.

A partial explanation of this problem may be found in the fact that the sophisticated diluvial theory of the 1830s was no mean opponent for a glacial theory to have to displace. It was a theory of great scientific status, scope and explanatory power. Although diluvial episodes were not strictly 'actual causes', in that none had been recorded in the short span of human history, this did not make them merely speculative. There was a constantly reiterated analogy with the known actualistic effects of tsunamis ('tidal waves'), such as the disastrous one that had followed the famous Lisbon earthquake. A 'mega-tsunami', as it were, was considered to be capable of explaining the diluvium quite satisfactorily. This could not be brought within the very strict canons of Lyell's version of the actualistic method, which (in principle if not always in practice) tolerated no deviation from the observed *degree* of intensity of 'actual causes'. But it was perfectly consistent with the more flexible actualism advocated, and practised, by most other geologists in the period: it postulated a past event or events similar in *kind* to an observable present process, though of such greater magnitude as was required to explain the observed phenomena.

Mega-tsunamis or sudden transient 'diluvial currents' could thus explain the diluvium. If the further cause of these currents was often left without discussion, that is no reflection on the scientific status of the theory. It does not betray the smuggling-in of non-material or supra-natural causes. Instead it shows the legitimate reticence of empirically-minded naturalists when faced with a phenomenon they believed *had* taken place, but for which they could not offer any satisfactory causal explanation. This was not the last time that geologists have recognised that in 'palaetiological' sciences (to use Whewell's excellent term) it is perfectly proper to concentrate on elucidating the character of past events, even when the cause of those events remains completely obscure.[9]

In fact, however, by the 1830s the cause of diluvial currents was no longer regarded as completely obscure. For the diluvial theory was given firm causal foundations by its integration into Léonce Élie de Beaumont's theory of *époques de soulèvement*. According to this, a diluvial current would be caused naturally whenever a new mountain range was suddenly elevated in an area covered by sea. Occasional mega-tsunamis would thus be produced by corresponding mega-tectonic events; and these paroxysmal elevations of mountain ranges were in turn linked causally to fundamental geophysical processes in the earth's interior.[10] Élie de Beaumont's theory was of outstanding explanatory power, and served to establish still more firmly the scientific status of the diluvial theory.

Is this, then, enough to account for the difficulty with which the glacial theory displaced it? It is clearly an important factor, but there are some puzzling features that still need to be explained. The opposition of Élie de Beaumont to Agassiz's theory would be understandable on this interpretation, but why should that ardent diluvialist Buckland have become Agassiz's most whole-hearted glacialist supporter? (The ease with which he made this change, thereby abandoning the last possibility of even a tenuous geological link with the Flood, should once more caution us not to place too great emphasis on the influence of "the Mosaic tradition".) Conversely, why did that equally ardent anti-diluvialist Lyell become an opponent of all but a minor aspect of Agassiz's theory?

Lyell's position is at least partly explicable in that he already had his own fairly satisfactory alternative to the diluvial theory. In order to explain away the paroxysmal character of the supposed diluvial episode, he had to invoke some alternative agency for the transport of erratics, and he revived an earlier suggestion that icebergs might have been responsible. But although the efficacy of icebergs for this purpose was confirmed in the 1830s by the results of polar exploration, Lyell still had to account for the position of erratics on land and in fairly southerly latitudes. Their distribution on land could be explained easily in terms of his belief in the continual movements of elevation and depression in the earth's crust. This seemed even more plausible after his visit to Sweden in 1834 had convinced him that such elevation was occurring, even at the present day, insensibly slowly and without earthquakes.[11] The southerly distribution of erratics could likewise be explained in terms of his climatic theory. Pointing out that local climate depends more on the configuration of land, sea, winds and currents than on mere latitude, he could argue that the geographical changes caused by recent elevation and depression could be sufficient to account for the change in the southward limits of drift-ice from the polar regions.[12]

This was quite a satisfactory explanation of the northern diluvium; and indeed many diluvialists seem to have adopted a similar use of icebergs as a means of transporting the largest erratics, though without accepting the gradualistic element of Lyell's interpretation. But Lyell was obliged to modify his explanation rather unconvincingly when dealing with the similar erratics of the Alpine region; their movement across the Swiss plain was attributed to transport on icebergs, which had been detached from glaciers into glacial lakes and then released by the sudden ('catastrophic'!) collapse of the lake dams.[13]   The diluvialists, on the contrary, had no such difficulty in subsuming the northern and Alpine erratics under a single explanation, since the blocks on the Jura could readily be attributed to a diluvial current caused by the sudden elevation of the Alps themselves.

But that explanation brings us back to the glacial theory: why was the extension of the observable Alpine glaciers not utilised earlier as a way of accounting for the erratics within the Alps and on the Jura; and why did this not lead more easily to a full glacial theory, accounting likewise for the northern erratics by the ice-sheets of a general 'Ice-Age'?   The glacial theory did in fact develop in such a sequence of stages, spreading its scope outwards like the glaciers themselves; but this still leaves the puzzle of its curiously hesitant formulation.

Professor Carozzi's introduction is a useful summary of these stages in the development of the theory.[14]   It began, in effect, in the Alpine valleys among peasants whose lives were directly affected by the periodic minor extension and retreat of the glaciers, and who were in a position to observe at first hand the action of glaciers in scratching the bedrock over which they moved, in transporting huge angular boulders, and in piling up such boulders as moraines.   As von Buch and other travellers often reported, it was a common belief among such peasants that the glaciers had formerly extended much further than at any time within their memory. This was a natural inference from their observation of scratched 'striated pavements', perched blocks and moraines, all extending some miles beyond the present glaciers.

The inaccessibility of the Alpine valleys is hardly enough by itself to account for the delayed impact of this idea of glacial extension.   One unusually literate peasant, Jean-Pierre Perraudin, is said to have told Jean Charpentier of it as early as 1815, yet it was not until 1834 that Charpentier read the paper which brought the question to the attention of men of science generally.   Even then, Charpentier borrowed the main features of his theory from another Swiss, the civil engineer Ignace Venetz, whose own ideas show the same curiously slow development. Venetz spent the spring of 1818 in the Val de Bagnes, where Perraudin lived, dealing with the aftermath of the catastrophic bursting of a naturally dammed lake (ironically this famous disaster became one of the actualistic analogies cited in discussions of the diluvial theory).   Venetz had previously read a paper on present glaciers to the *Société helvetique des sciences naturelles*; but in 1821, perhaps as a result of meeting Perraudin, he wrote another paper suggesting that the Alpine glaciers had formerly extended a few miles beyond their present limits; this however remained unpublished for another twelve years.[15]   In 1828 he enlarged the scope of the theory beyond the limits of the Alps, and it thereby became for the first time a true glacial theory with more than local implications. To the *Société helvetique*, assembled appropriately at the Great St Bernard, he suggested that a former vast extension of glacial conditions would explain not only the Alpine moraines but also the erratic boulders on the Jura and in northern Europe.

This sensational hypothesis at last stimulated Charpentier to study the glacial problem more closely. But another six years passed before in 1834 he felt sufficiently convinced by Venetz's solution to propound it in a paper of his own. It was this paper of Charpentier's, published in the *Annales des mines* in 1835 and in Jameson's *Edinburgh journal* and Froriep's *Notizen* in 1836, that at last extended discussion of the theory beyond the limited circle of Swiss naturalists, and made it available to men of science throughout Europe.[16]

Charpentier's argument for the former extension of the Alpine glaciers was impeccably based on actualistic method. He had studied Venetz's evidence for himself and found it overwhelming: the ancient erratics, moraines and scratched bedrock surfaces, throughout the Alpine valleys and even as far as the Jura, were too similar to those associated with the present glaciers to be attributed to any different cause. Yet this conclusion was not accepted easily; and its extension to cover the northern diluvium was still more retarded, although glacial interpretations of the northern erratics were already available. Jens Esmark had proposed such an explanation for both the erratics and the fjord topography of Norway as early as 1824; and since his paper was translated in Jameson's *Journal* and abstracted in Férussac's *Bulletin*, it should have been widely known.[17] A more detailed theory of the same kind, postulating an extension of polar ice to explain the transport of erratic blocks from Scandinavia across the Baltic on to the north German plain, was published in 1832 by Bernhardi, not in some obscure periodical but in Leonhard's and Bronn's *Jahrbuch*.[18] But such papers seem to have been generally ignored. One possible reason lies in their speculative flavour. Esmark's is set in the context of a 'cosmogony' explicitly based on William Whiston's much earlier system, complete with comets and changes in the earth's orbit. This was hardly calculated to win support for his theory in the climate of anti-cosmogonic empiricism which characterised the geology of the 1820s. Bernhardi's paper, though much more closely argued from the evidence, might have been tarred with the same brush, owing to his comment that his theory was "a further step" beyond Esmark's. But the Venetz-Charpentier theory could not reasonably be dismissed on these grounds.

A much more substantial reason for the neglect of these papers, and for the reluctant acceptance of the glacialism of the Swiss naturalists, lies in the fact that the application of any glacial theory to the Jura, and even more to northern Europe, made it at once a rival to the established diluvial

XIV

theory. Here then, leaving aside any supposed influence of "the Mosaic tradition", there may be much truth in Professor Carozzi's suggestion that diluvialism was too deeply entrenched to yield easily to glacialism. This is indeed supported by Charpentier's explicit statement that he was aware that his hypothesis was opposed to the opinion of the best geologists of the time.

But Charpentier also stated what he seems to have regarded as an even more important objection to his theory: "Comment concilier une semblable hypothèse avec la masse de faits qui prouvent que jadis la température de nos climats a été bien plus élevée qu'elle ne l'est maintenant?"[19] Here, I believe, we touch on a possible solution to the historical problem I have stressed. A general glacial theory, of more than merely local effects, may have been hard to accept simply because it contradicted the dominant geological synthesis of the period.

The content, and even the existence, of this synthesis have been obscured by Lyell's persuasive interpretation of the state of geology at this period. It is all too easy to accept his insinuation that his opponents were fundamentally unscientific cosmogonists dealing in unexplained or supranatural catastrophes. I have already suggested the need for an historical examination of the most important single 'catastrophist' theory, namely the diluvial theory; but this is only part of a wider area of research that deserves detailed exploration. How accurate a label is "catastrophism"? How central were saltatory or paroxysmal events in the theoretical framework of Lyell's opponents? Were they introduced, as Lyell alleged, in order to explain observed effects within a too restricted time-scale, or because the effects themselves seemed to demand causal processes greater in intensity than those recorded in human history? Tentatively I would suggest that Whewell's label has had the unfortunate consequence of focusing attention on the saltatory element, to the neglect of a more fundamental feature in these theories, namely the overall *directional* nature of geological change.[20] It is well-known that the 'progressionist' interpretation of the history of life was one of Lyell's main targets for attack; but this interpretation was integrated with, and in some sense causally dependent on, a more fundamental belief in the directional pattern of events in the history of the earth itself. The overtones of improvement in the word 'progression' make it unsuitable as a descriptive label for the resultant synthesis of biology and geology; the relatively neutral term *directionalist* seems more appropriate, although even this has teleological overtones which are certainly not intrinsic to the synthesis.

The earth itself was believed to have changed directionally, approximating ever more closely to its present state of relative stability; and this directionally changing environment was regarded as having been the physical substrate for the directional or 'progressive' changes in the history of life.[21] This synthesis was based on a revival of earlier theories of a cooling earth, of which Buffon's *Époques de la nature* had been perhaps the most influential. Louis Cordier's rigorous demonstration of the universality of the geothermal gradient in mines had given greater scientific authority to the idea of a "central heat" in the earth, and Fourier's application of his heat theory to the problem of a cooling earth had lent new prestige to the interpretation of the central heat as a *residual* heat.[22] Fourier's work became the ultimate causal justification for directionalist interpretations of many other problems, ranging from the origin of granite and gneiss to the distribution of fossil plants. Empirical observations of all kinds seemed to confirm the view that the temperature of the earth's surface had fallen throughout geological time, at first relatively rapidly from its original fluid state, then more and more slowly, until in the present epoch the rate of heat-loss from the interior was masked by the effects of solar radiation, so that the earth's surface had attained a relatively stable state. Against this background of gradual directional change, occasional paroxysmal events could be generated without any suspension of the ordinary 'laws of nature'. For example Élie de Beaumont explained his *époques de soulèvement* as the occasional result of slowly accumulating compressional strains in the crust of a gradually shrinking earth.

Just how much, and in what ways, was this consensus of geological opinion modified by Lyell's work? Should we take Lyell's view of the debate at its face value, and assume that the publication of the *Principles of geology* changed the geological scene almost overnight?[23] This is another area where further research is needed. Provisionally I would suggest that the explanatory power and scope of the directionalist synthesis may not have been greatly affected. While Lyell's demonstration of the heuristic value of the actualistic method was generally admired, his extremely rigorous version of actualism was criticised for being aprioristic; and his use of 'actual causes' to argue for a steady-state pattern in earth-history was felt to involve serious straining of the evidence.[24] Indeed, far from being weakened by Lyell's system, the directionalist synthesis seems to have gained added strength in the 1830s from the accumulation of further evidence for a 'progressive' history of life—which Lyell had explained away rather unconvincingly; and Lyell's steady-state interpretation of the earth itself fared little better, since he was unable to point to any indefinitely renewable source for the earth's internal heat, on which his system of perpetually balanced forces ultimately depended.

Charpentier was therefore right to see in the theory of gradual refrigeration the chief obstacle to any glacial hypothesis. The former extension of the Alpine glaciers for a few miles beyond their present positions— Venetz's earlier hypothesis—might have been acceptable, as being a reasonable extrapolation from the minor fluctuations known in historic times. But to extend the glaciers of the Rhône basin, as Charpentier's theory did, right across the Swiss plain and up on to the slopes of the Jura, was a far more radical suggestion. And to apply a glacial explanation also to the northern erratics would involve reconstructing ice sheets over large areas of northern Europe. This would inescapably imply a much colder period in the recent past, whereas the whole trend of geological research favoured a warmer climate. This contradiction between the glacial theory and the expectations of the directionalist synthesis largely accounts, I believe, for the otherwise puzzling lack of enthusiasm for the idea of an Ice Age.

One reason for the neglect of Bernhardi's earlier glacial hypothesis for the northern erratics, in addition to those already mentioned, may be precisely his failure to reconcile these northern ice-sheets with the idea of a gradually cooling earth: he admitted the difficulty, but only tackled it by suggesting that a glacial episode was within the scope of a reasonable fluctuation of the earth's climate, which it clearly was not. Charpentier himself admitted that the same problem had made him reluctant to accept Venetz's hypothesis of a great extension of the glaciers. Having found the evidence compelling, however, he minimised the difficulty by pointing out that it was unnecessary to postulate the extension of the Alpine glaciers far beyond the Alps, or even to fill the whole of the Swiss plain with ice. But even this could not dispel the problem altogether. Significantly he tried to solve it without having to postulate any globally cold period at all. He attributed the great extension of the Alpine glaciers not to an Ice Age but to the greater elevation of the Alps themselves in the geologically recent past. Having calculated from the present temperature gradient that this would require Alps of Himalayan height, he then had to explain how such a great elevation, followed by a reduction to the present height, had all occurred within a geologically short period. For the elevation, he was able to invoke a recent *époque de soulèvement*, but for the subsequent reduction in height, and consequent retreat of the glaciers, he had to fall back on a rather vague process of 'stabilising' of the newly-elevated mountains. Despite this weak point, his 1834 paper was a well-argued and persuasive re-interpretation of the Alpine and circum-Alpine 'diluvial' phenomena. But of course it could not explain the diluvium of northern Europe.

Agassiz was among the sceptics when Charpentier read his paper, but he was converted to the idea of glacial extension in 1836, as a result of being shown the Alpine phenomena by Charpentier and Venetz. The following year, after only a limited amount of further fieldwork in the Jura, he made the theory the subject of his opening address as president of the *Societé helvetique*—having composed it, so it is said, only the night before its delivery. The *Discours de Neuchâtel* presented a theory that was not only more radical than Charpentier's, but also fundamentally different in character; its publication in the *Bibliothèque universelle de Genève* and translation in Jameson's *Journal* a few months later gave

Agassiz's theory a wide circulation even before the publication of the *Études*.[25] Agassiz adopted Charpentier's view of the former extension of glaciers, but only for the erratics within the Alpine valleys; he gave an entirely different explanation of the erratics on the Jura. He followed Charpentier in using a recent *époque de soulèvement* as part of his explanation, but he used it in a different way. He postulated that *before* the recent elevation of the Alps, an "Ice Age" of intense cold had covered the whole northern sector of the globe, as far south as the Mediterranean, with a huge *nappe* of ice (or more accurately, of *névé*). The upheaval of the Alps had tilted the surface of this ice-sheet in the vicinity of the newly-raised mountains; and blocks of rock had then slid down this slope until they came to rest on the slopes of the Jura. Thus the theory postulated two essentially separate stages for the Ice Age: first the covering of large areas of the earth's surface by a static *nappe* of ice; and only later, during the subsequent warming in global temperature, the movement of the glaciers, which had finally shrunk to their present size. Agassiz himself stressed the importance of distinguishing these stages, and without doing so it is impossible to assess the reception of the theory.

The introduction of this highly speculative notion of a severe Ice Age *preceding* the extended glaciers poses an historical problem. Its speculative non-actualistic character was bound to make Agassiz's theory more difficult for his contemporaries to accept. Moreover, at first sight it seems totally unnecessary for Agassiz's own purposes. Once he was prepared—unlike Charpentier—to consider the possibility that the whole earth had passed through a colder period in the recent past, he could have extended the Charpentier–Venetz thesis, without further difficulty, to account for all aspects of the diluvium. Moreover, in the light of Charpentier's use of scratched bedrock surfaces as a criterion of glacial extension, Sefstroem's recent description of extensive surfaces of this kind in Sweden should have made a glacial explanation of the northern diluvium more plausible than ever.[26] Agassiz, it might seem, was in possession of all the data needed in order to postulate the former existence of ice-sheets just as far as the northern diluvium and erratics extended— and no further. Such moving ice-sheets could have been responsible for transporting the erratics and for scratching the underlying bedrock, and their later retreat and disappearance could have been due to the same climatic amelioration that led to the retreat of the glaciers in the Alps.

This of course is—with some simplification—the modern interpretation of the evidence, but it differs significantly from Agassiz's theory. For Agassiz, the movement of Alpine glaciers and northern ice-sheets was a feature only of the waning phases of the Ice Age: the Ice Age proper had been characterised by a much more widespread static sheet of *névé*. It is this that seems at first sight a superfluous hypothesis.

But its significance can perhaps be seen in the use that Agassiz made of it: to this extremely severe Ice Age he attributed the extinction of the spectacular 'diluvial' fauna.   In the manner characteristic of the age, Agassiz was here trying to synthesise his geology with his biology; and it may be no accident that he borrowed the concept of a drastic *Eiszeit* from another biologist, the botanist Karl Schimper.   It is at this point, I believe, that Agassiz's glacial theory must be studied in the context of the rest of his scientific work; for it was of course as a biologist, or more precisely as a palaeontologist, that he made his name in the scientific community in the 1830s, with the publication of the first parts of his ambitious *Poissons fossiles.*[27]

The possibility of an essential connection between his biological work and his glacial theory is not discussed by Professor Carozzi, but it deserves serious consideration.   At the present time, any suggestions of this kind must be highly tentative, for we still lack a satisfactory analysis of Agassiz's scientific career.   What is needed is a full-scale study of his scientific development, from his early training under such *Naturphilosophen* as Döllinger, Oken and Schelling, through his short but important period as a protégé of Cuvier, into his scientific maturity in the 1840s (his contribution to the scientific scene after his departure from Europe in 1846 seems by contrast relatively peripheral, except within the American setting).[28]   But pending such a study, it is still worth considering the relation between his glacial work and his biology, if only because it seems unlikely that he undertook such a time-consuming research project merely as a healthy open-air diversion.

Agassiz, then, used the *nappe* hypothesis explicitly in order to explain the apparently sudden extinction of the 'diluvial' fauna.   Here surely we can trace Cuvier's influence.   Agassiz's *Poissons fossiles* was modelled consciously on his hero's great *Ossemens fossiles;* Cuvier had indeed handed over his own materials for the work shortly before his death.   But Agassiz inherited from Cuvier not only specimens but also a coherent set of biological methods, problems and attitudes.[29]   The method of reconstruction of fossils on the principle of the *convenance des parties* was put to good use in the *Poissons fossiles;* but Agassiz also took over Cuvier's concern with the problem of extinction.   If extinct faunas had been as well-adapted as the application of the anatomical principles suggested (or, more accurately, presupposed!), how could they have become extinct, unless by some drastic *révolution?*   Thus if Agassiz was to remove the diluvial *révolution* by reinterpreting the diluvium as glacial, he may have felt compelled to replace it by another *révolution* with equally drastic ecological effects.   The mere extension of glaciers and ice-sheets, even as widely as the diluvium extended, would hardly be sufficient for this purpose, for the pre-glacial fauna could simply have migrated southwards away from the ice.   Only a *sudden* Ice Age of global extent seemed to be adequate as a mechanism for the extinction of the mammoths and other great mammals of the diluvium. Even Agassiz, with his predilection for dramatic hypotheses, stopped short of suggesting that the ice *nappe* had covered the whole earth; but he did assert, both in the *Discours* and in the re-written version that forms the final chapter of the *Études*, that the cold epoch had been intense enough to destroy all life at the earth's surface.

How was such an epoch to be reconciled with the theory of a cooling earth? Agassiz admitted that his theory would seem "in direct contradiction with the well-known facts demonstrating a considerable cooling of the earth since the remotest times",[30] and like Charpentier he had to attempt some reconciliation. Indeed Agassiz's Ice Age made the problem far more difficult than for Charpentier. Professor Carozzi dismisses Agassiz's solution as "a wild biological explanation",[31] and attributes its re-appearance in the final pages of the *Études* to Agassiz's reluctance to admit he had been wrong. This seems unconvincing: Agassiz could easily have dropped the idea from what he intended as an impressive *magnum opus*, and his detailed analysis of present glacial action would have been unaffected. The fact that he did include his admittedly speculative solution as the conclusion—and culmination—of the *Études* suggests that it was more important to him than Professor Carozzi allows.

Agassiz introduced his solution to the problem by setting it in the context of his palaeontological work, and it is only in this context, I believe, that its importance can be understood. For he inherited from Cuvier not only methods and problems, but also the fundamental biological attitude that had underlain all Cuvier's work: the conception of the organism as a functionally integrated whole. If each species was a stable adaptive mechanism of intricate construction, well fitted for its appropriate mode of life, it could only survive during a certain period if the requisite environment also continued to exist on earth. That a given fauna had characterised the earth's surface during a certain geological period therefore implied that environmental conditions had remained stable within that period. But Cuvier's work had also demonstrated the apparent periodisation of successive faunas, and he had therefore introduced successive *révolutions* to account for their extinction. Agassiz's study of fossil fish seems to have convinced him of the validity of Cuvier's approach,[32] but he was then faced with a serious problem: how could the required succession of periods of stable conditions be reconciled with the theory of a gradually cooling earth? Cuvier himself may never have felt the force of this dilemma, since most of his creative work was completed before the directionalist synthesis had become a compelling pattern in geological research. But the final pages of the *Discours* show Agassiz triumphantly producing a solution which at the same time made the Ice Age part of the fundamental pattern of earth-history.

Agassiz accepted the long-term cooling of the globe; but he asserted (like Bernhardi) that "nothing has demonstrated to us that this cooling process was continuous and took place without oscillations".[33]   He then argued that on "physiological" grounds (*i.e.*, on the Cuvierian conception of adaptation) a sequence of periods of stable conditions, rather than a continuous gradual cooling, should be expected to have occurred.   Hence, borrowing a diagram, and probably the idea itself, from Schimper, he converted a gradual decline in temperature into a stepwise decline:

This reconciled an overall directional cooling of the globe to the Cuvierian requirement of periods of stable conditions; but it also provided an explanation for the *révolutions* that had separated these periods.   The Ice Age thus became merely the last of many similar sudden episodes of drastic cooling, each followed by a warming to the somewhat lower temperature of the next period.   The last of these episodes had been an Ice Age merely because the long-term cooling had by then proceeded far enough for the sudden cooling to go below freezing-point; but earlier episodes would have been equally drastic in their ecological effects, since no set of organisms could survive such a sudden and radical change in environment.[33a]

Agassiz's theory therefore postulated that the history of the earth had been divided into periods of stable conditions, within each of which an appropriately adapted set of organisms could flourish on earth; the episode of sudden cooling at the end of each period would then not only account for the mass extinction of those organisms, but also act as a prelude to the establishment of a new stable environment with a suitable new set of organisms adapted to it.   Although, as Professor Carozzi points out, the idea of repeated episodes of sudden cooling does not appear explicitly in Agassiz's work after the *Études*, it may have contributed to his continued insistence on the periodisation of organic life, and on the absolute nature of the faunal discontinuities between the successive periods[34]—and this at a time when most other palaeontologists were moving towards Lyell's concept of piecemeal faunal change.

It is hardly surprising that Agassiz's sensational theory of an extremely extensive *nappe* of ice, and his still more speculative integration of this into the general history of the earth, should have had no enthusiastic reception from the leading geologists of the time.   A common reaction to Agassiz's theory was one of embarrassment and irritation, that a young naturalist of outstanding promise in biological work should waste his time on such extravagant speculations.

But the formidable trio of first-rate scientists mentioned by Professor Carozzi—von Buch, Élie de Beaumont and Humboldt—had a more substantial reason for opposing the theory.   In the case of Élie de Beaumont there may have been personal reasons too: Agassiz's theory relegated his *époques de soulèvement* to a less important position in the history of the earth, depriving them of any connection with the successive periods of extinction. But for all three, the theory would have seemed fundamentally unsound because it failed to reconcile the *Eiszeit* concept satisfactorily to their conviction about a gradually cooling earth. Agassiz, as we have seen, attempted a reconciliation, yet left the causal mechanism for such drastic fluctuations in temperature unexplained.

It is not surprising that Buckland was one of the first converts to the theory, and indeed one of Agassiz's few wholehearted supporters. For Buckland shared with Agassiz not only a liking for sensational theorising but also the same concern with finding a sufficiently drastic mechanism to explain the extinction of the diluvial fauna, especially since he had ceased to believe in a single universal deluge.

A more typical reaction was the cautious acceptance of the less speculative aspects of the theory by geologists such as Bernard Studer. Here the recognition of traces of small valley glaciers in areas such as the Jura and the Vosges, far from any modern glaciers, served to confirm the general idea of glacial extension.[35] At the same time it threw doubt on Charpentier's earlier suggestion that the extension of Alpine glaciers could be explained simply by the greater altitude of the newly-elevated Alps, for according to Élie de Beaumont's work the Jura and the Vosges had *not* been elevated recently. The former existence of glaciers in such areas implied that there had been a genuinely colder period in Europe in the recent past; but this conclusion did not entail the acceptance of Agassiz's much more drastic hypothesis of a static *nappe* of ice preceding the glaciers. In tracing the 'progress' of the glacial theory we need to discover in more detail just what elements of Agassiz's ideas were being accepted. Generally it seems to have been only the idea of an extension of valley glaciers in highland areas, not their extension as ice-sheets into lowland areas, and still less Agassiz's hypothesis of an episode of drastic global cooling.

This kind of modest glacial theory can be seen in Charpentier's own reaction to the glacial controversy of the late 1830s. Charpentier had the misfortune to see his own substantial *Essai sur les glaciers*[36] beaten by a few months by the publication of Agassiz's *Études*. Although he was only able to add a brief passage on Agassiz's theory to his own already-completed manuscript, his criticisms were highly pertinent; like most other geologists he found the concept of an ice *nappe* unsatisfactory and indeed unnecessary. On the other hand, Charpentier now acknowledged that his earlier hypothesis of the higher elevation of the Alps was untenable, and that the glacial interpretation of diluvial phenomena beyond the Alpine region implied a generally colder period in the recent past. But unlike Agassiz he tried to minimise the degree of cooling necessary to explain the extension of the glaciers, and avoided postulating any reversal of the gradual long-term cooling of the globe. He suggested instead a meteorological explanation, pointing out that precipitation and cloud cover are much more important factors in producing glacial extension than mere cold. On this basis he was able to argue that the required extension of the Alpine and other glaciers could have been produced quite easily, simply by a few centuries of cold wet summers like those that had produced a minor extension only a few years before. Retaining his earlier postulate of the recent elevation of the Alps, he was able to propose a fairly cogent causal link between this *soulèvement* and the meteorological changes required to extend the glaciers. This hypothesis, which shows far greater understanding of glaciology than Agassiz's more sensational ideas, enabled Charpentier to extend a modest glacial theory beyond the Alps, without conflicting with his underlying and continuing belief in the gradual long-term cooling of the whole globe. Charpentier's theory was justly commended for its actualistic character: unlike Agassiz's theory it avoided postulating conditions to which the present offered no analogy, and it explained the phenomena simply by the prolonged action of processes that were still observable. But as the same reviewer pointed out regretfully, "En science, comme à table, *Tarde venientibus ossa*".[37] The previous year had seen the publication of no less then three books on the Alpine glaciers, and Agassiz's jumping the gun with a more dramatic theory deprived Charpentier's well-argued work of the attention it deserved.

In Britain[38] Agassiz's theory had the same ambivalent reception as on the Continent. The modest theory of former valley glaciers in highland areas gained wide support, especially after Agassiz and Buckland had discovered glacial traces throughout the Scottish Highlands; and the inference of a cool episode received unexpected support from the discovery of Arctic molluscs in geologically recent marine deposits in Scotland. But there was general scepticism about what William Conybeare dubbed "the Bucklando-Agassizean Universal Glacier",[39] i.e., the hypothesis of a *nappe* of ice covering wide areas of the earth's surface. It seemed much more plausible to continue to explain the lowland diluvium in terms of diluvial currents, utilising icebergs broken from the highland glaciers as a means of transporting the larger erratics.

When Buckland reported to Agassiz that Lyell had "adopted your theory in toto!!!",[40] he was probably indulging in wishful thinking: it is hard to imagine Lyell accepting the non-actualistic *nappe* concept, even temporarily, whereas he could and did accept enthusiastically the idea of local glaciers. Since he had already postulated a recent cooler period to account for the southerly drift of icebergs bearing the 'diluvial' erratics, Agassiz's discovery of former valley glaciers in Scotland fitted neatly into the same explanatory framework. Charles Darwin likewise could be enthusiastic about his own discovery of traces of valley glaciers in Wales, but without modifying his essentially Lyellian position. There is much further work to be done on the reception of Agassiz's theory and the later history of glacial explanations. But it is possible now to suggest reasons for the slow acceptance of glacialism. Agassiz's theory might be thought 'catastrophic' enough to have appealed to the diluvialists; but the sophisticated diluvial theory of the 1830s and 1840s was very different from its cruder earlier versions, and the speculative non-actualistic character of Agassiz's Ice Age made his theory as unacceptable to Lyell's opponents as it was to Lyell himself.

But its unacceptable character goes deeper than this. I have suggested that historical concentration on the 'catastrophic' aspect has hidden the more fundamental directional pattern that underlay the theories of Lyell's opponents. It was principally at this level that their theories conflicted radically with Lyell's steady-state interpretation of the same phenomena. But Agassiz's glacial theory was *equally* irreconcilable to both. Although Lyell and the directionalists differed fundamentally in their interpretation of the basic pattern of earth-history, they were united in being unable to accept a theory that involved a major climatic fluctuation in the recent past. A minor fluctuation could be accommodated within either framework, so that both groups were prepared to accept the former extension of the Alpine glaciers and the former existence of valley glaciers in Scotland. But an Ice Age was impossible to accept: for Lyell it was too great a departure from the steady state; for the directionalists it conflicted with the belief in the gradual cooling of the globe. Roderick Murchison was perhaps more prescient than he realised when he said in 1842 that he saw "no stopping place" once any part of Agassiz's glacial theory was accepted.[41] The very slow and reluctant acceptance of an 'Ice Age' during the following twenty years may perhaps reflect the slow transformation of the sharp distinction between directionalism and uniformitarianism: both systems were forced to assimilate a larger and more dramatic fluctuation of climate than the proponents of either had expected to find in the recent history of the earth.

This is a superb edition of the *Études*, and its usefulness is enhanced by the editor's inclusion of a new translation of the *Discours de Neuchâtel* as a prelude to the main work. Agassiz published the *Études* in two volumes, the text in octavo, the plates in folio. In the present edition the plates have been reduced in size, so that both text and plates have been included in a single volume. On the resultant ample pages, the text is set out with attractive typographical design and with usefully wide margins. Professor Carozzi is well qualified for the editorial task, being both a native of Switzerland and a professional geologist widely read in the history of his subject. His translation is clear and accurate. He provides several useful locality maps to assist the reader through the descriptive sections of the book, and he has converted Agassiz's sketchy and often inaccurate footnotes into full references to the works cited. (Agassiz corrected the proofs of the *Études* at his glacial field-station, the so-called Hôtel des Neuchâtelois, a shelter built beneath a huge boulder on the Aar glacier: an appropriate place, but hardly ideal for checking references). There is only one small feature of the editor's work that must be criticised. He has used his knowledge of modern glaciology to comment on, and explain, some of Agassiz's conclusions; but instead of placing all these comments in editorial footnotes, where they would have been valuable to readers without his scientific background, he has interpolated some of them in the main text, where they form rather irritating anachronistic interruptions to the flow of Agassiz's argument.

REFERENCES

1. L. Agassiz, *Études sur les glaciers* (Neuchâtel, 1840); translated as *Untersuchungen über die Gletscher* (Solothurn, 1841).
2. For an important exception, see Walter F. Cannon, "The Uniformitarian-Catastrophist debate", *Isis*, li (1960) 38–55. These terms were first used by [William Whewell], "*Principles of geology*, by Charles Lyell . . . Vol. II", *Quarterly Review*, xlvii (1832) 103–132.
3. Carozzi, *Studies*, p. xi. Compare the similar remark by Agassiz's earliest scientific biographer: "Even the Uniformitarians . . . with Charles Lyell as their leader, did not see the splendid opportunity to add a new crown of laurels to Uniformitarianism, or the doctrine of existing causes." Jules Marcou, *Life, letters, and works of Louis Agassiz* (New York, 1896, 2 vols), i, 167.
4. R. Hooykaas, *Natural law and divine miracle. A historical-critical study of the principle of uniformity in geology, biology and theology* (Leiden, 1959); 2nd impression under title *The principle of uniformity* (Leiden, 1963). For the distinction between actualism and uniformitarianism, see especially the introduction to the 2nd impression; also M. J. S. Rudwick, "The principle of uniformity", *History of science*, i (1962) 82–86. I have suggested possible lines for a further analysis in "Uniformity and progression: reflections on the structure of geological theory in the age of Lyell", to be published in the proceedings of the Symposium on the History of Science and Technology held at the University of Oklahoma in April 1969.
5. Carozzi, *Studies*, pp. 133, 147. Ch. 1 of the *Études* is an historical introduction; ch. 2–16 deal with present glaciers; ch. 17 discusses the evidence for their former extension and ch. 18 elaborates Agassiz's theory of an Ice Age.
6. Charles Lyell, *Principles of geology* . . . (London, 1830–33), 1st edition, subtitle.
7. William Buckland, *Vindiciae Geologiae: or the connexion of geology with religion explained* (Oxford, 1820); *Reliquiae Diluvianae: or, observations on the organic remains contained in caves, fissures, and diluvial gravel, and on other geological phenomena, attesting the action of an universal deluge* (London, 1823).
8. John Fleming, "The geological deluge, as interpreted by Baron Cuvier and Professor Buckland, inconsistent with the testimony of Moses and the phenomena of nature", *Edinburgh philosophical journal*, xiv (1826) 205–239. Compare [William Fitton], "Reliquiae Diluvianae . . .", *Edinburgh review*, xxxix (1823) 196–234. Fleming's coupling of Cuvier with Buckland under the same condemnation seems to have stemmed from his knowing Cuvier's *Discours* only in Jameson's English editions.
9. A good modern parallel is shown by the tenacity with which many twentieth-century geologists continued to believe that continental displacement ('continental drift') *had* occurred, and continued to study the evidence for it, notwithstanding the somewhat arrogant assertions of geophysicists that it could not have occurred because they could not think of a causal mechanism for it. More recently the latter have changed their minds and are now among the theory's most enthusiastic supporters. This would make an instructive study in the methodology of the geological sciences.

10. Élie de Beaumont, "Recherches sur quelque-unes des révolutions de la surface du globe, présentant différents exemples de coincidence entre le redressement des couches de certains systèmes de montagnes, et les changements soudains qui ont produit les lignes de démarcation qu'on observe entre certains étages consecutifs des terrains de sédiment", *Annales des sciences naturelles*, xviii (1829) 5–25, 284–416; xix (1830) 5–99, 177–240; also published separately (Paris 1830). See also the revised summary in Henry T. De La Beche, *Manuel géologique* (Paris, 1833), 616–665.

11. Charles Lyell, "On the proofs of a gradual rising of the land in certain parts of Sweden", *Philosophical transactions of the Royal Society of London*, 1835, 1–38.

12. Charles Lyell, "Address to the Geological Society, delivered at the anniversary, on 19th February, 1836", *Proceedings of the Geological Society of London*, ii (1836) 357–390.

13. Charles Lyell, *Principles of geology* . . . [e.g.,] 1st ed. (London, 1930–3), iii, 148–150; 5th ed. (London 1837), iv, 46–8.

14. Carozzi, *Studies*, xi-xl. See also Albert V. Carozzi, "Agassiz's amazing geological speculation: the Ice Age", *Studies in romanticism*, v (1966) 57–83. F. J. North, "The centenary of the glacial theory", *Proceedings of the Geologists' Association*, liv (1943) 1—28, is valuable for its use of otherwise unpublished material. There is another account in Richard J. Chorley, Antony J. Dunn, and Robert P. Beckinsale, *The history of the study of landforms, or the development of geomorphology*, i (London, 1964), ch. 13.

15. Venetz, "Memoire sur les variations de la température dans les Alpes de la Suisse (redigé en 1821)", *Denkschriften der allgemeinen Schweitzerischen Gesellschaft für die gesamnten Naturwissenschaften*, i (1833) 1–38.

16. J. de Charpentier, "Notice sur la cause probable du transport des blocs erratiques de la Suisse," *Annales des mines*, 3me sér., viii (1835) 219–236; "Account of one of the most important results of the investigations of M. Venetz, regarding the present and earlier condition of the glaciers of the Canton Vallais", *Edinburgh new philosophical journal*, xxi (1836) 210–220; "Bericht über eines der wichtigsten Resultate der von Hrn. Venetz rücksichtlich des jetzigen und vormaligen Zustandes der Gletscher des Cantons Wallis angestellten Untersuchungen", *Notizen aus dem Gebiete der Natur- und Heilkunde*, i (1836) 290–6.

17. Esmark's paper (*Nyt magazin for naturvidenskaperne*, i (1824) 28) was translated as "Remarks tending to explain the geological history of the earth", *Edinburgh new philosophical journal*, ii (1826) 107–121; and abstracted in *Bulletin des sciences naturalles et de géologie*, xi (1827) 402–3, xii (1827) 1.

18. A. Bernhardi, "Wie kamen aus dem Norden stammenden Felsbruchstücke und Geschiebe, welche man in Norddeutschland und den benachbarten Ländern findet, an ihre gegenwärtige Fundorte?" *Jahrbuch für Mineralogie, Geognosie, Geologie und Petrefaktenkunde*, iii (1832) 257–267.

19. Charpentier, *Annales des mines* (see note 16), 226.

20. Whewell's choice of the term 'catastrophe', when 'paroxysm' and 'revolution' were also in current use, obviously reflects his own belief in large-scale discontinuities in the organic world, coinciding with corresponding discontinuities in the inorganic environment. But it may also reflect his special interest in the most recent 'catastrophe', i.e., the diluvial episode, in relation to the history of Man, seen as the final discontinuity bringing the directional development of the whole terrestrial system to its teleological climax: as a classical scholar he would have been well aware that in Greek drama the καταστροφή is the episode (not necessarily 'disastrous') that brings the action to its final dénouement.

21. For an influential exposition of this directionalist synthesis, see H. T. De La Beche, *Researches in theoretical geology* (London, 1834).

22. L. Cordier, "Essai sur la température de l'interieur de la terre", *Mémoires du Muséum d'Histoire naturelle*, xv (1827), 161–244; also *Mémoires de l'Academie royale des Sciences de l'Institut de France*, vii (1827), 473–555. Fourier, "Mémoire sur les températures du globe terrestre et des éspaces planétaires", *ibid.*, vii (1827) 569–604.

23. For this 'revolutionary' view, see e.g., Leonard G. Wilson, "The origins of Charles Lyell's Uniformitarianism", *Geological Society of America, special papers*, lxxxix (1967) 35–62: "The year 1830 marked a great divide in the history of geology. Before that year there had been many great accomplishments, but there had been no critical assessment of the meaning of geological phenomena . . . the interpretation of the past history of the earth . . . remained fanciful and speculative. After 1830 geology became a science" (p. 35).

24. E.g., Adam Sedgwick, "Address to the Geological Society, delivered on the evening of 18th February 1831", *Proceedings of the Geological Society of London*, i (1831) 281–316. W. D. Conybeare, "On Mr Lyell's 'Principles of Geology'", *Philosophical magazine and annals*, new ser., viii (1830) 215–9; continued as "An examination of those phaenomena of geology, which seem to bear most directly on theoretical speculations", *ibid.*, viii (1830) 359–362, 401–6; ix (1831) 19–23, 111–7, 188–197, 258–270.

25. L. Agassiz, "Discours prononcé à l'ouverture des séances de la Société Helvetique des Sciences Naturelles, à Neuchâtel, le 24 juillet 1837", *Actes de la Société helvetique des Sciences naturelles*, 22me session, Neuchâtel, 24–26 juillet 1837, v–xxxii; reprinted in Marcou (see note 3), i, 89–108. Also published as "Des glaciers, des moraines et des blocs erratiques", *Bibliothèque universelle de Genève*, nouvelle série, xii (1837), 269–293; this was translated as "Upon glaciers, moraines and erratic blocks", *Edinburgh new philosophical journal*, xxiv (1838) 364–383. Professor Carozzi does not refer to this contemporary translation, and provides another—excellent—translation as a prelude to his edition of the *Études* (Carozzi, xli–lxi). The 1838 translation probably accounts for the fact, noted by Professor Carozzi (p. xxxviii, note 33), that Agassiz's theory was known in the U.S.A. by 1839.

26. Sefstroem himself gave them a diluvial interpretation: "Ueber die Spuren einer sehr grossen urweltlicher Fluth", *Annalen der Physik und Chemie*, (2), xxxviii (1836) 614–8; "On the traces of a vast ancient flood", *Edinburgh new philosophical journal*, xxiii (1837) 69–73. This was a preliminary account: by the time Agassiz re-wrote his theory for the *Études* the full memoir in Swedish (*Kongliche Vetenskagems-Academie, Handlungen 1836* (1838) 141–255) was available (in part) in translation as "Untersuchung über die auf den Felsen Skandinaviens in bestimmter Richtung vorhandenden Furchen und deren wahrscheinliche Entstehung", *Annalen der Physik und Chemie* (2), xliii (1838) 533–567.

27. L. Agassiz, *Recherches sur les poissons fossiles . . .* (Neuchâtel, 1833–43).

28. Edward Lurie, *Louis Agassiz: a life in science* (Chicago, 1960), is a valuable general biography, especially for the second, American, half of Agassiz's career, but does not analyse his scientific ideas at all closely. For Agassiz and Darwin, see E. Mayr, "Agassiz, Darwin and evolution", *Harvard Library bulletin*, xiii (1959) 165–194; and Edward Lurie, "Louis Agassiz and the idea of evolution", *Victorian studies*, iii (1959) 87–108. Primary source material is contained in Elizabeth Cary Agassiz, *Louis Agassiz, his life and correspondence* (Boston, 1885, 2 vols), but there is more scientific material in Marcou (see note 3). The chief collections of MS material are in the Houghton Library and the Museum of Comparative Zoology, Harvard University.

29. For the relation between the two, see Lurie's biography (note 28); also William Coleman, "A note on the early relationship between Georges Cuvier and Louis Agassiz", *Journal of the history of medicine and allied sciences*, xviii (1963) 51–63. For Cuvier's approach to biology, see William Coleman, *George Cuvier zoologist. A study in the history of evolution theory* (Cambridge, Mass., 1964).

30. Carozzi, *Studies*, p. lvii.

31. Carozzi, *Studies*, p. xviii.

32. See his early report "On a new classification of fishes, and on the geological distribution of fossil fishes", *Proceedings of the Geological Society of London*, ii (1834) 99–102; reprinted in *Philosophical magazine and annals*, v (1834) 459–462, and *Edinburgh new philosophical journal*, xviii (1835) 175–8.

33. Carozzi, *Studies*, p. lvii.

33a. It is worth quoting an early informal summary of the theory in Agassiz's own words, for it illustrates many of the points I have been making. It comes from a letter (or draft of a letter) to Buckland; the letter is undated, but internal evidence suggests that it was written early in 1838. It is one of Agassiz's earliest letters written in English, and is reproduced here exactly as written. (Quoted by kind permission of the Director of the Houghton Library of Harvard University: the letter is filed as bMS. Am. 1419, item 107.)

"Since I saw the glaciers I am of a quite snowy humour, and will have the whole surface of the earth covered with ice and the whole prior creations dead by cold, in fact I am quite satisfied that ice must be taken in every complete explanation of the last changes which occurred at the surface of Europe. In my opinion there is no doubt that ice covered the whole place where the Alps now stand before they were up heaved and that the organic remains of the Diluvium were dead and enclosed in ice like the Ma[mmoths] of Siberia from the north Pole to Italy and the South of France and that the ice melted after the elevation of the Alps and effected[,] by drawing back and alternatively [i.e., alternately] extending again[,] the curious dispersion of erratic blocs. Mr. Charpentier and V[enetz] in their papers only perceived the near relations of blocs and glaciers but overlooked the general relations with the whole process of cooling the earth undergo principaly at the end of each geological period whilst after each soulevement a new elevation of temperature took place and with it a new increase in organic life. I hope this view will be a new insight in the explanation of the organic progress [substituted, significantly, for "development"] of the earth, and enable us to come nearer the conditions of the first appearance of living beings. I am very anxious to know what my English friends will say to this result of my wanderings in the Alps last summer."

34. For an early example, see Louis Agassiz, "On the succession and development of organised beings at the surface of the terrestrial globe; being a discourse delivered at the inauguration of the Academy of Neuchatel", *Edinburgh new philosophical journal*, xxxiii (1842) 388–399. Also "A period in the history of our planet", *ibid.*, xxxv (1843) 1–29: "it is [the glacial period] which, like a sharp sword, has separated the totality of now living organisms from their predecessors" (p. 16). This popularly written essay contains one of the clearest statements of Agassiz's glacial theory in a general geological context.

35. Bernard Studer, "Notice sur quelque phénomênes de l'époque diluvienne", *Bullétin de la Société géologique de France*, xi (1840) 49–52; Renoir, "Note sur les glaciers qui ont recouvert anciennement le partie méridionale de la chaine des Vosges", *ibid.*, 53–65.

36. Jean de Charpentier, *Essai sur les glaciers et sur le terrain erratique du bassin du Rhône* (Lausanne, 1841).

37. Review in *Bibliothèque universelle de Genève*, n.sér., xxxvii (1842) 390–411; translated in *Edinburgh new philosophical journal*, xxxiii (1842) 104–124.

38. For a detailed study, see G. L. Davies, "The tour of the British Isles made by Louis Agassiz in 1840", *Annals of science*, xxiv (1968) 131–146. Also *The earth in decay: a history of British geomorphology 1578–1878* (London, 1969), ch. 8.

39. M. J. S. Rudwick, "A critique of uniformitarian geology: a letter from W. D. Conybeare to Charles Lyell, 1841", *Proceedings of the American Philosophical Society*, cxi (1967) 272–287.

40. E. C. Agassiz (see note 28), i, 309.

41. Roderick Impey Murchison, "Anniversary address of the President," *Proceedings of the Geological Society of London*, iii (1842) 673–867 (at p. 677).

INDEX

INDEX 3

Printed and bound by CPI Group (UK) Ltd, Croydon, CR0 4YY

21/10/2024

01777084-0004